Green Energy and Technology

Climate change, environmental impact and the limited natural resources urge scientific research and novel technical solutions. The monograph series Green Energy and Technology serves as a publishing platform for scientific and technological approaches to "green"—i.e. environmentally friendly and sustainable—technologies. While a focus lies on energy and power supply, it also covers "green" solutions in industrial engineering and engineering design. Green Energy and Technology addresses researchers, advanced students, technical consultants as well as decision makers in industries and politics. Hence, the level of presentation spans from instructional to highly technical.

Indexed in Scopus.

More information about this series at http://www.springer.com/series/8059

Chua Kian Jon · Md Raisul Islam ·
Ng Kim Choon · Muhammad Wakil Shahzad

Advances in Air Conditioning Technologies

Improving Energy Efficiency

 Springer

Chua Kian Jon
Department of Mechanical Engineering
National University of Singapore
Singapore, Singapore

Ng Kim Choon
Environmental Science and Engineering
King Abdullah University of Science
and Technology
Jeddah, Saudi Arabia

Md Raisul Islam
Department of Mechanical Engineering
National University of Singapore
Singapore, Singapore

Muhammad Wakil Shahzad
Environmental Science and Engineering
King Abdullah University of Science
and Technology
Jeddah, Saudi Arabia

ISSN 1865-3529   ISSN 1865-3537   (electronic)
Green Energy and Technology
ISBN 978-981-15-8476-3   ISBN 978-981-15-8477-0   (eBook)
https://doi.org/10.1007/978-981-15-8477-0

This Springer imprint is published by the registered company Springer Nature Singapore Pte Ltd.
The registered company address is: 152 Beach Road, #21-01/04 Gateway East, Singapore 189721, Singapore

Preface

Many cosmopolitan cities, particularly those in the tropics, are trapped in a vicious circle caused by urbanization, exacerbated by climate change, and locked in by their obsession with keeping their indoor cool and comfortable. Cities are getting hotter due to global warming. As they get hotter, their air-conditioning needs burgeon. Ironically, the more air-conditioning is used to cool the buildings' interior, the more heat is dissipated to the environment; forming undesirable heat zones.

First invented by Willis Carrier in 1902, vapour compression air-conditioning is the most widely used air-conditioning technology today. While, this technology has served us well for more than a century, it has also presented two key challenges— first, it is very energy-intensive and second, it is environmentally harmful. For instance, in the USA, about 90% of homes have one air-conditioning unit or more, and they account for close to 6% of the nation's total residential energy use. That alone contributes close to 100 million tons of carbon dioxide into the atmosphere every year [1]. In India, the International Energy Agency estimates that the peak electricity load from air conditioning could climb by 10% by 2050 if the technology does not modernize [2]. Further, these mechanical air conditioners employ hydrofluorocarbons (HFCs), a type of industrial chemical, as their cooling agents to remove heat from their surroundings. Typically, these chemicals are not known to be harmful unless they leak from the air conditioners to the environment; but leaks are common. When these HFCs are released into the atmosphere, they are capable of trapping many times more heat in the atmosphere than carbon dioxide; markedly contributing to global warming.

To address these challenges, installing more efficient cooling systems is probably the best place to start. The more efficient the cooling system, the lesser the energy used coupled with reduced heat dissipated to the environment; easing the urban heat island effect and lowering cities' contribution to climate change.

Presently, based on the plethora of research activities conducted over the last few years, there is no shortage of innovative ideas to make air conditioning greener and better. Some of these ideas strive to enable current units to operate more efficiently, some attempt to combine old and new technologies, while others attempt to engineer new air-conditioning processes entirely. One concept toys with the idea of ditching the chemical coolants and using water as the only coolant to yield almost free cooling. Another idea uses membrane to effectively remove moisture from the air. The problems with large-scale employment of any of these ideas are cost, economic of scale production, and wide-scale technological deployment. It is noteworthy that current vapour compression units have had more than 100 years to become exceptionally cheap. Nevertheless, the future of air conditioning is really not in the traditional compression technologies as we know them today if we are to mitigate issues related to energy consumption and environmental well-being.

This book highlights the key recent developments in air-conditioning technologies for cooling and dehumidification with the specific objectives to improve energy efficiency and to minimize environmental impacts. Key technologies related to cooling include heat-driven absorption and adsorption cooling and water-based dew-point evaporative cooling. Technologies concerned with dehumidification involve new generations of adsorbent-desiccant dehumidifiers, liquid-based desiccants, and membranes that capable of sieving out water vapour from the air. Losses in cooling cycles and thermo-economic analysis for a sustainable economy are also presented. Since each of the individual area constitutes a broad and independent branch of thermal science, an in-depth coverage of basic heat and mass transfer principles is avoided. Instead references are provided for readers to consult standard textbooks or other relevant sources for detail derivations. The materials have been judiciously selected to convey to readers interesting perspectives on recent technological developments pertaining to cooling and dehumidification. Each chapter further endeavours to provide the readers with the tools necessary to perform similar studies for other thermal systems or processes involving the transfer of heat and mass during different stages of cooling and dehumidification, i.e. air conditioning. Some fundamental works on modelling of the dew-point evaporative cooling transfer processes are presented to illustrate to readers the direct link between fundamental thermal science and practical applications. Having witnessed the transitory process from basic thermal engineering to air conditioning applications demonstrated in this book, the reader can help to promote better interaction and dialogue between researchers and engineering practitioners. It is worthy to note that, although this book has been divided into chapters with certain air conditioning themes, the presentation in each chapter does not necessarily fit into neat pigeonholes; there are times when an overlapping of information exits.

The introductory chapter presents an overview of the present state of cooling, comparative energy consumption and sustainability of different existing and new cooling technologies. Several recently actively pursued research topics on air conditioning are introduced in Chap. 2 to provide readers with a holistic perspective on the potential of new air conditioning technologies and processes that are capable of replacing vapour compression cooling systems. Chapter 3 provides an extensive coverage on dew-point evaporative coolers. This chapter covers the fundamental development of several counter-flow dew-point evaporative cooling processes including several computational models during the transient and steady-state cooling phases. It also includes several key applications of this unique evaporative cooler under several industrial settings. Chapters 4–6, respectively, present recent developments on air dehumidification, namely, desiccant-coated, liquid desiccant, and membrane-based dehumidification. In desiccant-coated dehumidification, several polymeric desiccants that can be coated on metallic heat exchangers are portrayed. In the chapter on liquid desiccant, the focus is on how novel moisture removal cycles can be developed to promote energy-efficient dehumidification. The fundamental principle of membrane dehumidification, the importance of material selection, and the rubric to assess membrane dehumidification process are also included. Chapter 7 describes dissipative losses in different cooling cycles and how these losses can be quantified and minimised. The focus of Chap. 8 is on the efficacy comparison for cooling cycles and how efficiently these cycles convert energy to produce cooling effects. The last chapter, Chap. 9, will be of great interest to technology investors and entrepreneurs. It focuses on the thermo-economic analysis for cooling needs particularly life-cycle costing of different cooling technologies; benchmarking their economic potential with the conventional vapour compression air conditioner.

Due to the broadness of recent scientific developments on air conditioning, the selection of the materials and their balance has been a most difficult task. Pertinent materials have been selected from literature and our published works. These are judiciously put together in an easily digestible format. Credits should belong to the original sources. Finally, we would like to add that the technical content presented in this book has all been done in the spirit of contributing to the knowledge pool of the existing resources on air conditioning—a subject that is mature yet still has significant room to study and is certainly worthy to pursue.

In the spirit of expressing gratitude, the authors like to extend their heartfelt thanks and appreciation to some team members who have assisted and contributed to the documentation of some of the technical content presented in the various chapters. Some of these people include research staff and ex-PhD students, namely, Cui Xin, Lin Jie, Vivekh Prabakaran, Bui Duc Thuan, M Kum Ja, Oh Seung Jin, Chen Qian, Muhammad Burhan, and other graduate students who have worked in our laboratories.

Singapore, Singapore   Chua Kian Jon
Singapore, Singapore   Md Raisul Islam
Jeddah, Saudi Arabia   Ng Kim Choon
Jeddah, Saudi Arabia   Muhammad Wakil Shahzad

References

1. How Bad Is Your Air-Conditioner for the Planet? https://www.nytimes.com/2016/08/10/science/air-conditioner-global-warming.html
2. India is the epicenter of rethinking air conditioning: https://qz.com/1675017/the-next-big-disruption-in-air-conditioning-will-be-tested-in-india/

Contents

Chapter 1
Present State of Cooling, Energy Consumption and Sustainability

1.1 Introduction

The inexorable pace in economic growth of many countries globally has serious implications for the energy–food–environment nexus. The global energy demand is predicted to increase up to 850 quadrillion British thermal units (Quads) in 2050 as compared with 500 Quads in 2010, nearly 50% increase in three decades [1]. It is estimated that Non-Organization for Economic Cooperation and Development (non-OECD) countries will be leading with 71% increase in energy demand as compared with 18% in developed countries by 2040 [1]. In terms of energy utilization, the buildings, industry and the transport sectors are the three major energy consumers [2–3]. Energy is consumed in buildings for a variety of services, namely, comfort and hygiene, food preparation and preservation, entertainment and digital communications. Generally, the type and level of services as well as the quantity of energy demand differ from a country to another, depending on many parameters such as the culture, technological advances and development as well as the behaviour of populace of the nation. Global trends have shown that more economies have embarked upon electrification and urbanization trends. Consequently, large variations in cultural attitudes, populace behaviours, the selection of construction materials and practices, fuel-mix and technologies have a significant influence on a wide range of energy use as shown in Fig. 1.1 [4].

Within the building sector, the five major services that contributed 86% of primary energy use are: (1) the heating and cooling (thermal comfort) 36%, (2) lighting 18%, (3) sanitation and hygiene, including, washing, drying and water heating 13%; (4) communication, entertainment and office equipment 10%, and (5) cooking and food refrigeration 9% and others 14% [5–7]. It is worthy to note that air-conditioning systems consume a major portion of building energy.

© Springer Nature Singapore Pte Ltd. 2021
C. Kian Jon et al., *Advances in Air Conditioning Technologies*, Green Energy and Technology, https://doi.org/10.1007/978-981-15-8477-0_1

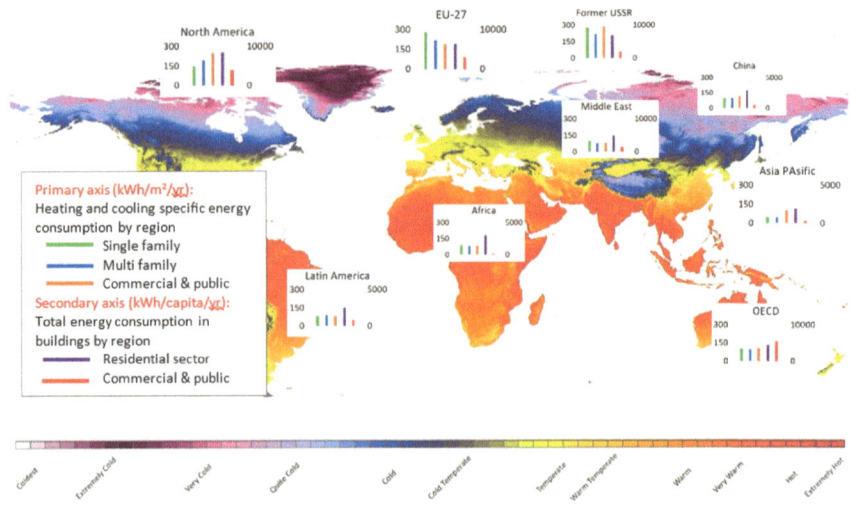

Fig. 1.1 Specific energy consumption for heating, cooling and total energy consumed by building sectors in the different regions of the World

1.2   Building Comfortable Zone

In most of the developed and developing countries, air-conditioning systems became very common features due to significant benefits to productivity, human health and comfort. The ANSI/ASHRAE Standard 55 (Thermal Environmental Conditions for Human Occupancy) is an international standard published by the American Society of Heating, Refrigerating, and Air-Conditioning Engineers. It was started in 1966 and revises after every 3 years and most recent version of the standard was published in 2017. It is to specify the various combinations of indoor thermal environmental factors as well as personal factors that will produce thermal environmental conditions acceptable to a majority of the occupants within a space [8–11].

Thermodynamically, the human body can be viewed as a heat engine where food can be input energy and work activities considered as output. During activities, body generates and release heat to ambient to fulfil thermodynamic conditions to continue process. The quantity of heat released depends on environmental conditions. For example, in winter or cold environment, body release more heat as compared with hot environment, both lead to discomfort situation. So, maintaining thermal comfort conditions for occupants of buildings is one of the important goals of standard and design engineers.

Thermal comfort is essential to continue the life cycle because anything out of the envelop conditions can be life threatening. For example, hyperthermia conditions when core body temperature reaches above 37.5–38.3 C and hypothermia conditions below 35.0 C are not comfortable for stable body operation. The air-conditioning systems modify these conditions within building environment in order to reduce

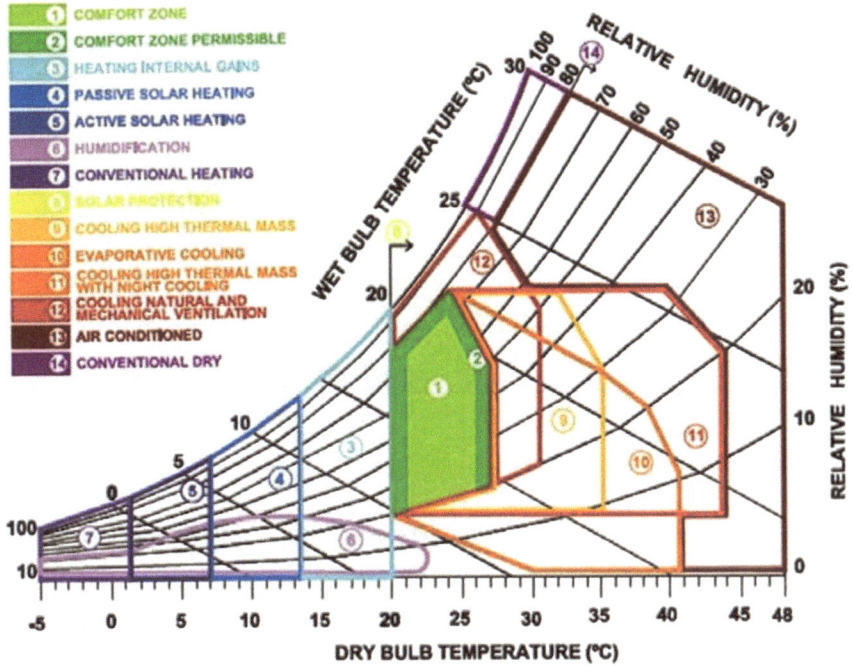

Fig. 1.2 Givoni's different level of comfortable zone on the psychrometric chart

hyperthermia and hypothermia to provide normal human body temperature. It is important to achieve stable physiological processes of the human body [12–14].

In 1963, Victor Olgyay introduced Bioclimatic Chart of thermal comfort representations and later in 1969 Givoni presented the psychrometric chart including Olgyay chart and additional strategies for heating and cooling as shown in Fig. 1.2 [15–18].

1.3 Air-Conditioning Market

Presently, the air-conditioning market is estimated over US$103 billion and 141 million units sold annually, which include new installation in new building, equipment for old building and replacement of the existing units. The number of units in different parts of the world is presented in Fig. 1.3 and data related to residential and packaged commercial units are shown in Fig. 1.4 and it does not include additional 400,000 commercial chillers units that cost around $8.5 billion annually [19–22]. At global level, the split air-conditioner product percentage is highest due to flexibility in installations followed by water-cooled and air-cooled chillers. The global air conditioner market by product and breakdown is shown in Fig. 1.5 [23].

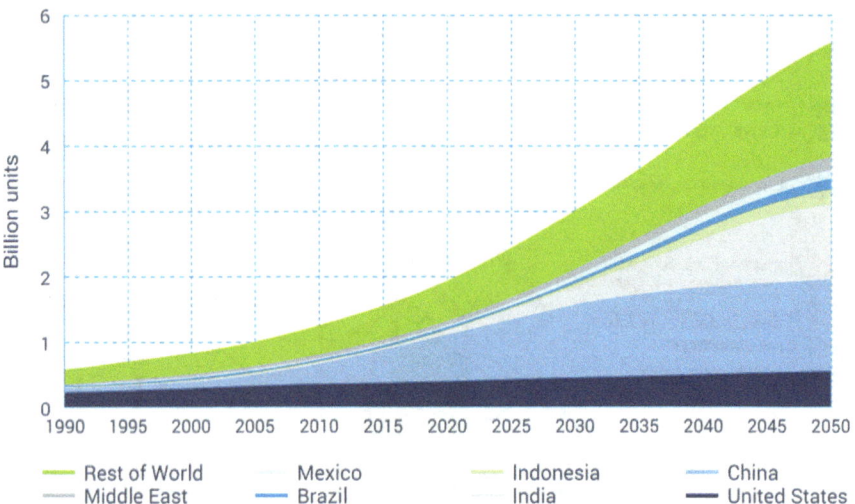

Fig. 1.3 The number of air-conditioner units (in billions) in different parts of the world from 1990 to 2050

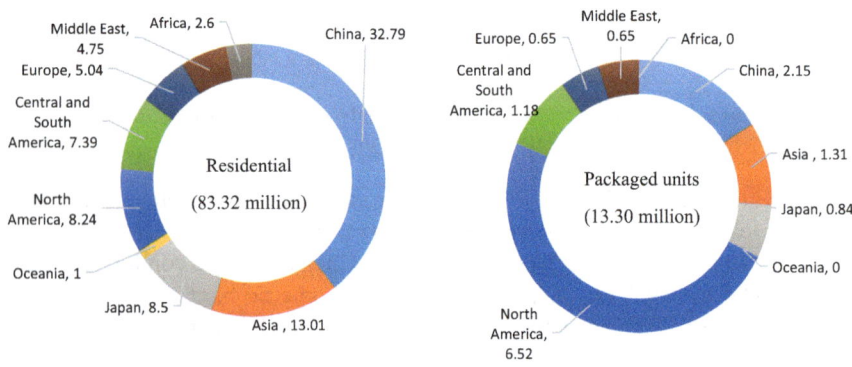

Fig. 1.4 Global air-conditioners sale in 2014 by region and by unit type

Fig. 1.5 Global air-conditioners market by product in 2018 (by volume)

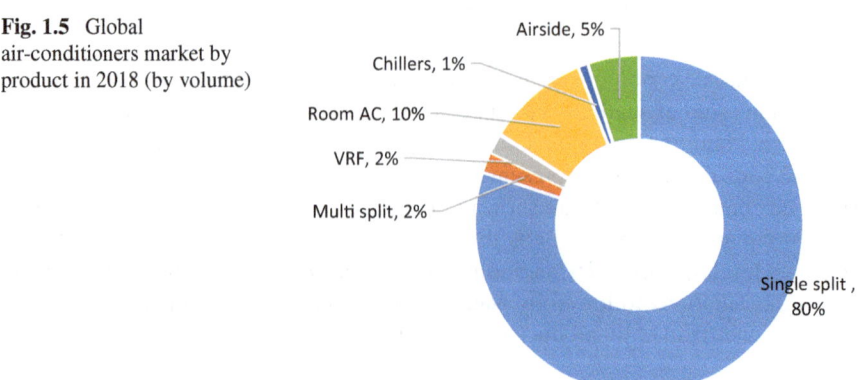

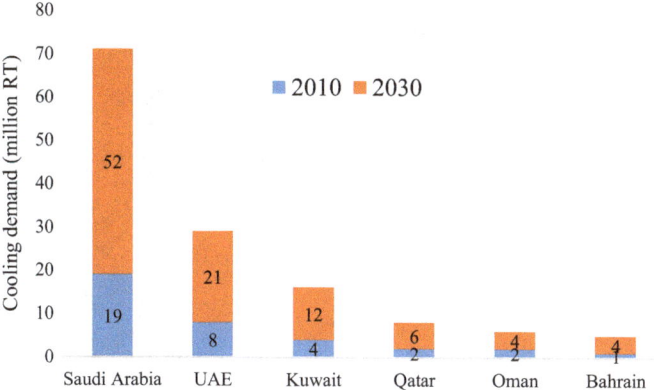

Fig. 1.6 The cooling capacity demand of GCC countries between the year 2010 and 2030, which is expected to increase by more than 3-fold

The global air-conditioner sales are growing rapidly, especially in the Middle East and developing countries. The data show that its potential is higher in China, India, Pakistan, Indonesia, Bangladesh Brazil and Philippines as compared with the USA [24]. In India, Turkey, China and Indonesia, the electricity consumption for air conditioning is growing over 20% annually and other countries such as Saudi Arabia, Australia, Canada, Brazil and EU facing 6–10% increase. In the gulf cooperation countries, the cooling capacity is expected to increase over 3-fold compared with 2010 due to high temperature coupled with humidity as shown in Fig. 1.6 [25–28].

1.4 Energy Consumed by the Air-Conditioning Systems

Based on the energy distribution within commercial and residential buildings, it is apparent that heating and cooling services formed the major shares of energy consumption. In commercial buildings, the air-conditioning services utilize around 40% of the total energy. On the other hand, in residential buildings, it consumes around 36% of the supplied energy. The share of energy consumed by different end-users in commercial and residential buildings is presented in Fig. 1.7 [29]. It is estimated that the air-conditioning energy demand will be tripled by 2050 and new power capacity equivalent to combined Japan, EU and USA will be required. It will also increase global building's air-conditioner stack to 5.6 billion by 2050 and compared with 1.6 billion in 2019, it means 10 new air conditioners will be sold in every second in next 30 years [30].

The large economies such as the USA and Japan air-conditioner market are saturated and progress is slow. Overall, cooling represents over 60% of residential load in the Middle East and 25–30% in the USA. Fig. 1.8a shows the percentage of

Fig. 1.7 a Commercial
buildings energy
consumption by different
sectors. **b** Residential
buildings energy
consumption by different
sectors

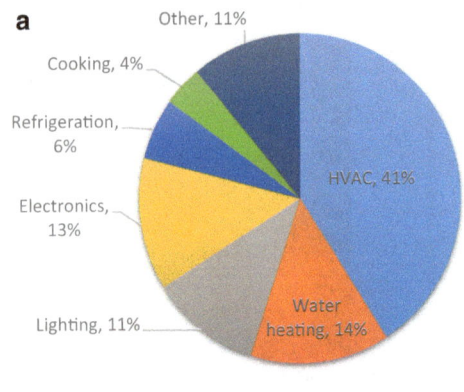

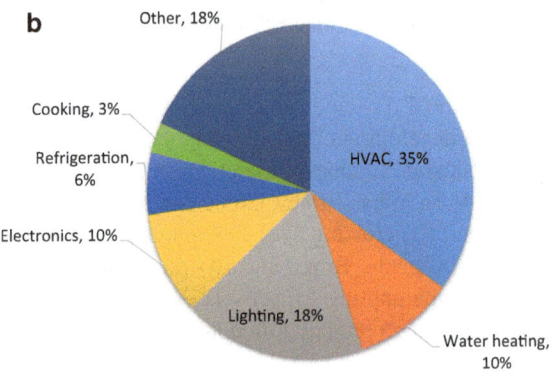

residential electricity consumed for air conditioners and Fig. 1.8b presents the total
electricity utilized in TWh [31].

1.5   Cooling Degree Days (CDDs)

The NASA record shows that the global temperature is increasing slowly but steadily.
The world is getting warmer and temperature readings around the globe have been
rising since industrial revolution coupled with human activities as shown in Fig. 1.9
[32]. Since 1880s, the average global space temperature has increased by a little
more than 1 °C (2°F) according to the scientists at NASA's Goddard Institute for
Space Studies (GISS). This increment in space temperature causing uncomfortable
environment in many parts of the world and cooling demand is increasing sharply
[33]. Cooling degree days (CDDs) is widely parameter to measure cooling demand.
It is the sum of daily mean temperatures above 18.3 °C (65°F). The CDD around
the world over 4500 weather monitoring stations was presented recently as shown

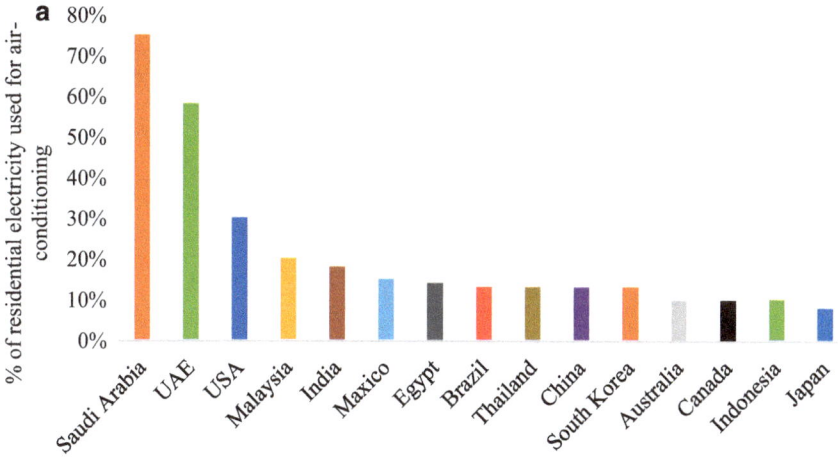

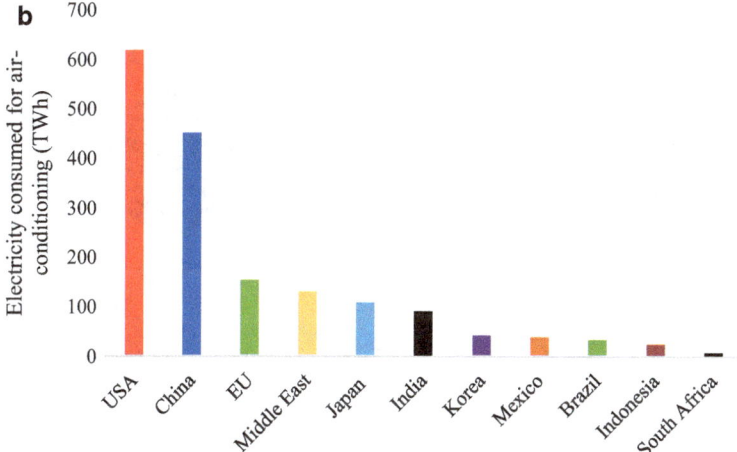

Fig. 1.8 **a** Percentage of residential electricity consumed for air conditioning globally. **b** Total electricity consumed for air conditioning globally

in Fig. 1.10. It can be observed that most of our planet area is exposed to over 4000 CDD annually. The desert areas such as Africa, Middle East and Southern Asia are exposed to maximum CDDs due to highest temperature throughout the year [34].

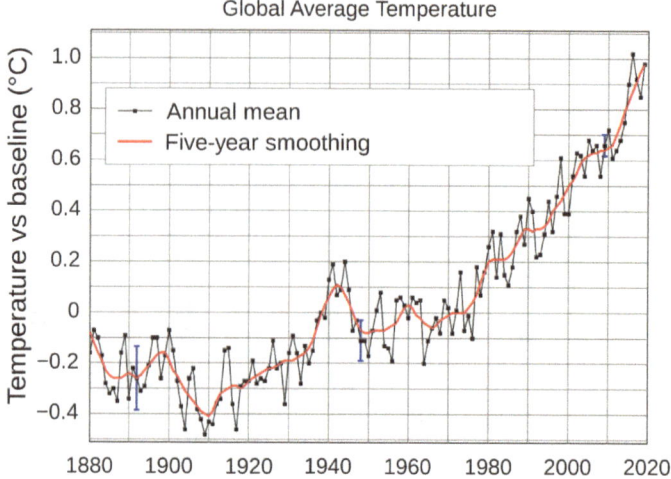

Fig. 1.9 Global average temperature increase from 1880 to 2020 measured by NASA

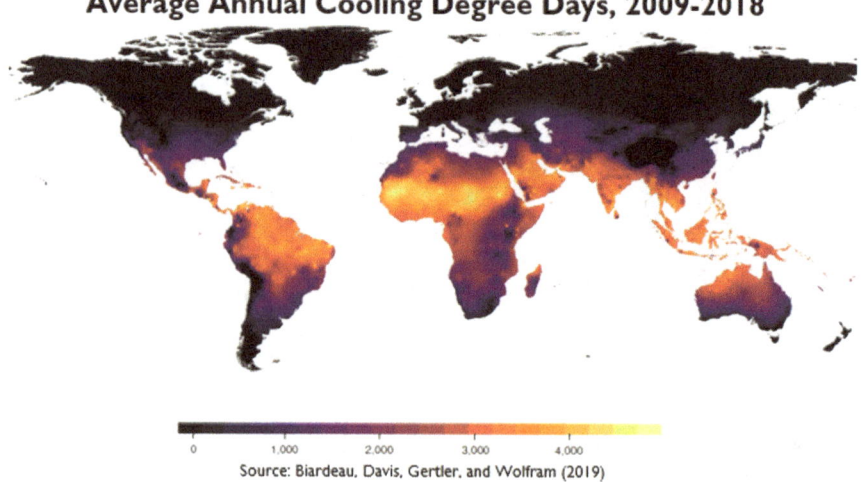

Fig. 1.10 Annual average cooling degree days from 2009 to 2018

1.6   CO₂ and GHG Emission from Air-Conditioning Systems

A recent report by the Global Alliance for Buildings and Construction highlighted that there are currently 1.6 billion air-conditioning units installed worldwide (35% in China, 23% in USA) and the figure could rise to 2.6 billion in the next decade [35].

Almost 90% air-conditioner market is covered by mechanical vapour compression (MVC) chillers, a more than 100-year-old technology. The MVC system uses refrigerants such as hydrofluorocarbons or HFCs that absorb and release heat to produce cooling by utilizing large amounts of electricity. The conventional air conditioner has two types of pollution, CO_2 emission from electricity generation system and greenhouse gases from chemical-based refrigerants [36]. Over 70% of GHG emissions from air-conditioning systems are due to indirect emissions from electricity generation and remaining 30% from direct chemical-based refrigerants [37].

Since 1900s when William Carrier invented 1st chiller system, the compressor technology at the heart of most AC units has barely reached 14% of its theoretical maximum efficiency [38] and the corresponding coefficient of performance ranges from 4 to 5 as shown in ASHARE classification in Fig. 1.11.

Due to the limited performance of refrigerant compressors, the overall power consumption is growing very fast with capacity and the corresponding CO_2 emission is also increasing. Since 1990, the air-conditioning CO_2 emissions have triple, reaching 1600 million metric tons and this is equivalent to the entire CO_2 emissions of Japan. The increase in the number of residential air-conditioner units and the corresponding CO_2 emission since 1990 is presented in Fig. 1.12 [39]. It can be seen that the trend is sharper after 2015 and it is due to improved lifestyle, more development activities and GDP increase. The global warming also has a great impact on the number of unit's installations due to temperature rise. The HFCs and other refrigerants used by MVC systems also produce greenhouse gases that can be up to 4,000 times more potent than carbon dioxide.

In 1985, the Vienna convention provided a framework for the Montreal protocol that mitigated the ozone-depleting substances being used from 1987. The original protocol has been amended several times for developed and developing countries, between 1990 (London) and 2007 (Montreal). Montreal Protocol and Europe Council Directive (3093/94) limited the use of HCFC and CFC gases to reduce the environmental impact. The protocol eradicated the production of CFCs in the developed countries from 1995 and planned to phase-out HCFCs completely by the year 2030, as shown in Fig 1.13 [40].

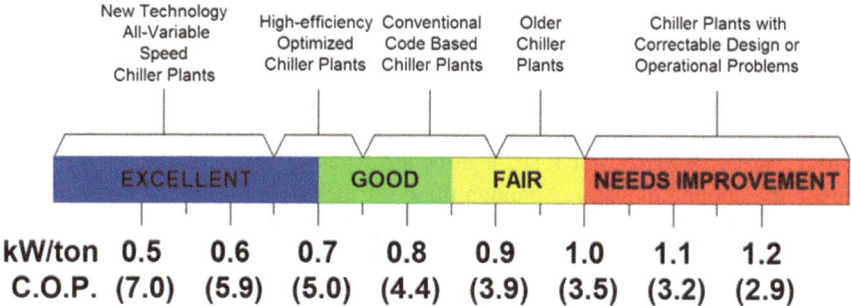

Fig. 1.11 ASHRAE air-conditioner performance (COP) classifications

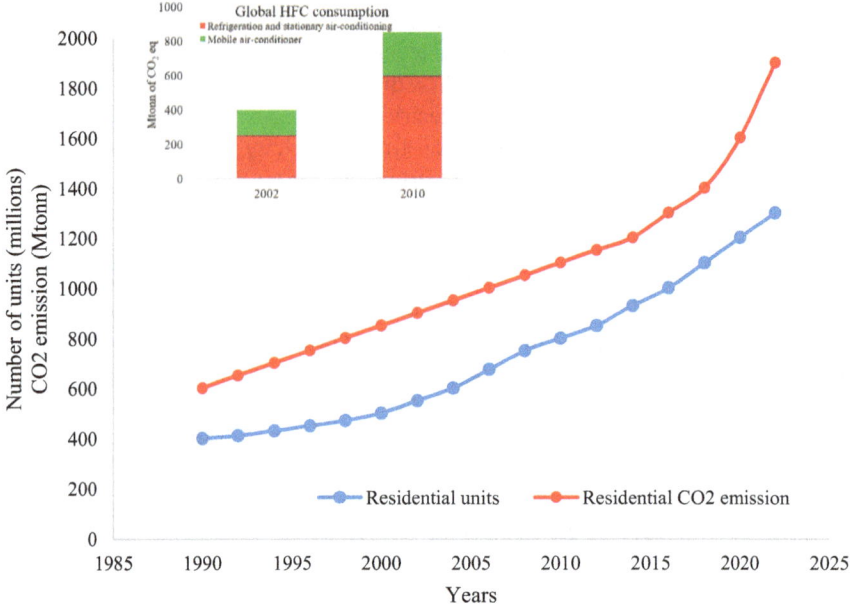

Fig. 1.12 Number of residential air-conditioner units and corresponding CO_2 emission

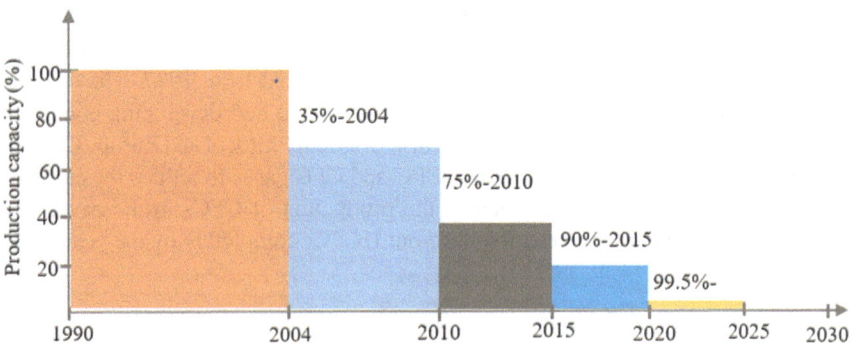

Fig. 1.13 The phase-out plan of HCFCs as mandated by the Montreal Protocol from 1990 to 2030

Currently, Hydrofluorocarbon (HFC) fluids and their mixtures such as R134a, R404A, R507, R407C, and R410A gases are widely used as alternative refrigerants in the cooling industry. Although these refrigerants have low ozone depletion potential (ODP) but their high global warming potentials (GWP) have significant environmental impact [18–21]. Consequently, cooling scientists and engineers are greatly motivated to investigate the alternate cooling processes and technologies that will further reduce energy consumption and environmental impact in order to meet the desired future goals of sustainable cooling. Some improvement in cooling systems, as reported by the International Energy Agency (IEA), is the application of energy mix (especially from renewable energies) and development of innovative cooling cycles such as heat-driven sorption cycles. The application of proposed energy mix and innovative cycles can reduce CO$_2$ emissions by up to 2 Giga-tons (Gt) by 2050—around 25% of today's emissions from buildings was estimated by IEA [41, 25].

1.7   Future Roadmap

According to the International Energy Agency (IEA), the air-conditioning demand will be tripled by 2050 and it will be one of the major electricity consumers. The major limitations of conventional MVC systems are the levelling-off conventional chiller's efficiency at 0.85 ± 0.03 kW/Rton due to pairing of dehumidification and cooling processes in one machine. So innovative solutions are required to overcome this 100 year's limitation. In addition, chlorofluorocarbons (CFCs) have already been phased out and hydrochlorofluorocarbons (HCFCs) will likely be phased out completely by 2030, so alternative non-chemical-based refrigerants need to develop to utilize for future cooling.

Out-of-box solutions are required to overcome the limitations of conventional systems to maintain the cooling demand trend. The present technologies level and future roadmap are presented in Fig. 1.14. Recently, researchers have attempted to decouple the dehumidification and cooling process by adsorbent coated evaporator and condenser heat exchangers. The adsorbent helps to remove moisture before

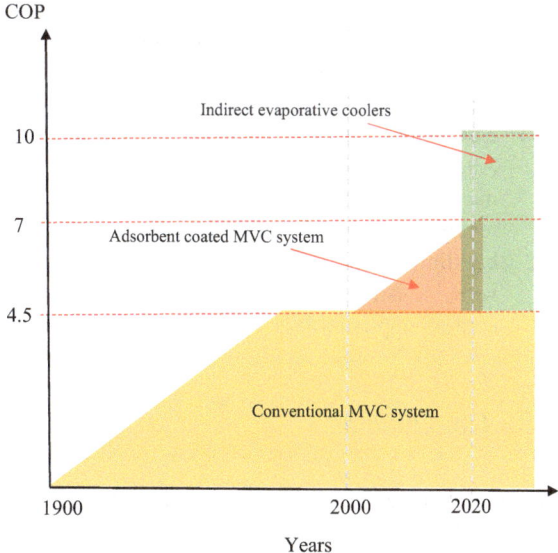

Fig. 1.14 Future roadmap for sustainable cooling processes

cooling or air and it eliminates the need for cooling air below dew point temperature and hence improved COP to 6–7 [42]. The other solution is to propose solid desiccant dehumidifier coupled with indirect evaporator coolers (IEC) but still the bottleneck of this hybrid system is that most solid desiccants are capable performing with COPs spanning 0.5–0.7. Recently, researchers [43–45] have shown that the IEC cooling COP can achieve 25–30 and if dehumidifiers can be developed with COP levels spanning 2–3, the overall attainable COP of the air-conditioning system is able to approach 10. This will possibly be the best achievement of any known air conditioning technology and will help to meet future cooling demand sustainably.

The following recommendations can also help to reduce the energy demand for air conditioning:

(i) **Encourage the development of alternative cooling solutions**
     The government plays an important role to reduce the country's cooling demand and its impact on energy by supporting improved building envelope solutions such as low emissivity window panels, integrated building cooling system and alternative renewable energy utilization for cooling.
(ii) **Encourage for efficient systems**
     The proper standard implementations can dampen the energy demand by improving air-conditioning systems efficiency. The government can issue energy bands for commercial systems and provide incentives to operate in high efficient bands. This will encourage all users to purchase and operate highly efficient systems.
(iii) **Encourage for smart control systems**
     Improved control can also help to reduce the energy consumption of cooling systems by operating them more efficiently. The government can promote innovative business models and provide incentives to encourage digital technologies applications. This includes smart temperature control thermostats and optimized load distribution. The government should also support small manufacturers to access to digital technologies solutions to develop smart control systems to improve efficiency.
(iv) **Encourage for alternative energy utilization**
     Alternative energy utilization to meet growing cooling demand can be instrumental and easily implementable. Government can promote solar PV and other renewable energy for cooling systems operation. It can also be integrated with building system to level-off the peak cooling demand. Government can work with industry to promote renewable energy and its integration for possible cooling applications.

1.8 Conclusions

A comprehensive literature review on cooling demand its energy consumption and direct and indirect emission has been conducted. It is estimated that by 2050, global energy demand will increase by 50% up to 850 Quads. Residential and commercial building consume major share of overall electricity produced.

Air conditioning is required to maintain comfortable condition for stable operation of human body and to enhance productivity. International standard ASHRAE 55 provides detailed guidelines for comfortable indoor conditions in terms of temperature and humidity.

The air-conditioning market is over US$100 billion and 141 million units sold every year. The split air conditioners are dominating in the market due to flexibility in installation and operation.

Air conditioners consume about 40% of the energy of commercial buildings and 36% of residential buildings. It is estimated that this energy will be tripled by 2050 due to growing cooling demand. The global air conditioner units will also increase to 5.6 billion and 10 units will be sold every second. The increase in cooling demand is due to industrialization and increase in ambient temperature. The degree cooling day increased to 4000 in most of the countries around the globe. The conventional air conditioners cause 70% indirect CO_2 emission and 30% direct GHG emissions. Since 1900, the air-conditioners emission has been tripled and it is expected to increase many folds by 2050.

Montreal Protocol was established to phase out chemical-based refrigerants to prevent GHG emissions. It plans to phase out CFCs and HCFCs by 2030 and 2047 in developed and developing countries.

For future cooling supplies, out-of-box solutions are required. The conventional chiller achieved only up to 0.85 ± 0.03 kW/Rton energy efficiency since 1900s due to operational limitations of combined latent and sensible load. The decoupled solutions are introduced and it is expected that innovative solution will be able to achieve 0.60 ± 0.03 kW/Rton energy efficiency. This will be sustainable solutions for future applications using natural refrigerants.

References

1. International Energy Outlook 2019 with projections to 2050, U.S. Department of Energy Washington, DC 20585 (2019). https://www.eia.gov/outlooks/ieo/pdf/ieo2019.pdf (2019)
2. Ürge-Vorsatz D, Cabeza LF, Serrano S, Barreneche C, Petrichenko K (2015) Heating and cooling energy trends and drivers in buildings. Renew Sustain Energy Rev 41:85–98. https://doi.org/10.1016/j.rser.2014.08.039
3. Lombard PL, Ortiz J, Pout C (2008) A review in building energy consumption information. Energy and Building. 40:394–398. https://doi.org/10.1016/j.enbuild.2007.03.007
4. Global energy assessment—toward a sustainable future, Chapter: 10, Cambridge University Press and IIASA, pp 649–760 (2012)
5. Akbari H, Menon S, Rosenfeld A (2009) Global cooling: Increasing worldwide urban albedos to offset CO_2. Climatic Change. 94:275–286. https://doi.org/10.1007/s10584-008-9515-9
6. Energy technology perspectives 2012. https://www.iea.org/reports/energy-technology-perspectives-2012 (2012)
7. World energy balances and statistics. IEA Online Data Services. https://data.iea.org/ieastore/statslisting.asp (2019)

8. ANSI/ASHRAE standard 55-2017. Thermal Environmental Conditions for Human Occupancy
9. Toftum J (2005) Thermal comfort indices. In: Handbook of human factors and ergonomics methods. CRC Press, Boca Raton, FL, USA
10. Lenzuni P, Freda D, Del Gaudio M (2009) Classification of thermal environments for comfort assessment. Ann Occup Hyg 53:325–332. https://doi.org/10.1093/annhyg/mep012
11. Lomas KJ, Giridharan R (2012) Thermal comfort standards, measured internal temperatures and thermal resilience to climate change of free-running buildings: a case-study of hospital wards. Build Environ 55:57–72. https://doi.org/10.1016/j.buildenv.2011.12.006
12. Axelrod YK, Diringer MN (2008) Temperature management in acute neurologic disorders. Neurol Clin 26:585–603. https://doi.org/10.1016/j.ncl.2008.02.005
13. Laupland KB (2009) Fever in the critically ill medical patient. Crit Care Med 37(Supplement):S273–S278. https://doi.org/10.1097/ccm.0b013e3181aa6117
14. Brown DJA, Brugger H, Boyd J, Paal P (2012) Accidental hypothermia. N Engl J Med 367:1930–1938. https://doi.org/10.1056/nejmra1114208
15. Givoni B (1969) Man. Applied Science Publishers, Climate and architecture
16. Givoni B (1992) Comfort, climate analysis and building design guidelines. Energy and Buildings. 18:11–23. https://doi.org/10.1016/0378-7788(92)90047-K
17. Olgyay V (2015) Design with climate: bioclimatic approach to architectural regionalism. Princeton University Press
18. Kiyan V, Sandrine M (2018) Development of psychrometric diagram for the energy efficiency of air handling units. Int J Vent 3:491
19. Milnes J (2014) Global A/C market starting to warm up. ACHR News. August 18, 2014. Citing BSRIA data. https://www.achrnews.com/articles/127385-global-ac-market-starting-to-warm-up
20. JARN (2015) BSRIA world air conditioning market study. https://www.ejarn.com/news.aspx?ID=34847
21. Goetzler W, Guernsey M, Young J, Fuhrman J, Abdelaziz O (2016) The future of air conditioning for buildings.https://www.energy.gov/sites/prod/files/2016/07/f33/The%2OF uture%20of%20AC%20Report%20-%20Full%20Report_0.pdf
22. The Importance of Energy Efficiency in the Refrigeration, Air-conditioning and Heat Pump Sectors, UN Environment (2018) https://conf.montreal-protocol.org/meeting/workshops/ene rgy-efficiency/presession/breifingnotes/briefingnote-a_importance-of-energy-efficiency-in-the-refrigeration-air-conditioning-and-heat-pump-sectors.pdf
23. BSRIA (2019) Annual global air conditioning study shows growth.https://www.bsria.com/uk/news/article/bsria-annual-global-air-conditioning-study-shows-growth/
24. Davis L (2019) Predicting global air conditioning demand, by nation. https://energypost.eu/predicting-global-air-conditioning-demand-by-nation/
25. How much can HVAC technology slow carbon emissions? https://www.buildings.com/article-details/articleid/12424/title/how-much-can-hvac-technology-slow-carbon-emissions (2011)
26. Sayed TE, Fayad W, Monette SP, Sarraf G (2012) The need for GCC governments to take action: unlocking the potential of district cooling. Booz & Company Report
27. Al-Faris AR (2002) The demand for electricity in the GCC countries. Energy Policy. 30:117–124. https://doi.org/10.1016/S0301-4215(01)00064-7
28. Patlitzianas KD, Doukas H, Askounis DT (2007) An assessment of the sustainable energy investments in the framework of the EU-GCC cooperation. Renew Energy 32:1689–1704. https://doi.org/10.1016/j.renene.2006.08.002
29. Real prospects for energy efficiency in the United States (2010) National Academy of Sciences, National Academies Press. https://doi.org/10.17226/12621
30. Air conditioning use emerges as one of the key drivers of global electricity-demand growth, IEA report (2018). https://www.iea.org/news/air-conditioning-use-emerges-as-one-of-the-key-drivers-of-global-electricity-demand-growth
31. EnerDemand: The global efficiency and demand database (2019) https://www.enerdata.net/publications/executive-briefing/the-future-air-conditioning-global-demand.html
32. https://earthobservatory.nasa.gov/world-of-change/decadaltemp.php

33. National Research Council (U.S.) (2006) Committee on surface temperature reconstructions for the last 2,000 years surface temperature reconstructions for the last 2,000 years (2006). National Academies Press ISBN 978-0-309-10225-4.
34. Biardeau LT, Davis LW, Gertler P, Wolfram C (2020) Heat exposure and global air conditioning. Nat Sustain. 3:25–28. https://doi.org/10.1038/s41893-019-0441-9
35. Global alliance for buildings and construction. UN Environment Program (2020) https://globalabc.org/
36. Calm JM (2020) Emissions and environmental impacts from air-conditioning and refrigeration systems. Int J Refrig 25:293–305. https://doi.org/10.1016/S0140-7007(01)00067-6
37. Air conditioning is threatening our ability to tackle climate change. Here's what we need to do (2019) https://climatechange-theneweconomy.com/air-conditioning-is-threatening-our-ability-to-tackle-climate-change-heres-what-we-need-to-do/
38. Gloël J, Oppelt D, Becker C, Heubes J (2015) Green Cooling Technologies, Market trends in selected refrigeration and air conditioning subsectors. Published by Deutsche Gesellschaftfür Internationale Zusammenarbeit (GIZ) GmbH
39. Paupardin SE (2019) Is the world facing a looming cold crunch? https://www.sageglass.com/eu/visionary-insights/world-facing-looming-cold-crunch
40. Söğüt MZ (2015) Developing CO_2 emission parameters to measure the environmental impact on cooling applications. Int J Green Energy 12:65–72. https://doi.org/10.1080/15435075.2014.889008
41. Munzinger P, Gessner A (2015) Climate-friendly Refrigeration and air conditioning: a key mitigation option for INDCs. Published by Internationale Zusammenarbeit (GIZ) GmbH
42. Tu YD, Wang RZ, Ge TS, Zheng X (2017) Comfortable, high-efficiency heat pump with desiccant-coated, water-sorbing heat exchangers. Sci Rep. 7:40437. https://doi.org/10.1038/srep40437
43. Ng KC, Shahzad MW, Burhan M, Ybyraiymkul D, Oh SJ (2020) Combined direct and indirect evapourative cooling systems and methods. WO2020/058778 A1
44. Oh SJ, Shahzad MW, Burhan M, Chun W, Jon CK, KumJa M, Ng KC (2019) Approaches to energy efficiency in air conditioning: a comparative study on purge configurations for indirect evaporative cooling. Energy. 168:505–519. https://doi.org/10.1016/j.energy.2018.11.077
45. Shahzad MW, Burhan M, Ybyraiymkul D, Oh SJ, Ng KC (2019) An improved indirect evaporative cooler experimental investigation. Appl Energy 256:113934. https://doi.org/10.1016/j.apenergy.2019.113934

Chapter 2
Future of Air Conditioning

2.1 Introduction

Increasing population growth and rising energy demand in many countries are imposing great challenges in water, energy and environment sustainability nexus. Global energy and environmental scenarios are closely intertwined. The problems of supply and use of energy are intimately related to both global warming and climate change [1]. For instance, the electrical energy in Asia countries accounts for about 30% of the global energy demand [2]. In Singapore, electricity consumption rose by 3.3% to 46 TWh in 2014 [3]. Most of this was attributed to consumption by the industrial-related sector (43%), followed by the commerce and services-related sector (37%). Households accounted for 15% of the total electricity consumption.

Globally, the demand for HVAC (Heating Ventilation Air Conditioning) products is primary driven by key factors such as rising population, buzzing housing markets, and better consumers' standard of living in emerging markets, improved development of reliable energy resources, growing commercial and industrial units, among others. Asia–Pacific is leading the global HVAC market chiefly due to the high demand coming from fast emerging economies like China, India and South-East countries like Indonesia, Malaysia and Thailand. Air conditioning is widely used in Asia primarily due to its hot and humid climatic conditions. In Singapore, the energy consumed for Heating, Ventilation and Air-Conditioning (HVAC) applications typically comprises up to 50% [4] of the total energy consumption in a building. Figure 2.1 displays the energy consumption breakdown in a typical building in Singapore [5]. It is noteworthy that this figure can be considered to be quite representative of most buildings experiencing tropical climates.

In conventional vapour–compression air-conditioning systems, the compressor has made a marked improvement in terms of energy efficiency, chronologically from 1.2 kW/RT in 1990s to 0.85 ± 0.05 kW/RT. To achieve better energy efficiency for air-conditioning applications, manufacturers have introduced many individual technologies integrated to, collectively, improve overall system efficiency. These include multi- and variable-speed drives, novel compressor, fan, motor, and heat

© Springer Nature Singapore Pte Ltd. 2021
C. Kian Jon et al., *Advances in Air Conditioning Technologies*, Green Energy and Technology, https://doi.org/10.1007/978-981-15-8477-0_2

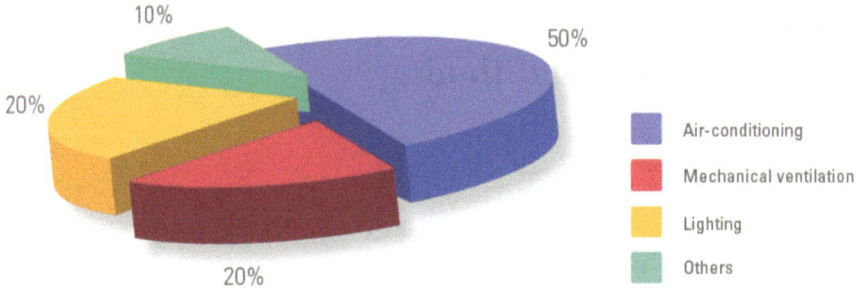

Fig. 2.1 Breakdown of energy consumption within a building in Singapore [5]

exchanger designs, electronic expansion valves and advanced controls. Since 2000, however, this improvement trend has begun to level off asymptotically as portrayed in Fig. 2.2. This implies that the efficiency improvement for compressor stages and heat exchangers has almost peaked. Despite the massive investment on R&D for cooling technologies by chillers' manufacturers, there exist physical and material limits to which efficiencies of the major components in the cooling cycle can continue to improve. Therefore, if the pursuit to improve the cooling efficiency of conventional chillers (with CFC/HCFC refrigerants) continues, the improvement in kW/Rton may only be marginal of less than 5% and no quantum leap in energy efficiency improvement can be realized. Therefore, there is an exacting need for an out-of-box solution for cooling where the consumption of energy can be markedly reduced. Conservatively, that number can be as much as 30% in order to achieve a sustainable cooling

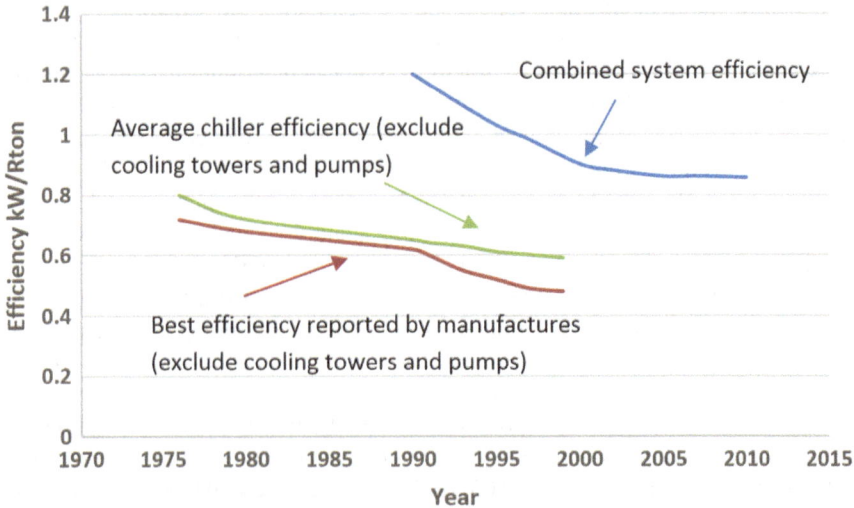

Fig. 2.2 The efficiency of the mechanical chiller improved from 1.2 kW/RT in 1990s to 0.85 kW/RT in 2010

solution for the future that addresses the space cooling needs of present and future generations.

It is the intent of this chapter to review some of the most recent developments concerning innovative cooling and dehumidification technologies that can potentially impact our environment and also lower the kW/Rton of cooling systems. It is commonplace knowledge that the literature on this subject is highly voluminous and expanding rapidly, therefore, it is simply impossible to produce a completely extensive review. However, this chapter intends to examine a specific number of articles, particularly recent ones, to present a comprehensive overview of key research developments and directions in this growing, multidisciplinary field. This review chapter is judiciously divided into two key categories that include novel cooling technologies that handle sensible load and innovative dehumidifiers that take care of latent load. The individual section details and highlights the recent progresses made in their respective areas. It is noteworthy that several of these energy-efficient cooling and dehumidification technologies described in the chapter are at the research stage or are on-going studies. Some have been implemented on a larger scale and, therefore, are prime examples of practical solutions that can be readily applied to suit specific needs.

2.2 Absorption/Adsorption Chillers

The technology of thermally driven absorption and adsorption chillers has a long history. The term 'adsorption' was first coined by Kayser in 1881, and the very first adsorption cooling system was invented by Michael Faraday in 1848, utilizing ammonia and silver chloride as a working pair (adsorbent and adsorbate) [6]. Similar to the adsorption chiller development, the earliest absorption chiller, using water and sulphuric acid as a working pair (refrigerant and absorption solution), was invented by the French scientist Ferdinand Carré in 1858 [7].

Although thermally driven absorption and adsorption chillers have long technological history, they are still unable to compete with electrically driven chillers due to their poor thermal performance. An electrically driven chiller is capable of providing four to seven folds of cooling capacity in comparison to the thermally driven chillers with the same equivalent amount of energy input [8].

It is noteworthy that thermally driven chillers have their own advantages of having benign effects on the environment with their zero Ozone Depletion Potential (ODP) and zero Global Warming Potential (GWP) [9–11] particularly when compared with their electrically operated counterpart. Further, the implementation of the Montreal Protocol to phase-out CFC refrigerants has promoted attention to focus on thermally driven chillers. The chief advantage of thermally driven chillers is that they are known to have lesser moving parts and so that it has lesser electricity demand resulting higher electrical COP (Cooling Load RT/Electrical Power Input kW). These advantages that the thermally driven chillers possessed make them highly suitable for waste heat recovery applications through a tri-gen system (electricity, heat and chilled water)

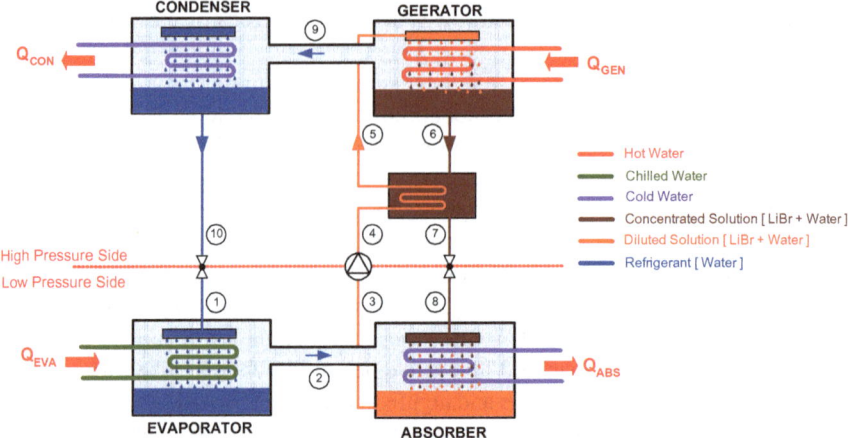

Fig. 2.3 Schematic illustrating the fundamental principle behind absorption cooling system

or a quad-gen system (electricity, heat, potable and chilled water) and solar-driven chiller system.

The basic working principle of an absorption cooling system is depicted in Fig. 2.3. The major components of the system are evaporator, absorber, generator and condenser. The liquid refrigerant (water) evaporates in the evaporator at low pressure by taking heat from its surroundings resulting in the production of chilled water. The vaporized refrigerant is then absorbed by the absorbent (LiBr solution) in the absorber resulting in a diluted solution. The solution is then pumped to the generator where it gets heated to vaporize the refrigerant from the solution. The diluted solution is preheated before it enters the generator by the hot concentrated solution from the generator. The vaporized refrigerant gets condensed by transferring heat to the surroundings in the condenser before returning to the evaporator. The cycle repeats itself. The system's higher and lower operating pressures of the system are determined by the condenser and the evaporator temperatures. The concentration of the solution at each point can be determined from the Dühring's plot as illustrated in Fig. 2.4 based on each point's pressure and temperature.

Absorption chillers are specifically designed based on their application requirements and, most importantly, availability of thermal energy sources. They include single-effect, indirect-fired using steam or hot water with a COP spanning 0.5–0.78, double-effect, indirect-fired with a COP ranging from 1.1 to 1.32, and double-effect, direct-fired using gas/oil burner with a COP of 1.02–1.19. The COPs of the above-mentioned chillers are portrayed in Fig. 2.5 for various percentages of chillers loading (8). The graph shows that the chiller's part load is a key parameter that influences the performance of chiller. It also demonstrates that all types of absorption chillers perform optimally at around 50% of part load. For high-temperature thermal energy

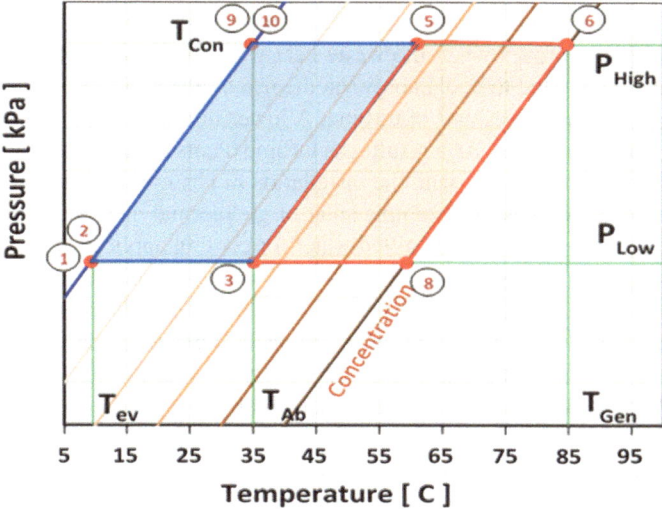

Fig. 2.4 Dühring's equilibrium chart illustrating the process flow and conditions of the working solution

Fig. 2.5 Illustrating the effect of the chiller's part load operation performance of absorption chiller

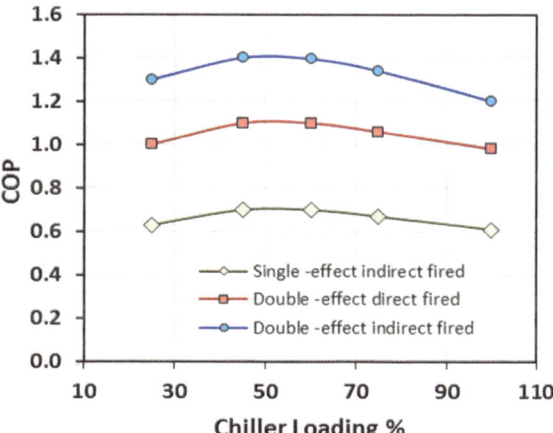

sources, double-effect indirect-fired absorption chillers are preferred. Most absorption chillers rely on the ARI Standard 560 for testing. This standard involves the integrated part load value (IPLV) testing standard for absorption chillers.

Albeit absorption chillers having lower thermal energy performance efficiency, they possess key merits such as having fewer moving parts, ability to be driven by low-grade thermal energy source, and employing an environmental-friendly working medium. Absorption chillers are essential for some industrial applications. First, waste heat recovery system from tri-gen and quad-gen systems, petroleum and chemical industries, palm oil production, and incinerating systems, etc., acts as the thermal fuel that feds the absorption chiller to produce cooling effect. Second, the efficiency of a gas turbine improves when the waste heat from the exhaust of the turbine is recovered and converted to cool energy via the use of absorption chillers. The cool energy cools the intake fresh air of turbine. A lower inlet air temperature increases air density and airflow rate. As a result, gas turbine plants claim an 11.3% improvement in average power output with the installation of absorption chiller [12]. Third, renewable energy sources, for example solar or geothermal means, can be coupled to absorption chillers to produce chilled water for cooling applications.

The absorption cooling system is already a well-established technology. Many research activities have focused on system components such as the use of heat transfer enhancements like falling film heat transfer, tube design, etc. However, the development of using ionic liquid as an absorbent is still at its infancy state [13], therefore there is potential for further fundamental research to yield better system improvement.

Conventional adsorption cooling system, shown in Fig. 2.6, comprises a condenser, an evaporator and a pair of sorption beds (adsorber and desorber based on operating mode). The sorption reactors are packed with the adsorbent material that adsorbs or desorbs the adsorbate. There are interconnecting valves to control the flow of the refrigerant. Adsorption is an exothermic process and, therefore, requires

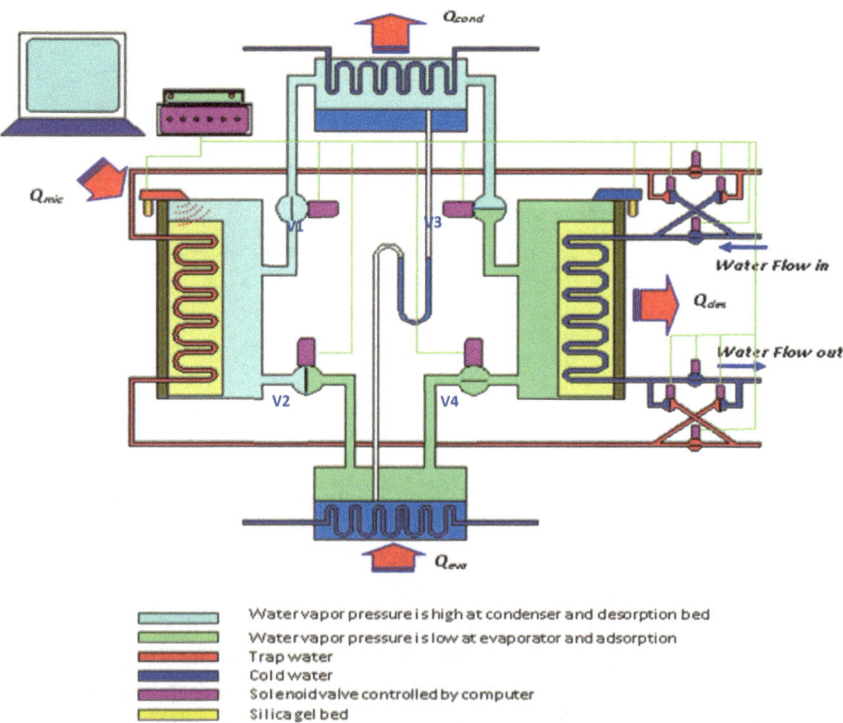

Fig. 2.6 Schematic diagram on the working mechanism of a two-bed adsorption chiller

external cooling while desorption is an endothermic process requiring external heating. In order to achieve continuous cooling, two sorption beds operate in parallel but under different desorption and adsorption modes. In the first desorption process, the adsorbent such as silica gel is heated by hot water to desorb the adsorbate (water vapour) from the adsorbent while valve 1 between the desorber bed and the condenser is opened and valve 2 between the desorber and the evaporator is closed. The generated adsorbate vapour enters the condenser and condenses by rejecting heat to the surrounding. The condensate, comprising the adsorbent solution, then flows down to the evaporator through the U-tube. Concurrently, the water cools the adsorbent in the absorber bed so that it has high affinity to absorb the adsorbate vapour while Valve 4 between the adsorber bed and the evaporator is opened and Valve 3 between the adsorber bed and the condenser is closed. Hence, the adsorbent solution from the evaporator is absorbed by the adsorbent from the absorber bed. This process causes the evaporation process in the evaporator by taking the heat from the surrounding. Chilled water is subsequently generated. The end of this cycle is seamlessly followed by the beginning of the next cycle. Each bed alternates between its roles as an adsorber and a desorber in the bath-operated cycle. The switching process, when both beds are isolated from the evaporator and condenser, takes place before alternating. This chain of events completes one adsorption cycle of the chiller. The temperature profiles of each unit in a two bath-operated cycle of adsorption chiller are illustrated in Fig. 2.7.

Most recently, at the National University of Singapore, Department of Mechanical Engineering, a silica gel-water, four-bed, two-evaporator adsorption chiller has designed, commissioned and studied [14]. Figure 2.8 portrays a schematic diagram of the four-bed, two-evaporator adsorption chiller, while Fig. 2.9 presents the picto-

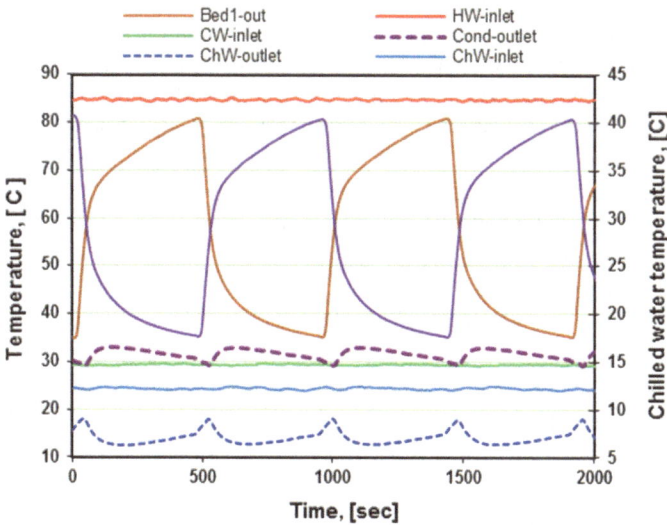

Fig. 2.7 Cyclic variation of the condenser, hot water and chilled water temperatures during the adsorption chiller operation

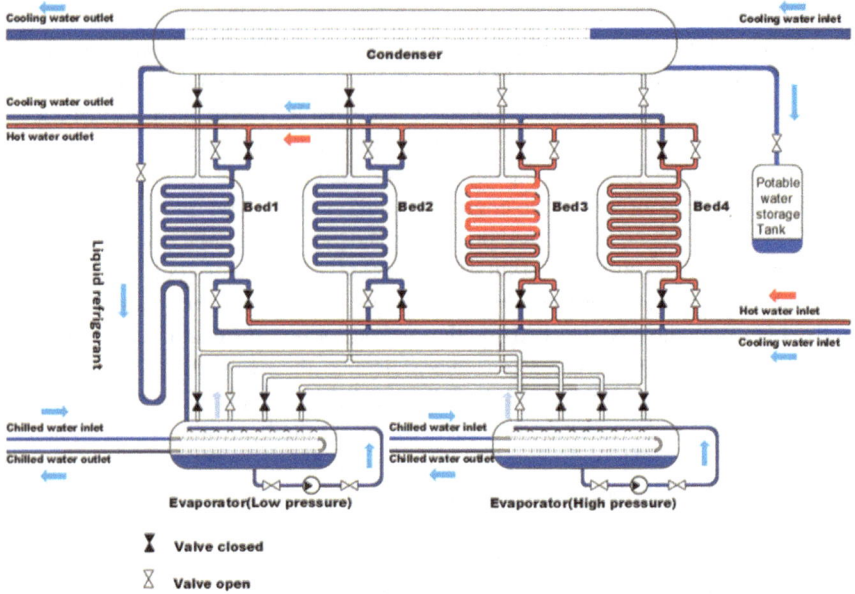

Fig. 2.8 A schematic diagram of the four-bed, two-evaporator adsorption chiller

rial perspective. The input hot water temperature ranges from 60 to 85 °C while the cooling water temperature in the condenser is around 28.1 °C and the pressure is about 4 kPa. Two evaporators, one condenser and four beds constitute the key components of the chiller. The low-pressure evaporator is employed for cooling while the high-pressure one conducts simultaneous thermal desalination and cooling. A system with two evaporators is able to produce two kinds of chilled water temperatures for different cooling requirement temperatures. During the adsorption process, the high-pressure evaporator provides the vapour for beds 2 and 4, while the low-pressure evaporator provides vapour for beds 1 and 3. As far as the high-pressure evaporator is concerned, wastewater, brackish water or seawater needs to be constantly supplied in order for the high-pressure evaporator to conduct thermal distillation so that potable water is continuously produced.

A hybrid adsorption system has been proposed to promote better system's COP. The integration of an adsorption with a thermoelectric cooling cycle evolves an electro-adsorption chiller (EAC) which is a novel technology that can markedly increase the system's COP [15, 16]. The heating and cooling processes during both desorption and adsorption are accomplished via the use of thermoelectric cooler. The block diagram for the theoretical COP calculation of EAC is shown in Fig. 2.10.

$$\text{COP}_{\text{NET}} = \frac{Q_{\text{EVAP}}}{P_{\text{IN,TE}}} = \text{COP}_{\text{ADS}}(1 + \text{COP}_{\text{TE}}) \tag{2.1}$$

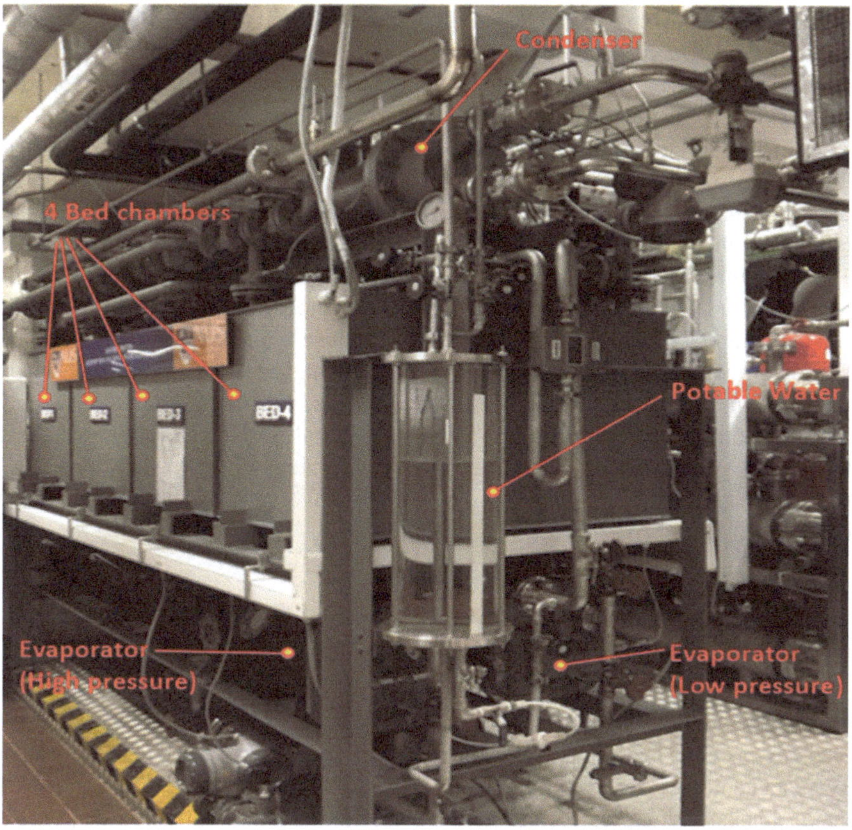

Fig. 2.9   A pictorial view of the four-bed, two-evaporator adsorption chiller with producing potable water

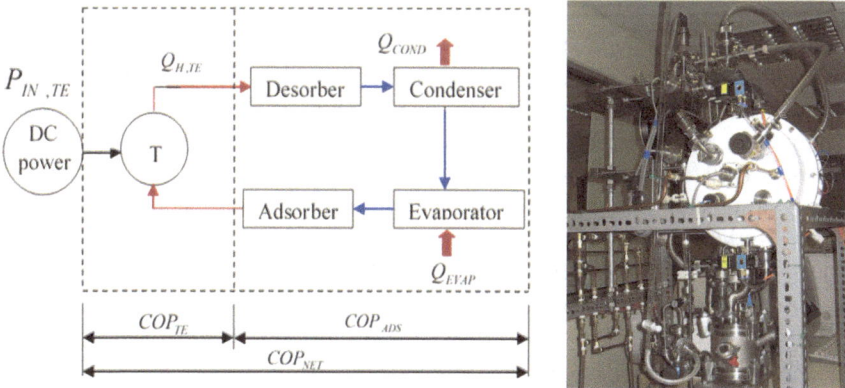

Fig. 2.10   a Experimental set up of the hybrid thermoelectric cooler integrated adsorption chiller; and **b** a block diagram that highlights its coefficient of performance of this hybrid system [35]

Having less moving parts while providing excellent heat recovery, the thermoelectric cooler enables the adsorption system to realize better COP during the early part of the EAC's switching interval. The maximum theoretical COP of EACs is found to be 1.105 while practically one can achieve nearly 0.86 at prototype design stage. The experimental set up of EAC is shown in Fig. 2.9. The key motivation of this hybrid system is to miniaturize the system for electronic cooling application [16]. Albeit the fact that EAC is capable of promoting COP, further research needs to be conducted to optimize the trade-off between performance and better miniaturization.

Key features/advantages of thermally driven adsorption and absorption cooling systems include marginal environmental impact on global warming, ability to utilize low-temperature thermal energy sources and reduced consumption of electricity. The major disadvantages associated with them are higher initial costs, larger physical footprint area especially for adsorption systems, the need to maintain vacuum for operation and the potential of absorbent crystallization. Nevertheless, these heat-driven systems key play roles in recovering energy from waste heat as well as harnessing renewable energy to provide cooling for domestic and industrial applications.

2.3 Dew-Point Evaporative Cooling Systems

A new revolutionary approach for sensible air cooling has evolved in recent years to provide occupants' thermal comfort in the most sustainable manner. By sustainable, we refer to a highly energy-efficient process without the use of energy-intensive compressors nor with the use of environmentally unfriendly CFC (chlorofluorocarbon)/HCFC (hydrochlorofluorocarbons).

The proposed "disruptive" technology rides on the wave of new fundamental developments in evaporative cooling theories to develop state-of-the-art direct/indirect evaporative heat and mass transfer (HMX) for highly effective sensible cooling. It is noteworthy that the invention of this technology was based on a fundamental breakthrough cycle envisioned by Dr. Maisotsenko [17]. The technology was, therefore, adeptly known as the M-cycle cooling technology. The highly novel technology was evolved by pushing the boundary of engineering science and innovating analysis of a thermodynamic cycle. Fundamentally, the M-cycle has the capability to cool air to below wet-bulb while approaching dew-point temperature [18, 19].

The working mechanism of the M-cycle HMX is essentially documented as follows [17]. The creative aspect of the dew-point evaporative cooling involves the pre-cooling of the working air before it enters the wet channels to pick up the moisture from water evaporation. This process reduces the inlet wet-bulb temperature of the working air, resulting in a lower product air temperature through heat transfer. Owing to its pre-cooling effect, the working air on the wet side (working air wet channel) has a much lower temperature and therefore, is able to absorb more heat from its two adjacent sides. As a result, the cooling (wet-bulb) effectiveness of the

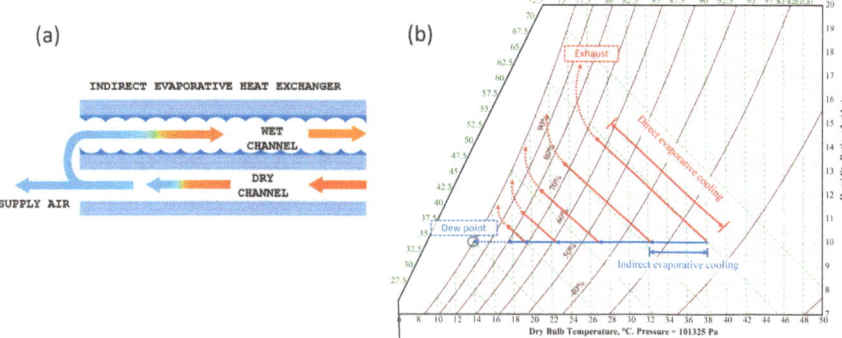

Fig. 2.11 a Air-flow configuration of the dew-point HMX; and **b** the airflow characteristics on the psychrometric chart after undergoing multi-stage dew-point evaporative cooling

new structure is higher than that of the traditional cross-flow exchanger. The interaction between air and water in channels is shown in Fig. 2.11a while the cooling process is depicted on a psychometric chart in Fig. 2.11b.

The process of dew-point evaporative cooling allows the cooling of the air to below its wet bulb (WB) temperature, towards the dew-point (DP) temperature. With a counter-flow cooler configuration, depicted in Fig. 2.12, process cooling is achieved by branching the working air from the supply air stream. The working air is directed into the wet channel to affect water evaporation and thereby exchange heat with the incoming supply air. The rest of the supply air is purged out as the product air, for cooling purposes.

The cross-flow HMX, shown in Fig. 2.13, was developed and commercialized by Coolerado Corporation™ for its low-pressure drop and ease of manufacturing [20]. The cooling performance of the product was evaluated under different weather conditions [19]. Testing was conducted to evaluate the performance of an early-2005 model Coolerado Cooler™ indirect evaporative cooling unit designed for the residential and small commercial market. A testing schedule was developed based on ASHRAE test standards for evaporative coolers, which focused on the arrangement of the test apparatus and determining the supply airflow [19]. Compared with conventional evaporative coolers (both direct and indirect), the commercial system is able

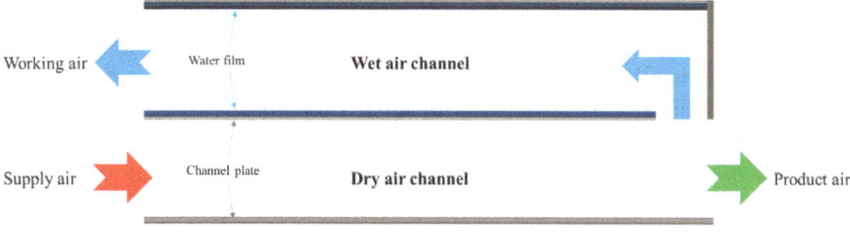

Fig. 2.12 The schematic diagram of a counter-flow dew-point evaporative cooler

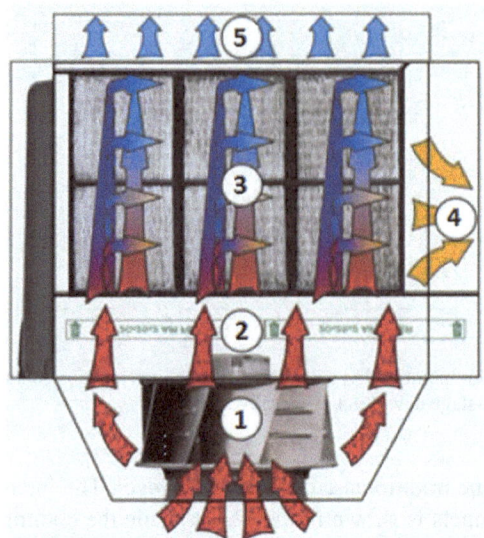

HOW IT WORKS

1. 100% Fresh air enters the system;

2. The air is filtered of dust/allergens;

3. Working air removes heat;

4. Heat and moisture is exhausted from the systems;

5. Cool product air enters the building with no added humidity;

Fig. 2.13 The working process of a commercial cross-flow HMX that is commercialized by Coolerado CorporationTM [39]

to produce air conditions that would keep a space within the ASHRAE comfort zone over a wider range of outdoor conditions. Quantitative results documented the wet-bulb effectiveness over the test conditions ranging from 81 to 91% with an average of 86% [38]. In another test evaluation of a more recent model Coolerado H80 to replace an old 7-ton vapour compression cooler at the University House in Davis, CA, showed Coolerado's energy efficiency to be markedly greater even though it has to operate with a significantly larger volume of ventilation air [20]. Most importantly, this new model has enabled energy reduction of up to 25% during peak hours operating.

The interest concerning dew-point HMX (heat and mass exchange) indirect evaporative cooler from the scientific community has increased dramatically over the past few years through both simulation modelling as well as essential experiments to demonstrate its efficacy under varying conditions. Figure 2.14 illustrates two different system configurations, namely, cross-flow and counter-flow configurations while Fig. 2.15 portrays a more detailed schematic of the counter-flow dew-point evaporative cooling. The ensuing sessions will focus on some recent numerical and experimental studies that highlight the potential of the dew-point evaporative cooling systems for varying air-conditioning applications.

Different mathematical models have been developed to predict the steady-state performance of dew-point evaporative cooling systems. Riangvilaikul and Kumar [21] performed a numerical study on their previous experimental setup, and their model predicted the product air temperature within 5% discrepancies. Zhan et al.

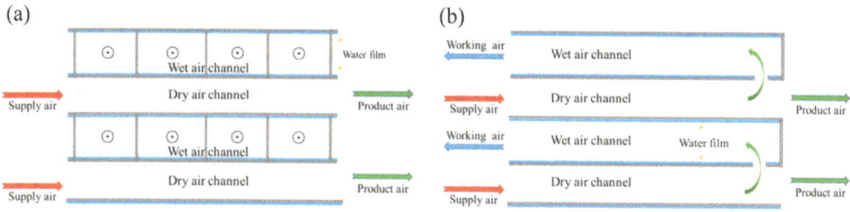

Fig. 2.14 Different system configurations of dew-point evaporative cooling: **a** cross-flow; and **b** counter-flow

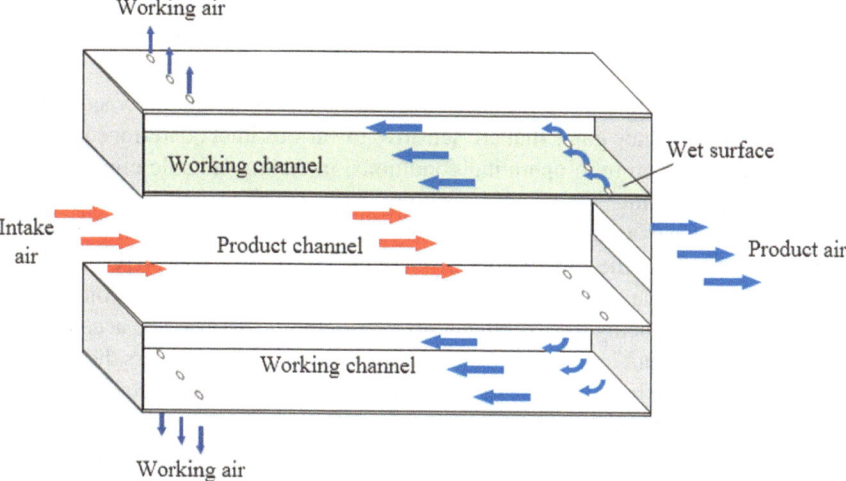

Fig. 2.15 A more detailed schematic of the counter-flow dew-point evaporative cooling

[22] carried out a numerical study, comparing the performance of M-cycle cross-flow and counter-flow configurations. Their model was validated with three sets of experimental data within an acceptable error range of 2–10%. Hasan [23, 24] proposed both numerical and analytical models to study the sub-wet-bulb indirect evaporative cooling. The performance of four different cooler configurations was studied, i.e. single-stage and two-stage counter-flow, two-stage parallel-flow and combined parallel-regenerative evaporative coolers. It showed that their respective wet-bulb effectiveness was 1.16, 1.26, 1.09 and 1.31 at inlet air conditions that are less than 30 °C and 34% RH. Heidarinejad and Moshari [25] presented a novel mathematical model for cross-flow indirect evaporative cooler. The longitudinal heat conduction along the wall and spray water temperature variation were considered. Their model is applicable to direct, indirect and regenerative evaporative coolers with high precision.

Anisimov et al. [26, 27] and Pandelidis et al. [28] used a modified ε-NTU method to study the M-cycle HMX. Their numerical models, validated with experimental data, employed a modified ε-NTU method to perform thermal calculations of the indirect evaporative cooling process and eventually quantifying the overall performance of the heat exchangers. Key results from their simulation showed that high-efficiency gains are sensitive to various inlet conditions, and allow for estimation of optimum operating conditions, including suitable climatic zones for the proposed units. Most importantly, results also highlighted that cross-flow heat and mass exchanger achieves the highest cooling capacity while the counter-flow configuration produced the lowest outlet temperatures. Anisimov et al. [29], in another work, describe numerical modelling of heat and mass transfer in five different exchangers utilizing the Maisotsenko cycle for indirect evaporative cooling. Again, the modified ε-NTU method was relied on to perform thermal calculations of the indirect evaporative cooling process, thus quantifying the overall performance of considered heat exchangers. Numerical simulation reveals many unique features of considered devices, enabling an accurate prediction of their performance. The results from computer simulations showed high-efficiency gains that are sensitive to various inlet conditions and allow for estimation of optimum operating conditions, including suitable climatic zones for their proposed unit.

Chen et al. [30] examined a hybrid cooling system, an indirect evaporative cooling (IEC) unit to enable the pre-cooling of incoming fresh air in an AHU. The main objective is to shave off some of the cooling load on the main air-conditioning system, thereby reducing the overall energy consumption. To evaluate accurately its energy-saving potential, they establish a short-cut model that considers the potential of condensation to predict the annual performance of the hybrid cooling system considering condensation in IEC unit. The model was subsequently validated with published data from literatures. Incorporating with a system model developed using TRNSYS, the performance of the proposed IEC hybrid cooling system operating in a Hong Kong's wet market is evaluated and the amount of energy saving compared with conventional systems is judiciously quantified.

The performance of a typical evaporative cooling system is largely dependent on the structure and design of the heat and mass exchanger. Jardi and Riffat [31] conducted a numerical analysis for a modified dew-point cooling system based on a proposed psychrometric energy core employing a cross-flow heat and mass exchanger as the key component for air-conditioning applications. Key results from their numerical study pointed out that with an intake air of 30 °C temperature and 50% relative humidity and a working-to-intake airflow ratio of 0.33, the evaporative cooling system is capable of achieving a wet-bulb effectiveness of 112% and a dew-point effectiveness of 78% with 5 mm channel height and 500 mm channel length.

Besides numerical studies to verify the performance of dew-point evaporative coolers, experimental studies have also been conducted to quantify their physical performance based on different supply air conditions. For example, Hsu et al. [32] compared the cooling effectiveness of the dew-point evaporative cooler with that of the conventional indirect evaporative cooler. They reported that the maximum WB effectiveness of a dew-point evaporative cooler approached 1.30, with either counter-flow or cross-flow configuration. Riangvilaikul et al. [33] developed a counter-flow dew-point evaporative cooler prototype with channel dimensions of 1200 mm $\times$ 80 mm $\times$ 5 mm ($L \times W \times H$). The cooler was tested under varying inlet air temperatures, humidity and velocities. The WB and DP effectiveness of the cooler spanned 0.92–1.14 and 0.58–0.84, respectively. Lee et al. [34] conducted an experimental study on a counter-flow dew-point evaporative cooler with finned air channels. The cooler was able to deliver a product air temperature of 22.0 °C when the supply air was 32 °C and 50% RH. The WB effectiveness during testing was between 1.18 and 1.22. Duan et al. [35] proposed a counter-flow dew-point evaporative cooler with triangular air channels. The cooler was fabricated and tested under various operational conditions. The researchers concluded that the effectiveness and efficiency were markedly influenced by the inlet wet-bulb depression, inlet air velocity and working air ratio. The corresponding WB effectiveness and COP were in the ranges of 0.55–1.06 and 2.8–15.5, respectively. Bruno [36] performed a long-term test of the counter-flow dew-point evaporative cooler for both commercial and residential applications. The product air temperature from the cooler was as low as 10.2 °C, with a cooling capacity of above 10 kW. The average COP in the 5-year test was between 7.0 and 8.0, leading to an energy saving of more than 50% compared with a packaged unit air conditioner. Woods et al. [37] proposed a desiccant-enhanced evaporative (DEVAP) air conditioner, which combines air dehumidification and evaporative cooling to improve the capability of handling humid air. Individual tests were carried out for their hybrid liquid desiccant dehumidifier and indirect evaporative cooler. The results showed that the dehumidification process almost reached its adiabatic potential and the dew-point effectiveness for air cooling was above 0.8 in the presented tests. Gao et al. [38] also carried out an experimental investigation on liquid desiccant indirect evaporative cooling. The energy balance in their dehumidifier and M-cycle cooler was within $\pm 20\%$, while the impacts of key parameters, namely, inlet airflow rate, temperature and humidity on the cooler's performance were examined. Moisture and temperature reduction, as well as sensible heat ratio, were used to evaluate the dehumidification and cooling ability of the hybrid system.

To sum it up for this section on dew-point evaporative coolers, existing studies have demonstrated that the counter-flow dew-point evaporative cooler is able to achieve wet-bulb effectiveness of above 1.20. The power consumption of a practical cooling system can be as low as 20% of a conventional air conditioner. Also, the performance of the dew-point evaporative cooler is strongly influenced by its operating and geometric parameters, as well as the ambient air conditions.

2.4 Solid-Based Desiccant Dehumidification

Solid-based desiccant dehumidification has gained increasing interest as a sustainable cooling technology, especially in high humid regions. Without the use of energy-intensive compressors, it can achieve substantial energy saving when integrated with an appropriate sensible cooler such as a direct/indirect evaporative cooler. With such an integrated system, there is no need for the temperature of process air to drop below

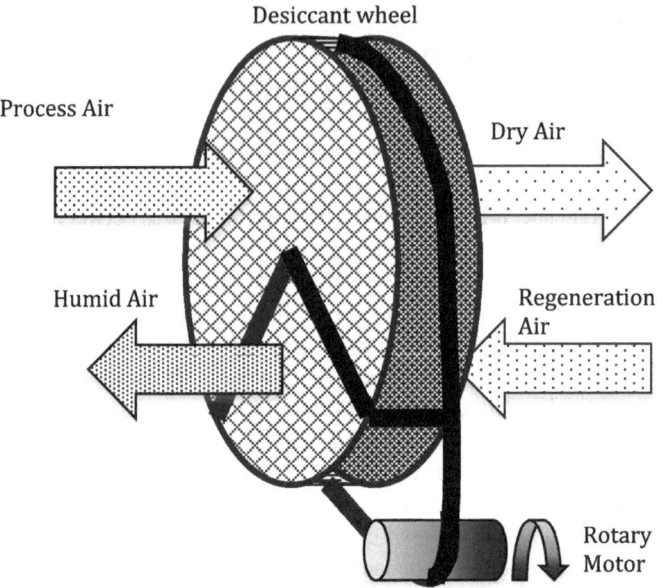

Process Air

Dry Air

Humid Air

Regeneration Air

Rotary Motor

Fig. 2.16 The basic operating principle of rotary desiccant dehumidifier for moisture absorption

its dew-point temperature in order to remove water vapour (latent cooling load) from the air. A typical desiccant wheel mechanical system is shown in Fig. 2.16. The air to be dehumidified enters the system, comes into contact with the desiccant wheel, and exits the dehumidifier hot and dry. The wheel is then rotated so that the desiccant portion that has picked up moisture is exposed to the hot reactivation air so that its moisture can be removed. The wheel rotation mechanism is repeated once moisture regeneration is completed.

Recently, studies have been focused on the development of desiccant materials to improve adsorption capacity and thus the performance of the solid-based desiccant dehumidification system. Commonly used solid desiccants include silica gel, activated carbon and molecular sieves such as zeolites [39, 40]. Silica gel has relatively high moisture adsorption capacity and low regenerative temperature (50–90 °C) while activated carbon and alumina require high regeneration temperature (250 °C) while providing reduced adsorption capacity. Molecular sieves such as zeolites enable the highest adsorption capacity due to their large pore size and highly polar surface [41].

Composite desiccants, made by impregnating hygroscopic substance in the pores of solid adsorbents, have also been found to be effective in increasing the moisture adsorption capacity [42, 43]. Dehumidifying air with a relative humidity of 84%, activated carbon impregnated with lithium chloride is able to absorb moisture as much as 194% of its mass when compared with the one with calcium chloride where its moisture adsorption has been observed to be 170% of its mass. Experiments have been conducted for composite desiccants consisting of calcium chloride being contained in the pores of silica gel [44–46]. The moisture adsorption capacity of this composite desiccant is better than that of pure silica gel. Combination of silica gel and lithium chloride has been experimentally studied [47, 48]. Similar to calcium chloride, the lithium chloride content in composite desiccants plays a significant role in enhancing moisture adsorption capacity. It is commonplace knowledge that lithium chloride is corrosive in nature, and this halts the progress of lithium chloride as a liquid desiccant. However, by impregnating it in the silica gel, the corrosive effect is kept to a minimum. In addition, the incorporating of lithium chloride to silica gel affects moisture regeneration. The regenerative temperature of this desiccant is found to be in the range of 60–100 °C. Accordingly, low-grade heat is sufficient for moisture regeneration.

Despite the advancement made on the desiccant rotary wheel and novel composite desiccant materials, the removal of adsorption heat generated during dehumidification remains a hurdle in improving solid-based desiccant dehumidification [49]. The adsorption heat attenuates the adsorption capacity of the desiccant; presenting a deviation from the ideal isothermal process. Greater irreversibility losses translating to higher entropy that eventually impacts on the system COP is anticipated. Some researchers [50–54] have proposed the concept of adsorbent-coated cross-cooled heat exchanger to promote air dehumidification as portrayed in Fig. 2.17. Fin tube heat exchangers are predominantly used in DCHEs. The fins are made up of thin Aluminium (Al) sheets while the tubes are made up of Copper (Cu). Figure 2.18a shows an uncoated heat exchanger, and Fig. 2.18b shows the result of coating the heat exchanger with silica gel [55]. Desiccants coated on the fins of these heat exchangers are capable of instantly adsorbing the water vapour molecules from the moist air. When these gaseous molecules are adsorbed, they release their kinetic energy into the desiccants as heat of adsorption, which results in reduced adsorption capacity and higher air temperature [56, 57]. By supplying the cooling fluid in the tubes of heat

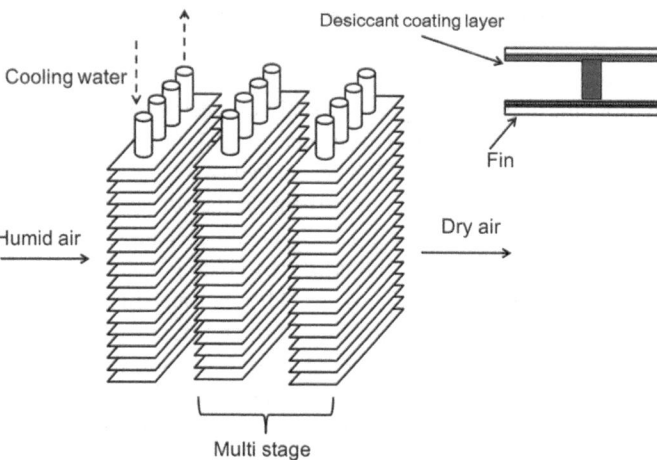

Fig. 2.17 The concept of an adsorbent-coated heat exchanger to realize air dehumidiciation

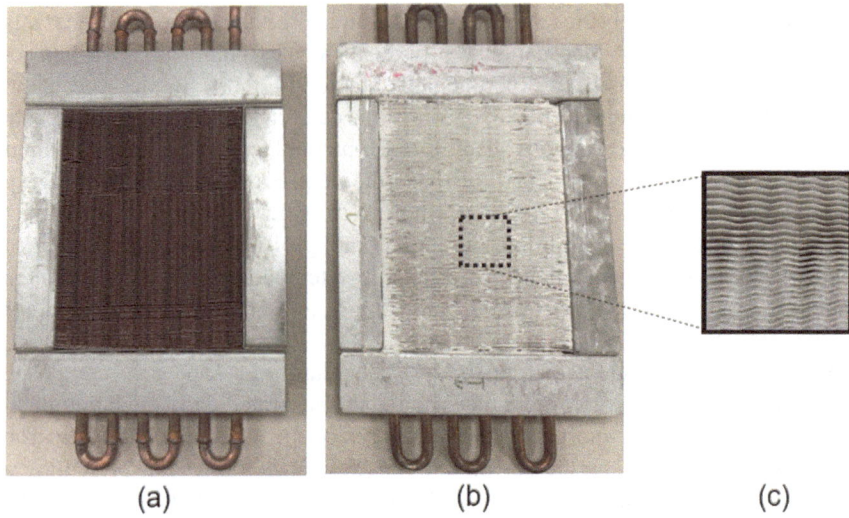

Fig. 2.18 Pictorial perspective of a single finned-tube heat exchanger DCHE system

exchanger, the released heat of adsorption is effectively taken away by the cooling fluid. Low manufacturing costs and simplicity of its structure serve as an added advantage to the production of DCHEs in bulk.

As the dehumidification performance of the heat exchanger depends largely on the sorption properties of the coated desiccant material, the selection of desiccant is carried out based on its adsorption capacity, and the desired regeneration temperature. Determination of the surface area characteristics of the desiccant is also important in analysing its adsorption capacity. Moreover, to attain uniform adhesion on the fin surface, high-quality binders must be incorporated. An exhaustive review of various solid desiccants used for adsorption cooling technology has been carried out by Zheng et al. [58].

Silica gels and zeolites have been widely used as desiccants because of their excellent hydrophilic nature and higher adsorption capacities. With the advancements in material science, many composite desiccants, bio-desiccants and polymers have been used in the DCHEs. Li et al. [59] have studied two types of silica gel powders: Type 3A and type RD powder by coating them on the surface of a heat exchanger, and compared their adsorption capacities. On the other hand, Zhao et al. [56] have investigated the performance of a dehumidifier by coating the heat exchangers using column chromatography silica gel particles. Mesoporous silica gel particles were preferred because of their high porosity, low cost, and stability [57].

Recently, one of the easier ways to enrich the water sorption characteristics of desiccant materials is by impregnating hygroscopic salts such as Lithium Chloride (LiCl) and Calcium Chloride ($CaCl_2$) in traditional desiccant materials such as silica gels and activated carbons [60]. Jiang et al. have infused LiCl salt solution in porous silica gels under different mass fractions [61]. Hu et al. have developed composite silica gel-based desiccants by impregnating LiCl and $CaCl_2$ solutions. Their findings indicate that both composite desiccants displayed an increase in sorption capacity by 30–45% [62]. Comparatively, LiCl impregnated silica gels had larger water uptakes when compared with the $CaCl_2$-based ones. However, the use of such liquid desiccants has also resulted in spill-over phenomenon. Suitable precautions have to be taken into consideration in order to preserve the uniform coating of desiccants on the metal surfaces [63]. Zheng et al. have impregnated activated carbon (AC) and activated carbon fibre (ACF) with LiCl and studied their textural properties [64]. It was observed that the surface area and pore volume of these composite desiccants decreased after the impregnation of LiCl salt particles. Equilibrium water sorption isotherms indicated that the composite carbons desiccants had shown three to four times rise in water sorption capacity over the pure silica gel desiccants. Between the two carbon-based desiccants, ACF had the largest water uptake.

Albeit the adsorption capacity of desiccants being improved via the introduction of hygroscopic salts, the presence of chlorides induced corrosion on the fins of heat exchangers. To resolve this issue, polystyrene cation exchange resin (IER) was applied to the adsorption dehumidification process. Fang et al. have modified the IER by replacing Sodium (Na) ions of IER with the cations of hygroscopic salts such as LiCl and $MgCl_2$ [65]. The modification of IER resin was carried out to enhance the dehumidification performance. Experiments were conducted to compare the adsorption performance of IER and modified-IER (MIER). Results highlighted an increase in the adsorption capacity of Li-modified and Mg-modified coating by 83.64% and 73.40%, respectively.

Experimental systems, designed and built to analyze the dehumidification performance of DCHEs, can be classified into two types depending on whether one or two exchangers are employed for the analysis. Single heat exchanger systems comprise a DCHE, one or two heating and cooling water sources, and many air ducts. An axial flow fan is installed at the outlet of heat exchanger to drive the air-flow. To simulate different indoor/outdoor temperature and humidity conditions for testing, a humidity generator is also used in some setups. A schematic diagram of the single heat exchanger testing system is shown in Fig. 2.19 [61].

One of the major disadvantages of the single heat exchanger system is that the dehumidification and regeneration processes are not continuous. Therefore, the analysis of both dehumidification and regeneration is not simultaneous. To address this difficulty, Zhao et al. have employed two heat exchangers, which help in achieving continuous dehumidification [56]. The schematic diagram of a two parallel DCHE system used or testing dehumidification performance is portrayed in Fig. 2.20.

To further support the possibility of using DCHEs in winter, T. S. Ge et al. [66] carried out a performance analysis of an air-conditioning system. Figure 2.21 describes the operating principle of DCHE air-conditioning system in winter. A mathematical model was developed using heat and mass transfer resistance approach. The analysis was carried out for regeneration temperatures from 20 to 40 °C. The results indicated that the average supply air cannot be used for meeting the indoor load in winter when the required regeneration temperature was 30 °C. Even though the

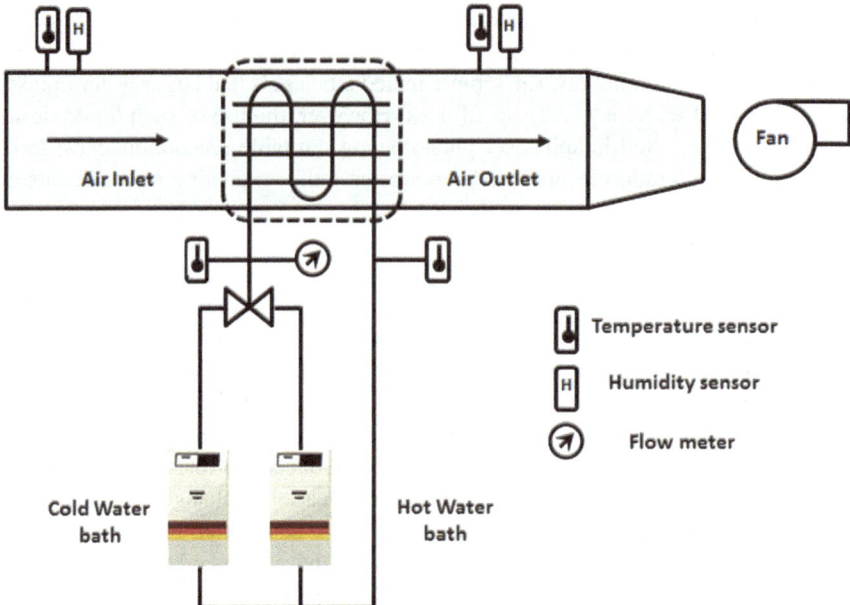

Fig. 2.19 Schematic diagram of single heat exchanger DCHE testing system with controlled conditions

humidification capacity was acceptable, the temperature of outlet air was not suitable for supplying into indoor space. Hence, a higher regeneration temperature was recommended. However, when the switch time spanned between 70 and 240 s, the supply air could meet the thermal requirements of indoor space.

DCHE systems are showing great potential in effective air dehumidification particularly with the incorporation of solar thermal and waste heat. Accordingly, DCHEs can be regarded as potential alternatives to conventional vapour compression air conditioners.

2.5   Liquid-Based Desiccant Dehumidification

Major studies have been conducted on liquid-desiccant dehumidification systems, technological and theoretical developments, configurations of related systems, and experimental and analytical studies to optimize the system performance are summarized [67–70]. Salam and Simonson reported that the use of a solution-to-solution heat exchanger with an effectiveness of 0.9 improves the total COP by around 23% while decreasing the capacities of solution heating and cooling equipment by 26% and 21%, respectively [67]. Mei and Dai [71] demonstrated that the mixture of lithium

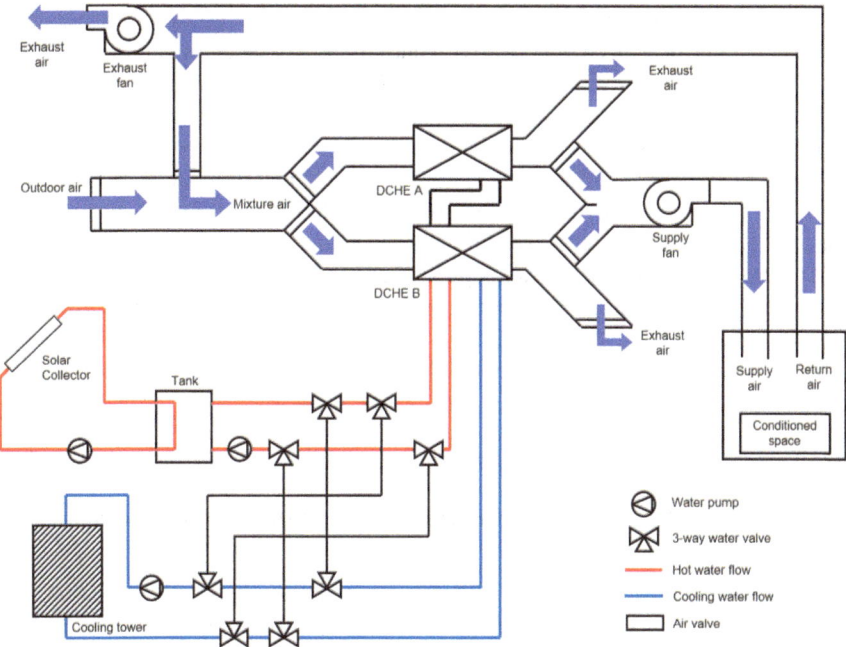

Fig. 2.20 Schematic diagram of two heat exchanger DCHE system for continuous dehumidification (adapted from reference [75])

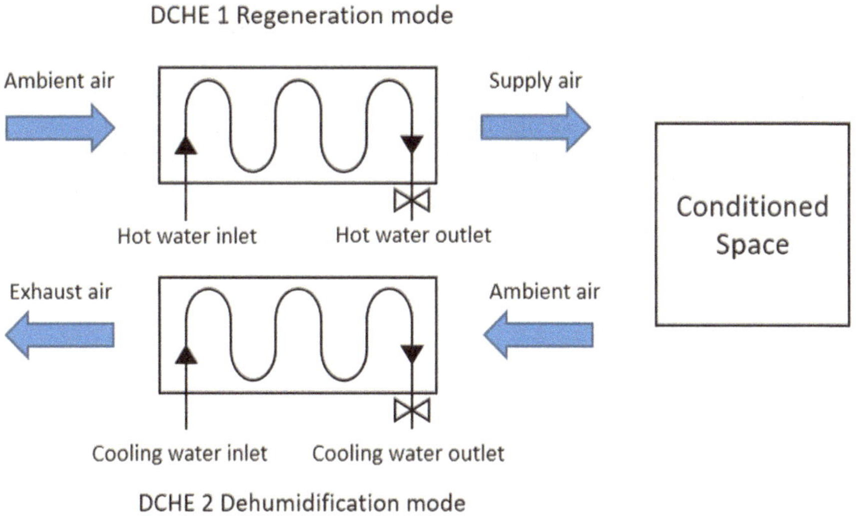

DCHE 1 Regeneration mode

Fig. 2.21 Schematic diagram explaining the operation of a DCHE air-conditioner in winter (adapted from reference [85])

chloride of 99.3% purity and calcium chloride of 90% purity resulted in higher dehumidification performance compared with pure lithium or calcium chloride solutions which is attributed to lower viscosity and surface vapour pressure.

Fundamental heat and mass transfer processes in the dehumidifier and regenerator of dehumidification systems have been intensively investigated both experimentally and numerically [71–76]. Conde [77] developed calculation models for computing the thermophysical properties of aqueous lithium chloride (LiCl) and calcium chloride ($CaCl_2$) solutions of mass fractions ranging from 0 to 0.6 and temperatures from -40 to 100 °C. The dehumidification performance of triethylene glycol (TEG), LiCl, LiBr and $CaCl_2$ as desiccant materials has also been studied comprehensively [78, 79]. Jain and Bansal [78] reported that a maximum dehumidification effectiveness in the range of 0.8–0.95 can be achieved for the inlet air temperature of 15–30 °C and the ratio of desiccant solution to airflow rate of 0–16.

2.5.1   Adiabatic Liquid-Desiccant Dehumidification Systems

Adiabatic liquid-desiccant dehumidification systems comprised of structured or randomly packed adiabatic bed dehumidifier and regenerator as shown in Fig. 2.22. Concentrated solution of liquid-desiccant (also called strong solution) is cooled and sprayed from the top of the adiabatic bed of dehumidifier. Distribution of the wall flow and channel flow on the structured and random packing materials (pall ring, rosette ring, ladder ring, etc.) and their corresponding impacts on the heat and mass transfer performance have been analysed. The effects of different design and operating parameters such as the temperature and flow rate of the solution, the concentration of solution, the height of the packed bed and the wetted area were studied for three flow patterns namely parallel flow, counter flow and cross flow [80].

Zhang et al. [74] experimentally investigated the mass-transfer characteristics of structured adiabatic packing dehumidifier and regenerator. Experimental results

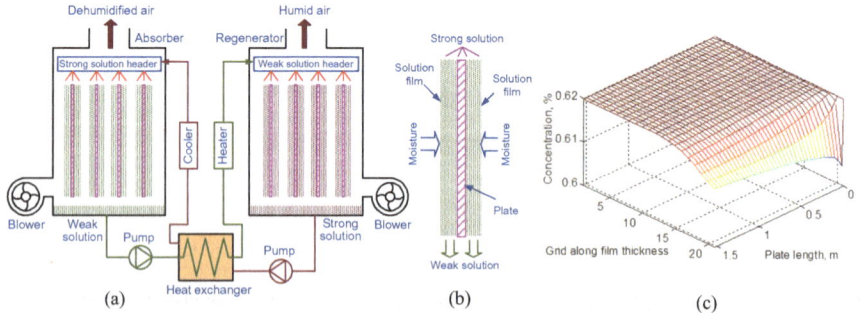

Fig. 2.22 a Schematic diagram of an adiabatic liquid-desiccant dehumidification system, **b** absorption of moisture in the adiabatic film and **c** variation of desiccant concentration across the adiabatic film of liquid-desiccant

showed that when the air velocity progresses from 0.5 to 1.5 m/s, the overall mass-transfer coefficient in the dehumidifier and regenerator increases from 4.0 to 8.5 g/m^2 s and from 2.0 to 4.5 g/m^2 s, respectively. Liu et al. [75] developed analytical solutions of heat and mass transfer between air and liquid desiccant in a cross-flow packed bed dehumidifier. The model was validated with experimental data for both enthalpy efficiency and moisture efficiency within an acceptable error range of 6.3% and 5.6%, respectively. The analytical solutions can be used in the optimization of the cross-flow dehumidifier. Wang et al. [81] experimentally studied the dehumidification performance of a counter flow liquid-desiccant dehumidifier using polypropylene gauze, CELdek and CELdek 5090 with a high specific surface area of 650 m^2/m^3 as the packing materials of height 0.3–0.5 m. Comparatively, CELdek 5090 showed better dehumidification performance.

2.5.2 Internally Cooled Liquid-Desiccant Dehumidification Systems

Jain and Bansal [78] illustrated that the dehumidification effectiveness of an internally cooled dehumidifier could be greater than 1. This is primarily due to a fast solution cooling using the cooling water inside the absorber which decreases its vapour pressure to a value lower than at the inlet. The moisture absorption process of an internally cooled liquid-desiccant dehumidification system and the changing desiccant concentration are shown in Fig. 2.23. Moisture absorption rate for internally cooled dehumidifiers improves when compared with adiabatic dehumidifiers. The attributing factor is the heat of absorption transfer to the coolant in order to maintain the temperature of the film at lower values. However, due to the poor mass diffusivity of water in the desiccant solution, the solution concentration change remains confined to a relatively small part of the film that is close to the exposed surface. Khan and Martinez [82] studied the performance of internally cooled dehumidifier

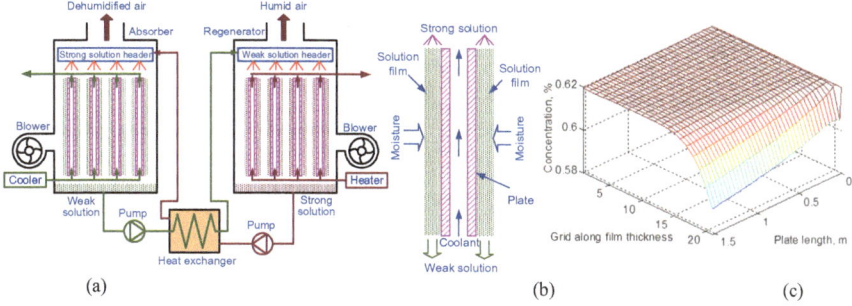

Fig. 2.23 a Schematic diagram of an internally cooled liquid-desiccant dehumidification system; **b** absorption of moisture in the internally cooled film; and **c** variation of desiccant concentration across the internally cooled film of liquid-desiccant

with LiCl as the liquid desiccant. The parameters they covered included humidity and
enthalpy effectiveness and these parameters were employed to predict the thermal
performance of the unit. They observed that both humidity and enthalpy effectiveness are greatly influenced by the number of mass transfer units. Xiong et al. [83]
observed an increase in the system COP from 0.23 to 0.72 by using a two-stage novel
liquid desiccant cooling system which is shown in Fig. 2.24. On the process side,
the air is first dehumidified in the first dehumidifier (1stDEH) using the solution of
CaCl$_2$ and then, in the second dehumidifier (2ndDEH) air is further dehumidified by
the solution of LiCl to the desired humidity of air. As far as capital cost is concerned,
it can be reduced by about 53% as compared to single-staged dehumidification using
LiCl because CaCl$_2$ is much cheaper than LiCl [83].

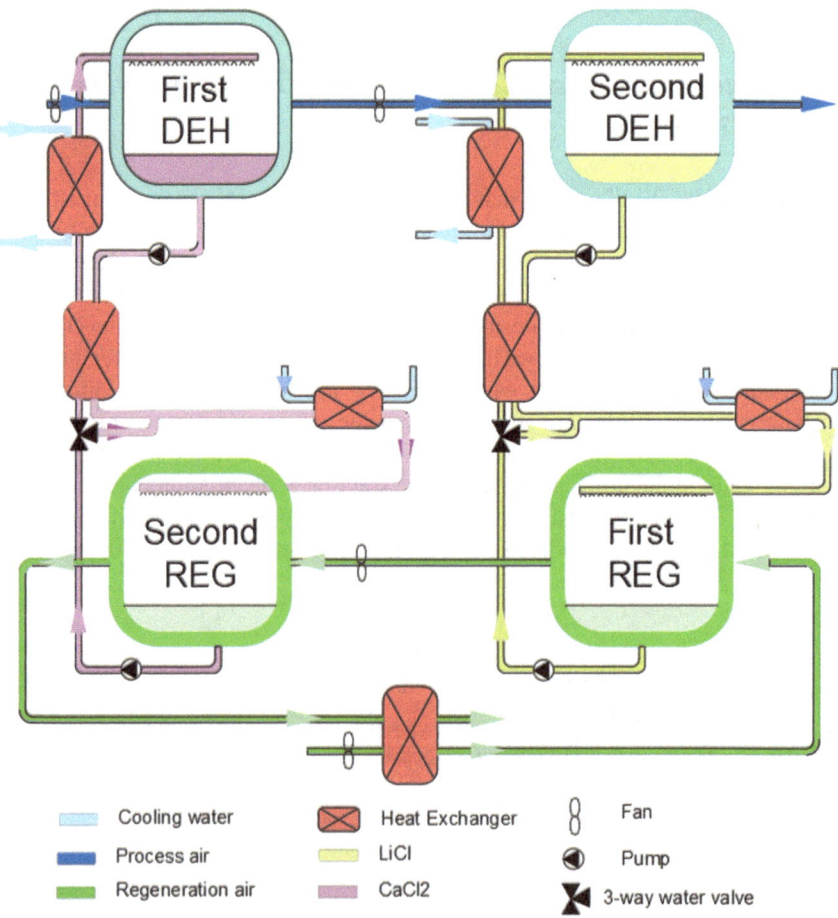

Fig. 2.24 Schematic explaining the operation of a two-stage liquid desiccant dehumidification
system

Cheng [84] developed a liquid desiccant dehumidifier using fin-plastic tube heat exchanger. The developed system has great potential for corrosion prevention and can be used as internally cooled/heated dehumidifier/regenerator. Gao et al. [85] compared the performance of two types of desiccant dehumidifiers—one internally cooled while the other is purely an adiabatic system. Their results indicated that the internally cooled dehumidifiers can significantly improve performance compared with the adiabatic dehumidifiers, especially when the solution is at a high temperature and/or low concentration in the downstream section of the liquid desiccant flow. It was also concluded that lower temperatures of the cooling water can markedly enhance dehumidification effectiveness and moisture removal rate in an internally cooled dehumidifier.

2.6 Membrane-Based Air Dehumidification

Recently, membrane-based dehumidification has gained significant research traction primarily due to its ease of applications and versatility [86–88]. In this process, the air is passed over a membrane surface at normal pressure. Vacuum pressure is applied on the opposite side of the membrane to create a driving force for water to permeate through the membrane. Such an air conditioning process is illustrated as A–B–C pathway on the psychrometric chart in Fig. 2.25. First, humid air is dehu-

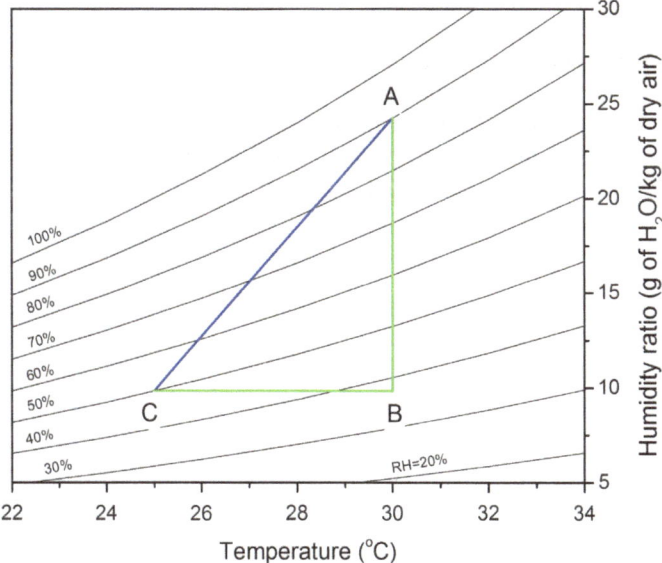

Fig. 2.25 The air-conditioning processes from input condition A (at temperature $T_A = 30\,°C$ and relative humidity $RH_A = 90\%$) to desired condition C (at temperature $T_B = 25\,°C$ and relative humidity $RH_C = 50\%$)

midified without any temperature change from A to B. Then, the dried air is cooled down to the thermal comfort level with minimal energy consumption from F to C. This isothermal dehumidification is considered to be a 'green' process because no thermal source is needed to facilitate thermal regeneration, resulting in minimal environmental emission [89].

A typical cross-flow membrane dehumidification setup is shown in Fig. 2.26a. In this setup, the humid air is passed over the feed side of a working membrane at ambient pressure. Vacuum pressure is applied on the permeate side of the membrane to create a driving force (Δp_w) for the water vapour permeation to take place through the membrane. Water vapour is selectively sieved out of the air stream, pumped and discharged to the ambient pressure by mechanical means such as a vacuum pump.

When humid air passes over the membrane, the water vapour partial pressure of the air stream (p_w) gradually lowers, as shown in Fig. 2.26b. The work required in this dehumidification process is mainly due to the electrical power to compress the water vapour from the vacuum pressure to the ambient pressure. The energy efficiency (normally expressed as coefficient of performance COP) of the dehumidification process is appropriately ratio of latent heat removed to the pump's work. It is apparent that without involving any phase change, membrane-based dehumidification potentially has higher COP than condensing water vapour by conventional processes. A dehumidification COP range of 2–3.5 has been reported for certain membranes and operation conditions [89–91].

Beside energy efficiency, the dehumidification ratio is also considered an essential parameter in evaluating the performance of a membrane system. The dehumidification ratio is expressed as the ratio of water vapour removed to the water vapour in inlet air. Recently, D. T. Bui et al. have reported that performance of a membrane system strongly depends on permeate pressure, velocity and RH of the feed air, permeability and selectivity of the working membrane [91].

Water vapour permeability and selectivity are two important mass transport characteristics that determine the performance of a membrane. Water permeability (P) is defined as the permeation rate through a unit of thickness and a unit of area of the

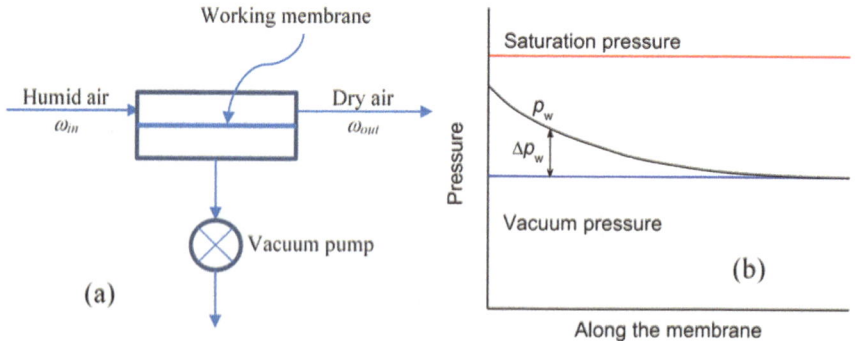

Fig. 2.26 **a** Typical membrane-based dehumidification setup; and **b** water vapour pressure profile (p_w) in air stream along an ideal working membrane

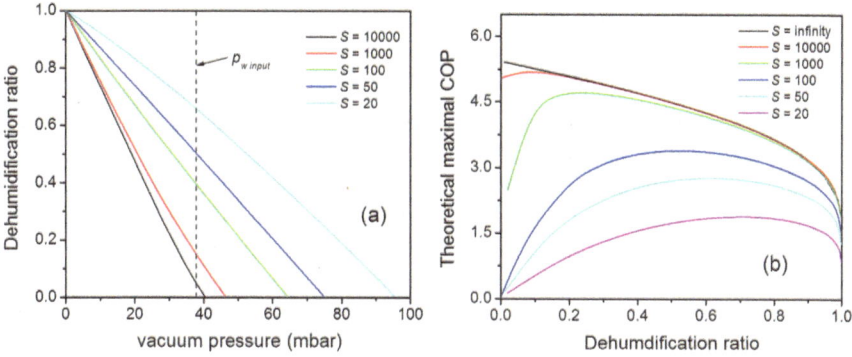

Fig. 2.27 **a** Recovery ratio versus applied vacuum pressure with different membrane selectivity; and **b** effect of membrane selectivity on the COP limit under isothermal compression (input air is at $T = 30\ °C$ and RH $= 90\%$)

membrane, under a unit transmembrane pressure. It is usually expressed in Barrer, (1 Barrer $= 10^{-10}$ cm^3 (STP) cm/cm^2 s cmHg $= 3.348.10^{-16}$ mol m/m^2 s Pa). Water vapour selectivity (S) towards another permeant is well defined as the ratio of their permeabilities.

It is apparent that membranes with high water vapour permeability are desired for dehumidification. The higher permeability a membrane has, the higher water vapour removal rate is obtained, the drier output air becomes, and the less membrane area is required. However, permeability does not affect the energy efficiency of a membrane system.

Figure 2.27a shows the effect of membrane selectivity on the dehumidification ratio of a membrane dehumidification system. Principally, the dehumidification ratio increases with lower vacuum pressure. When a low selectivity membrane is employed, there is a large portion of air in the permeate stream. Therefore, the driving force for the water permeation, which is the water vapour pressure difference between the two sides of the membrane, can be achieved with a permeate pressure even higher than the input water vapour pressure. For a specific permeate pressure applied, the membrane with a low selectivity has higher dehumidification ratio than a membrane with high selectivity. In a similar manner, Scovazzo et al. experimentally obtained high dehumidification ratio at a high permeate pressure by introducing air into the permeate stream as a sweep gas [90]

Although a lower membrane selectivity results in higher dehumidification ratio, it does not necessarily imply that higher energy efficiency will be achieved. Figure 2.27b shows the theoretical COP with assumption that the pump efficiency is 100% and the compression is isothermal process. The energy efficiency decreases with lowered membrane selectivity. It is because more energy is required to pump the permeated air when a membrane. When the membrane selectivity deviates from infinity, there exists a certain range of recovery ratio, at which COP peaks [91]. In term of energy efficiency, membranes with selectivity higher than 1000 are desirable.

2.6.1 Membrane Materials

The development of membranes for the purpose of air dehumidification closely resembles the membranes used in various dehydration applications such as dehydration of compressed air, natural gas, flue gas and pervaporation. Many types of organic and inorganic membranes have been synthesized and investigated for various industrial processes. Among them, the polymeric membranes have received close attention due to their low capital cost, lightness, physical robustness and ease of fabrication and modification. For low temperature and gentle working conditions in ventilation and air conditioning, polymeric membranes are capable of providing consistent and stable performances [92]. Figure 2.28 illustrates the fabrication of testing of a flat-sheet membrane module and the typical experimental setup employed to test and evaluate its dehumidification performance characteristics.

Mass transport in polymer materials is based on the solution diffusion mechanism. Gas molecules absorb on the membrane surface, diffuse through the membrane and then desorb at the opposite surface [93, 94]. Gas permeability through a membrane is a result of the gas dissolution (or sorption) and diffusion in the membrane material.

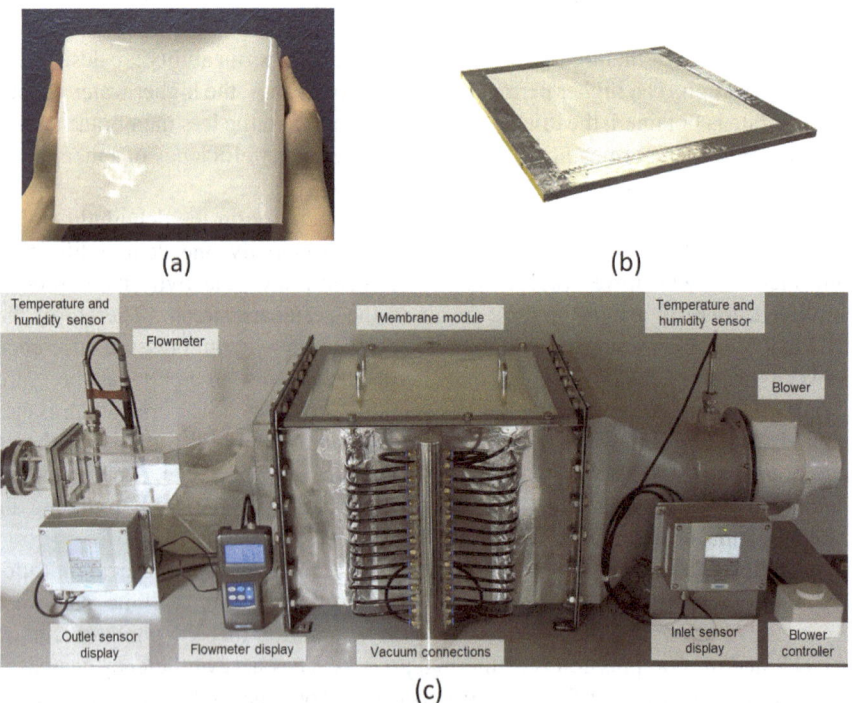

Fig. 2.28 **a** A single nanofiltration membrane sheet ready to be synthesized to a membrane dehumidification module; **b** a coated membrane frame; and **c** a membrane performance testing module setup

The molecular interaction between membrane material and permeant determines the permeation rate. Separation takes place at varying degrees depending on the components are that dissolved in and diffuse through the polymer.

Majority of polymers having high water vapour permeability and selectivity are hydrophilic materials. Water molecule preferentially permeates through a polymer membrane due to its smaller kinetic diameter and higher condensability than the other gases. It is noteworthy that the diffusivity of water greatly depends on its activity in air and its content in a polymer matrix. Higher water activity in air can result in higher amount of adsorbed water in polymer matric and higher water permeability due to the plasticization. Highest increase of 42 times was reported for PVA membrane when its water vapour activity was increased from 0.5 to 0.9 [93]. Although polymers have high permeability and selectivity, they are rarely used as single-layer membranes in vacuum-based membrane dehumidification because of their low mechanical strength. Thus far, the membranes used in air dehumidification are constructed based on a composite structure. Composite membranes are multi-layer membranes comprising support layers and an active layer. The support layers provide high mechanical strength and chemical stability to the membrane. They are usually made of mesoporous or/and microporous materials. The active layer is a thin permselective layer determining permeability and selectivity of the composite membrane. A desired composite membrane comprises of a highly microporous substrate and a thin, dense selective layer. Using the same material, permeation increases with the decrease in the membrane thickness. Therefore, the composite membrane has much higher water permeation rate and mechanical strength than a single-layer membrane having the same thickness.

Composite membranes often comprised of two primary forms: flat-sheet and hollow-fibre membranes. Flat-sheet membranes can be packed in modules similar to plate-and-frame heat exchanger. In these modules, vacuum and air channels are arranged alternately. Hollow fibre membranes usually packed in modules similar to shell-and-tube heat exchangers. In these modules, vacuum is applied on the lumen side and air flows on the shell side inside, or vice versa. Hollow structure enables fibre membrane to withstand a large transmembrane pressure difference. However, supplying the air to the lumen side of the hollow fibres can lead to large pressure losses, while supplying the air to the shell side may result in poor flow distribution. Xu et al. [95] synthesized membrane dehumidifier from composite materials comprising nonwoven fabrics, sodium alginate and lithium chloride (LiCl). Nonwoven fabrics were adopted as supporting substrate followed by the layering of sodium alginate and LiCl as the respective corresponding active layer and hydrophilic additive. Membranes with different LiCl mass fractions were tested and analysed to investigate the composite membrane performance. Results showed that the highest vapour permeance of the composite membrane is 29.8×10^{-8} kg/m^2 s Pa which is approximately 10 times higher than that of composite one without LiCl. Additionally, its water solubility declined by 20.6%; indicating a certain degree of membrane waterproofness.

In one of the very recent works on membrane dehumidification, Petukhov et al. designed and synthesized thin graphene oxide (GO) membranes supported on porous anodic aluminium oxide (AAO) substrates [96]. The GO membranes were prepared via a spin-coating process and employed pressure-assisted filtration to test for dehumidification. Key results highlighted that the GO membranes performed best when they were prepared from medium-flake-sized GO nanosheets which provided high enough defect density at the surface to allow water entrance to the interlayer space while enabling large enough nanosheet size to cover AAO nanochannels to facilitate continuous barrier coating towards permanent gases on the surface.

As far as the dependence of the membrane performance on its module design and configuration, Scovazzo and MacNeill evaluated the performance of four membrane module design geometries, counter-current flow, cross-current flow, and combined counter/cross current, for an air conditioning (AC) membrane module [97]. The key objective of their work was to focus on reducing the feed/retentate boundary layer resistance to mass transport. Testing and evaluation of the module geometries, using an industry-standard feed humidity for designing air conditioning systems, showed that membrane module design can produce vacuum sweep dehumidification membrane systems with better performance and energy efficiencies than standard cooling coil systems. The best module (cross-current flow) tested had an overall module water-vapour permeance of 4700 GPU for a feed humidity of 11 g/kg; producing a 26% reduction in absolute humidity with less than 2.5% product loss and requiring less than 6.2 compression ratio for the vacuum pump.

The pathway for membrane dehumidification is an exciting one. However, new research directions are required to pave new studies on membranes' performance at scaled-up levels; considering the fact that membranes can perform differently when they are expanded from lab scale to industrial size applications. By decoupling air dehumidification process from air cooling, the membrane presents a strong proposition to enable the energy efficiency of buildings' cooling plants to be dramatically improved.

2.7 Conclusions

An extensive literature review on the potential benefits of novel cooling and dehumidification technologies has been conducted. Key features of a few of the most promising coolers and dehumidifiers are judiciously selected and presented to identify their potential and merits in improving cooling energy efficiency.

As far as innovative cooling methods are concerned, two technologies stand out, namely, heat-driven adsorption/absorption chillers and dew-point water-based evaporative coolers. The use of thermally driven chillers presents a lower cost alternative to electrically driven air conditioners. Several key advantages of adsorption/absorption chillers include lower energy costs than electrically driven mechanical chillers, greater reliability as they have fewer moving parts than electrically driven mechanical chillers, and also they can be operated using waste heat that are easily harvested from many industrial processes. The dew-point evaporative cooler differs from the conventional swarm cooler because it exploits the evaporative potential of the dehumidified air enabling it to approach its dew-point temperature and not the wet-bulb temperature of ambient air. Operating without compressor and free of HCFC/CFC, the electricity-saving potential is highly significant while being an environmentally friendly sensible cooler.

To decouple sensible and latent loads, the use of an effective dehumidification technology is key in reducing energy consumption of cooling systems. Increasing interest and attention have been paid to the membrane-based technology, which is based on water vapour transmembrane transport driven by mass transfer potential. The virtue of engineering membranes for dehumidification is its non-regenerative nature and close control of water vapour permeation. In addition, liquid-based desiccant dehumidification has proven to be an effective method to extract the moisture of air particularly when low humidity is desired. However, the problem of potential carryover of the liquid desiccant to the air has to be judiciously addressed.

Considering the amount of recent development on innovative cooling and dehumidification technologies, readers should be mindful that there is no one single encompassing technology that can arrive at the lowest kW per refrigeration ton. Often than not, a combination of several of these technologies or strategies is necessary. A holistic approach is necessary towards achieving new cooling energy-efficient targets by considering the needs, practicality and constraints of the weather, environmental and available support resources.

References

1. Schiermeier Q (2011) Climate and weather: extreme measures. Nature 477:148–149. https://doi.org/10.1038/477148a
2. Energy consumption: The Asian experience, https://ejap.org/environmental-issues-in-asia/energy-consumption.html (2013)
3. Energy market authority: Singapore energy statistics. https://www.ema.gov.sg/Singapore_Energy_Statistics.aspx (2014)
4. National climate change secretariat—air-con system efficiency primer: a summary. https://www.nccs.gov.sg/sites/nccs/files/Aircon%20Primer.pdf (2011)
5. National Environment Agency (2010) https://www.nea.gov.sg/docs/default-source/weather-and-climate/second-nc.pdf
6. Rezk ARM (2012) Theoretical and experimental investigation of silica gel/water adsorption refrigeration systems. Presented at the University of Birmingham (United Kingdom)
7. Wikipedia (online). https://en.wikipedia.org/wiki/Absorption_refrigerator
8. Sakraida VA (2009) Basics for absorption chillers. Eng Syst. 26:36–48
9. Loh WS, Kumja M, Rahman KA, Ng KC, Saha BB, Koyama S, El-Sharkawy II (2010) Adsorption parameter and heat of adsorption of activated carbon/HFC-134a pair. Heat Transfer Eng. 31:910–916. https://doi.org/10.1080/01457631003603949
10. Ng KC, Wang X, Lim YS, Saha BB, Chakarborty A, Koyama S, Akisawa A, Kashiwagi T (2006) Experimental study on performance improvement of a four-bed adsorption chiller by using heat and mass recovery. Int J Heat Mass Transf 49:3343–3348. https://doi.org/10.1016/j.ijheatmasstransfer.2006.01.053

11. Chua HT, Ng KC, Wang W, Yap C, Wang XL (2004) Transient modeling of a two-bed silica gel-water adsorption chiller. Int J Heat Mass Transf. 47:659–669. https://doi.org/10.1016/j.ijh eatmasstransfer.2003.08.010

12. Ameri M, Hejazi SH (2004) The study of capacity enhancement of the Chabahar gas turbine installation using an absorption chiller. Appl Therm Eng. 24:59–68. https://doi.org/10.1016/ S1359-4311(03)00239-4

13. Preißinger M, Pöllinger S, Brüggemann D (2013) Ionic liquid based absorption chillers for usage of low grade waste heat in industry. Int J Energy Res. 37:1382–1388. https://doi.org/10. 1002/er.2997

14. Chen WD, Chua KJ (2020) Parameter analysis and energy optimization of a four-bed, two-evaporator adsorption system. Appl Energy 265:114842

15. Chua HT, Ng KC, Gordon JM (2002) US Patent No. 6434955

16. Ng KC, Chakraborty A, Wang XL, Gao LZ, Sai MA (2005) Experimental investigation of an electro-adsorption chiller. In: Proceedings of ASME heat transfer conference. Presented at the ASME heat transfer conference: 2005 summer heat transfer conference, San Francisco, CA, USA. https://doi.org/10.1115/HT2005-72226

17. M-Cycle—Valeriy MAISOTSENKO. https://www.rexresearch.com/maisotsenko/maisot senko.html (2007)

18. Riangvilaikul B, Kumar S (2010a) An experimental study of a novel dew point evaporative cooling system. Energy Build. 42:637–644. https://doi.org/10.1016/j.enbuild.2009.10.034

19. Elberling L (2006) Laboratory evaluation of the Coolerado cooler-indirect evaporative cooling unit. Pacific Gas and Electric Company

20. Grupp D, Wolley J (2014) Coolerado H80 rooftop unit filed test. Cooling Efficiency Center, UC Davis. Presented at the (2014)

21. Riangvilaikul B, Kumar S (2010b) Numerical study of a novel dew point evaporative cooling system. Energy and Buildings. 42:2241–2250. https://doi.org/10.1016/j.enbuild.2010.07.020

22. Zhan C, Duan Z, Zhao X, Smith S, Jin H, Riffat S (2011) Comparative study of the performance of the M-cycle counter-flow and cross-flow heat exchangers for indirect evaporative cooling—paving the path toward sustainable cooling of buildings. Energy. 36:6790–6805. https://doi. org/10.1016/j.energy.2011.10.019

23. Hasan A (2012) Going below the wet-bulb temperature by indirect evaporative cooling: analysis using a modified ε-NTU method. Appl Energy 89:237–245. https://doi.org/10.1016/j.apenergy. 2011.07.005

24. Hasan A (2010) Indirect evaporative cooling of air to a sub-wet bulb temperature. Appl Therm Eng 30:2460–2468. https://doi.org/10.1016/j.applthermaleng.2010.06.017

25. Heidarinejad G, Moshari S (2015) Novel modeling of an indirect evaporative cooling system with cross-flow configuration. Energy and Buildings. 92:351–362. https://doi.org/10.1016/j. enbuild.2015.01.034

26. Anisimov S, Pandelidis D, Danielewicz J (2015) Numerical study and optimization of the combined indirect evaporative air cooler for air-conditioning systems. Energy. 80:452–464. https://doi.org/10.1016/j.energy.2014.11.086

27. Anisimov S, Pandelidis D (2015) Theoretical study of the basic cycles for indirect evaporative air cooling. Int J Heat Mass Transf 84:974–989. https://doi.org/10.1016/j.ijheatmasstransfer. 2015.01.087

28. Pandelidis D, Anisimov S, Worek WM (2015) Comparison study of the counter-flow regen-erative evaporative heat exchangers with numerical methods. Appl Therm Eng. 84:211–224. https://doi.org/10.1016/j.applthermaleng.2015.03.058

29. Anisimov S, Pandelidis D, Danielewicz J (2014) Numerical analysis of selected evaporative exchangers with the Maisotsenko cycle. Energy Convers. Manage. 88:426–441. https://doi.org/ 10.1016/j.enconman.2014.08.055

30. Chen Y, Luo Y, Yang H (2015) A simplified analytical model for indirect evaporative cooling considering condensation from fresh air: development and application. Energy Build. 108:387–400. https://doi.org/10.1016/j.enbuild.2015.09.054

31. Jradi M, Riffat S (2014) Experimental and numerical investigation of a dew-point cooling system for thermal comfort in buildings. Appl Energy. 132:524–535. https://doi.org/10.1016/j.apenergy.2014.07.040
32. Hsu ST, Lavan Z, Worek WM (1989) Optimization of wet-surface heat exchangers. Energy. 14:757–770. https://doi.org/10.1016/0360-5442(89)90009-1
33. Riangvilaikul B, Kumar S (2010c) An experimental study of a novel dew point evaporative cooling system. Energy Build. 42:637–644. https://doi.org/10.1016/j.enbuild.2009.10.034
34. Lee J, Lee D-Y (2013) Experimental study of a counter flow regenerative evaporative cooler with finned channels. Int J Heat Mass Transf. 65:173–179. https://doi.org/10.1016/j.ijheatmasstransfer.2013.05.069
35. Bruno F (2011) On-site experimental testing of a novel dew point evaporative cooler. Energy Build. 43:3475–3483. https://doi.org/10.1016/j.enbuild.2011.09.013
36. Duan Z, Zhan C, Zhao X, Dong X (2016) Experimental study of a counter-flow regenerative evaporative cooler. Build Environ. 104:47–58. https://doi.org/10.1016/j.buildenv.2016.04.029
37. Woods J, Kozubal E (2013) A desiccant-enhanced evaporative air conditioner: Numerical model and experiments. Energy Convers Manage. 65:208–220. https://doi.org/10.1016/j.enconman.2012.08.007
38. Gao WZ, Cheng YP, Jiang AG, Liu T, Anderson K (2015) Experimental investigation on integrated liquid desiccant—indirect evaporative air cooling system utilizing the Maisotesenko—cycle. Appl Therm Eng. 88:288–296. https://doi.org/10.1016/j.applthermaleng.2014.08.066
39. Demir H, Mobedi M, Ülkü S (2011) Microcalorimetric investigation of water vapor adsorption on silica gel. J Therm Anal Calorim 105:375–382. https://doi.org/10.1007/s10973-011-1395-y
40. Li X, Li Z, Xia Q, Xi H (2007) Effects of pore sizes of porous silica gels on desorption activation energy of water vapour. https://doi.org/10.1016/j.applthermaleng.2006.09.010
41. Bonaccorsi L, Freni A, Proverbio E, Restuccia G, Russo F (2006) Zeolite coated copper foams for heat pumping applications. Microporous Mesoporous Mater. 91:7–14. https://doi.org/10.1016/j.micromeso.2005.10.045
42. Aristov YI, Restuccia G, Cacciola G, Parmon VN (2002) A family of new working materials for solid sorption air conditioning systems. Appl Therm Eng. 22:191–204. https://doi.org/10.1016/S1359-4311(01)00072-2
43. Smith DM, Lucky EA, Natividad V (2000) USA Patent 6559096
44. Zhang XJ, Sumathy K, Dai YJ, Wang RZ (2006) Dynamic hygroscopic effect of the composite material used in desiccant rotary wheel. Sol Energy 80:1058–1061. https://doi.org/10.1016/j.solener.2005.07.008
45. Zhang XJ, Qiu LM (2007) Moisture transport and adsorption on silica gel-calcium chloride composite adsorbents. Energy Convers. Manage. 48:320–326. https://doi.org/10.1016/j.enconman.2006.04.001
46. Jia C (2010) Study on adsorption mechanism and dehumidification property of composite desiccant. https://doi.org/10.4028/www.scientific.net/AMR.150-151.912
47. Gordeeva L, Grekova A, Krieger T, Aristov Y (2009) Adsorption properties of composite materials (LiCl + LiBr)/silica. https://doi.org/10.1016/j.micromeso.2009.06.015
48. Jia CX, Dai YJ, Wu JY, Wang RZ (2007a) Use of compound desiccant to develop high performance desiccant cooling system. Int J Refrig 30:345–353. https://doi.org/10.1016/j.ijrefrig.2006.04.001
49. Huan Z, Jianlei N (1999) Two-stage desiccant cooling system using low-temperature heat. Build Serv Eng Res Technol. 20:51–55. https://doi.org/10.1177/014362449902000202
50. Weixing Y, Yi Z, Xiaoru L, Yuan X (2008) Study of a new modified cross-cooled compact solid desiccant dehumidifier. https://doi.org/10.1016/j.applthermaleng.2008.01.006
51. Chang C, Luo CW, Lu YS, Cheng BY, Lin ZH (2017) Effects of process air conditions and switching cycle period on dehumidification performance of desiccant-coated heat exchangers. Sci & Techno for the Built Env 23:181–90
52. Fathalah K, Aly SE (1996) Study of a waste heat driven modified packed desiccant bed dehumidifier. Energy Convers Manage 37:457–471. https://doi.org/10.1016/0196-8904(95)00201-4

53. Li A, Thu K, Ismail AB, Shahzad MW, Ng KC (2016a) Performance of adsorbent-embedded heat exchangers using binder-coating method. Int J Heat Mass Transf. 92:149–157. https://doi.org/10.1016/j.ijheatmasstransfer.2015.08.097

54. Ge TS, Dai YJ, Wang RZ, Peng ZZ (2010) Experimental comparison and analysis on silica gel and polymer coated fin-tube heat exchangers. Energy. 35:2893–2900. https://doi.org/10.1016/j.energy.2010.03.020

55. Oh SJ, Choon Ng K, Thu K, Kum Ja M, Islam MR, Chun W, Chua KJE (2017) Studying the performance of a dehumidifier with adsorbent coated heat exchangers for tropical climate operations. Sci Tech Built Environ 23:127–135. https://doi.org/10.1080/23744731.2016.1218234

56. Zhao Y, Ge TS, Dai YJ, Wang RZ (2014) Experimental investigation on a desiccant dehumidification unit using fin-tube heat exchanger with silica gel coating. Appl Therm Eng. 63:52–58. https://doi.org/10.1016/j.applthermaleng.2013.10.018

57. Ismail AB, Li A, Thu K, Ng KC, Chun W (2013) On the thermodynamics of refrigerant + heterogeneous solid surfaces adsorption. Langmuir 29:14494–14502. https://doi.org/10.1021/la403330t

58. Zheng X, Ge TS, Wang RZ (2014) Recent progress on desiccant materials for solid desiccant cooling systems. Energy. 74:280–294. https://doi.org/10.1016/j.energy.2014.07.027

59. Li A, Thu K, Ismail AB, Shahzad MW, Ng KC (2016b) Performance of adsorbent-embedded heat exchangers using binder-coating method. Int J Heat Mass Transf. 92:149–157. https://doi.org/10.1016/j.ijheatmasstransfer.2015.08.097

60. Saha BB, Chakraborty A, Koyama S, Aristov YI (2009) A new generation cooling device employing $CaCl_2$-in-silica gel–water system. Int J Heat Mass Transf 52:516–524. https://doi.org/10.1016/j.ijheatmasstransfer.2008.06.018

61. Jiang Y, Ge TS, Wang RZ, Hu LM (2015) Experimental investigation and analysis of composite silica-gel coated fin-tube heat exchangers. Int J Refrig. 51:169–179. https://doi.org/10.1016/j.ijrefrig.2014.11.012

62. Hu L, Ge T, Jiang Y, Wang RZ (2014) Hygroscopic property of metal matrix composite desiccant

63. Jia CX, Dai YJ, Wu JY, Wang RZ (2007b) Use of compound desiccant to develop high performance desiccant cooling system. Int J Refrig. 30:345–353. https://doi.org/10.1016/j.ijrefrig.2006.04.001

64. Zheng X, Wang RZ, Ge TS (2016) Experimental study and performance predication of carbon based composite desiccants for desiccant coated heat exchangers. Int J Refrig. 72:124–131. https://doi.org/10.1016/j.ijrefrig.2016.03.013

65. Fang Y, Zuo S, Liang X, Cao Y, Gao X, Zhang Z (2016) Preparation and performance of desiccant coating with modified ion exchange resin on finned tube heat exchanger. Appl Therm Eng. 93:36–42. https://doi.org/10.1016/j.applthermaleng.2015.09.044

66. Ge TS, Dai YJ, Wang RZ (2016) Performance study of desiccant coated heat exchanger air conditioning system in winter. Energy Convers Manage. 123:559–568. https://doi.org/10.1016/j.enconman.2016.06.075

67. Abdel-Salam AH, Simonson CJ (2016) State-of-the-art in liquid desiccant air conditioning equipment and systems. Renew Sustain Energy Rev. 58:1152–1183. https://doi.org/10.1016/j.rser.2015.12.042

68. Mujahid Rafique M, Gandhidasan P, Rehman S, Al-Hadhrami LM (2015) A review on desiccant based evaporative cooling systems. Renew Sustain Energy Rev 45:145–159. https://doi.org/10.1016/j.rser.2015.01.051

69. Yin Y, Qian J, Zhang X (2014) Recent advancements in liquid desiccant dehumidification technology. Renew Sustain Energy Rev. 31:38–52. https://doi.org/10.1016/j.rser.2013.11.021

70. Mohammad AT, Mat SB, Sulaiman MY, Sopian K, Al-Abidi AA (2013) Survey of liquid desiccant dehumidification system based on integrated vapor compression technology for building applications. Energy Build. 62:1–14. https://doi.org/10.1016/j.enbuild.2013.03.001

71. Mei L, Dai YJ (2008) A technical review on use of liquid-desiccant dehumidification for air-conditioning application. Renew Sustain Energy Rev. 12:662–689. https://doi.org/10.1016/j.rser.2006.10.006

72. Park J-Y, Yoon D-S, Lee S-J, Jeong J-W (2016) Empirical model for predicting the dehumidi-
    fication effectiveness of a liquid desiccant system. Energy Build. 126:447–454. https://doi.org/
    10.1016/j.enbuild.2016.05.050
73. Wang X, Cai W, Lu J, Sun Y, Ding X (2013) A hybrid dehumidifier model for real-time
    performance monitoring, control and optimization in liquid desiccant dehumidification system.
    Appl Energy. 111:449–455. https://doi.org/10.1016/j.apenergy.2013.05.026
74. Zhang L, Hihara E, Matsuoka F, Dang C (2010) Experimental analysis of mass transfer in
    adiabatic structured packing dehumidifier/regenerator with liquid desiccant. Int J Heat Mass
    Transf. 53:2856–2863. https://doi.org/10.1016/j.ijheatmasstransfer.2010.02.012
75. Liu X-H, Jiang Y, Qu K-Y (2008) Analytical solution of combined heat and mass transfer
    performance in a cross-flow packed bed liquid desiccant air dehumidifier. Int J Heat Mass
    Transf. 51:4563–4572. https://doi.org/10.1016/j.ijheatmasstransfer.2007.11.059
76. Ren CQ (2008) Effectiveness–NTU relation for packed bed liquid desiccant–air contact systems
    with a double film model for heat and mass transfer. Int J Heat Mass Transf 51:1793–1803
77. Conde MR (2004) Properties of aqueous solutions of lithium and calcium chlorides: formu-
    lations for use in air conditioning equipment design. Int J Therm Sci 43:367–382. https://doi.
    org/10.1016/j.ijthermalsci.2003.09.003
78. Jain S, Bansal PK (2007) Performance analysis of liquid desiccant dehumidification systems.
    Int J Refrig 30:861–872. https://doi.org/10.1016/j.ijrefrig.2006.11.013
79. Koronaki IP, Christodoulaki RI, Papaefthimiou VD, Rogdakis ED (2013) Thermodynamic
    analysis of a counter flow adiabatic dehumidifier with different liquid desiccant materials.
    Appl Therm Eng. 50:361–373. https://doi.org/10.1016/j.applthermaleng.2012.06.043
80. Qi R, Lu L, Yang H, Qin F (2013) Investigation on wetted area and film thickness for falling
    film liquid desiccant regeneration system. Appl Energy. 112:93–101. https://doi.org/10.1016/
    j.apenergy.2013.05.083
81. Wang L, Xiao F, Zhang X, Kumar R (2016) An experimental study on the dehumidification
    performance of a counter flow liquid desiccant dehumidifier. Int J Refrig. 70:289–301. https://
    doi.org/10.1016/j.ijrefrig.2016.06.005
82. Khan AY, Martinez JL (1998) Modelling and parametric analysis of heat and mass transfer
    performance of a hybrid liquid desiccant absorber. Energy Convers Manage. 39:1095–1112.
    https://doi.org/10.1016/S0196-8904(97)00032-0
83. Xiong ZQ, Dai YJ, Wang RZ (2010) Development of a novel two-stage liquid desiccant dehu-
    midification system assisted by CaCl₂ solution using exergy analysis method. Appl Energy.
    87:1495–1504. https://doi.org/10.1016/j.apenergy.2009.08.048
84. The performance and the applied research of liquid desiccant fresh air conditioning system,
    Ph.D. thesis. Tsinghua University, Beijing, China (2009)
85. Gao WZ, Shi YR, Cheng YP, Sun WZ (2013) Experimental study on partially internally cooled
    dehumidification in liquid desiccant air conditioning system. Energy Build. 61:202–209. https://
    doi.org/10.1016/j.enbuild.2013.02.034
86. Woods J (2014) Membrane processes for heating, ventilation, and air conditioning. Renew
    Sustain Energy Rev 33:290–304. https://doi.org/10.1016/j.rser.2014.01.092
87. Bolto B, Hoang M, Xie Z (2012) A review of water recovery by vapour permeation through
    membranes. Water Res. 46:259–266. https://doi.org/10.1016/j.watres.2011.10.052
88. Yang B, Yuan W, Gao F, Guo B (2013) A review of membrane-based air dehumidification.
    Indoor Built Environ. 24:11–26. https://doi.org/10.1177/1420326X13500294
89. Xing R, Rao Y, TeGrotenhuis W, Canfield N, Zheng F, Winiarski DW, Liu W (2013)
    Advanced thin zeolite/metal flat sheet membrane for energy efficient air dehumidification and
    conditioning. Chem Eng Sci. 104:596–609. https://doi.org/10.1016/j.ces.2013.08.061
90. Scovazzo P, Scovazzo AJ (2013) Isothermal dehumidification or gas drying using vacuum
    sweep dehumidification. Appl Therm Eng. 50:225–233. https://doi.org/10.1016/j.appltherm
    aleng.2012.05.019
91. Bui TD, Chen F, Nida A, Chua KJ, Ng KC (2015) Experimental and modeling analysis of
    membrane-based air dehumidification. Sep Purif Technol. 144:114–122. https://doi.org/10.
    1016/j.seppur.2015.02.019

92. Wolińska-Grabczyk A, Jankowski A (2015) Woodhead publishing series in energy. Woodhead Publishing, Oxford. Presented at the (2015)
93. Frisch HL (1968) "Diffusion in polymers" edited by J. Crank and G. S. Park. Academic Press, London and New York, 452 pg. J Appl Poly Sci 14:1657–1657 (1970). https://doi.org/10.1002/app.1970.070140623
94. Shao P, Huang RYM (2007) Polymeric membrane pervaporation. J Membr Sci 287:162–179. https://doi.org/10.1016/j.memsci.2006.10.043
95. Xu J, Zhang C, Ge T, Dai Y, Wang R (2018) Performance study of sodium alginate-nonwoven fabric composite membranes for dehumidification. Appl Therm Eng 128:214–224. https://doi.org/10.1016/j.applthermaleng.2017.09.020
96. Petukhov DI, Chernova EA, Kapitanova OO, Boytsova OV, Valeev RG, Chumakov AP, Konovalov OV, Eliseev AA (2019) Thin graphene oxide membranes for gas dehumidification. J Membr Sci 577:184–194. https://doi.org/10.1016/j.memsci.2019.01.041
97. Scovazzo P, MacNeill R (2019) Membrane module design, construction, and testing for vacuum sweep dehumidification (VSD): Part I, prototype development and module design. J Membr Sci 576:96–107. https://doi.org/10.1016/j.memsci.2018.12.076

Chapter 3
Dew-Point Evaporative Cooling Systems

List of Symbols

A	Area, m^2
c	Specific heat, J/(kg K)
c_p	Specific heat at constant pressure, J/(kg K)
c_v	Specific heat at constant volume, J/(kg K)
d	Diameter, m
D	Diffusion coefficient, m^2/s
D_h	Hydraulic diameter, m
ex	Specific flow exergy, J/kg
E	Relative error, %
$\dot{E}$	Energy transfer rate, W
$\dot{Ex}$	Exergy transfer rate, W
F	Correction factor
F_o	Fourier number
Gz	Graetz number
H	Height, m
H_t	Channel height, m
h	Heat transfer coefficient, W/(m^2 K)
h_{fg}	Latent heat evaporation, J/kg
h_m	Mass transfer coefficient, m/s
i	Specific enthalpy, J/kg
j	Diffusive mass flux, kg/(m^2 s)
k	Thermal conductivity, W/(m K)
l_e	Characteristic length, m
L	Channel length, m
Le	Lewis number
LMTD	Log Mean Temperature Difference
m	Mass, kg
$\dot{m}$	Mass flow rate, kg/s

© Springer Nature Singapore Pte Ltd. 2021
C. Kian Jon et al., *Advances in Air Conditioning Technologies*, Green Energy
and Technology, https://doi.org/10.1007/978-981-15-8477-0_3

$\dot{M}$	Water evaporation rate, kg/s
n	Mass transfer rate, kg/s
$n^{''}$	Mass flux, kg/(m^2 s)
N	Number
Nu	Nusselt number
P	Pressure, Pa
Pr	Prandtl number
q	Heat transfer rate, W
$q^{''}$	Heat flux, W/(m^2 s)
$\dot{Q}$	Cooling capacity, W
r	Working air ratio/solution flow ratio
R	Specific gas constant, J/(kg K)
R_m	Membrane diffusion resistance, s/m
Re	Reynolds number
Sc	Schmidt number
Sh	Sherwood number
t	Time, s
T	Temperature, K
u	Specific internal energy, J/kg
U	Overall heat transfer coefficient, W/(m^2 K)
$\dot{U}$	Internal energy transfer rate, W
v	Velocity, m/s
$\dot{V}$	Volumetric flow rate, m^3/s
W	Width, m
$\dot{W}$	Power consumption, W
X	Concentration

Greek Symbols

α	Thermal diffusivity, m^2/s
δ	Thickness, m
ε	Effectiveness
ϕ	Relative humidity, %
η	Efficiency
ρ	Density, kg/m^3
π	Dimensionless number
μ	Dynamic viscosity, Pa s
ν	Kinematic viscosity, m^2/s
υ	Specific volume, m^3/kg
ω	Humidity ratio, kg/kg dry air
$\dot{\omega}$	Mole fraction ratio, mol/mol dry air
ξ	Ratio between the change of specific enthalpy and the change of wet-bulb temperature, kJ/(kg K)

Subscripts

0	Initial state/reference state
1	First stage
2	Second stage
A	Air
A	Area
CV	Control volume
c	Constant
d	Cry channel
D	Diameter
de	Dehumidified
dp	Dew point
e	Evaporation
ex	Exergy
f	Water film
i	In/inner
l	Length/liquid
lm	Log mean
lat	Latent
m	Mean/membrane
o	Out/observation/outer
p	Product
pl	Plate
r	Room
s	Supply
sa	Saturation
sf	Surface
sh	Shell
st	Steady-state
sen	Sensible
th	Thermal
v	Vapour
vac	Vacuum
w	Wet channel/working/water
wb	Wet bulb
COP	Coefficient of performance
DB	Dry bulb
DP	Dew point
DPEC	Dew point evaporative cooling
HMX	Heat and mass exchanger
HVAC	Heating, ventilation and air conditioning
MLDD	Membrane liquid desiccant dehumidification
RH	Relative humidity

SHR   Sensible heat ratio
WB   Wet bulb

Abbreviations

COP   Coefficient of performance
DB   Dry bulb
DP   Dew point
DPEC   Dew point evaporative cooling
HMX   Heat and mass exchanger
HVAC   Heating, ventilation and air conditioning
MLDD   Membrane liquid desiccant dehumidification
RH   Relative humidity
SHR   Sensible heat ratio
WB   Wet bulb

3.1 Introduction

It is reported that the world primary energy consumption has increased from 8.1 to 13.3 billion tons oil equivalent of energy from 1965 to 2016 translating with 63.1% increment [1]. According to U.S. Energy Information Administration (EIA) [2], the primary energy is used mainly for electricity production, transportation, industrial, residential, and commercial sectors, while around 41% of the primary energy consumed is used for residential and commercial buildings [3, 4].

The energy consumption in buildings is mainly incurred by air-conditioning systems [5, 6]. Ever since Wills Carrier invented the electricity-driven refrigeration cycle in 1902, it has been possible to artificially control the indoor air conditions against various climates, known as air conditioning [7]. Later, the mechanical vapour compression chiller was developed based on the refrigeration cycle, as an essential component in the Heating, Ventilation and Air Conditioning (HVAC) system. For more than a century, the HVAC system has grown from being a luxury item to a necessity in the modern society.

Since the 1970s, several global issues, such as oil and ecological crises, have surfaced and motivated people to reconsider the use of conventional vapour compression chiller [8]. The major associated problems can be summarized as (1) environmental issue. The global ozone depletion and greenhouse effect have significantly arisen since 1990s and which are attributed partly due to the utilization of environmentally unfriendly refrigerants, such as chlorofluorocarbons (CFCs) and hydrofluorocarbon (HFCs); (2) energy issue. In many developed and developing countries, the energy consumed in buildings accounts for 20–40% of the total energy consumption. Around 50% of the building energy consumption is contributed by HVAC systems, due to the intensive electricity demand of mechanical chillers [4, 5]; (3) safety issue. Some of the high pressure flammable refrigerants employed may lead to fire and explosion arising from electrical and leakage problems; (4) technical issue. To remove the excessive moisture from the air, the mechanical chiller has to cool the air to below its dew-point (DP) temperature. The coupled latent and sensible cooling usually creates a 'cold' or overcooled indoor condition, resulting in inefficient air conditioning. Consequently, the broad application of the chiller-based air conditioning system gradually became an obstacle to achieve the goal of a green and low-carbon society.

To resolve the drawbacks of the existing air-conditioning systems, one possible approach is to find potential alternatives to mechanical vapour compression chillers. Several researchers have proposed heat-driven cooling technologies, such as absorption and adsorption refrigeration. These cooling systems can be powered by low-grade waste heat, with water commonly adopted as the cooling refrigerant. However, both absorption and adsorption chillers are known to have low energy efficiencies. Further, they require a substantial amount of continuous heat supply, such as the harvested heat from gas engines, power plants and solar collectors, etc. In addition, by replacing the compact compressor with bulky absorber/adsorber/desorber to achieve similar cooling capacity, the chiller design can potentially become more complex; requiring larger physical space to install and operate.

As mechanical vapour compression chillers continue to dominate the global air-conditioning market, some researchers are devoted to re-visiting a conventional cooling technology, namely, evaporative cooling. This technology absorbs heat from the air via the evaporation process of water, so that the air temperature is reduced. Due to the large latent heat of evaporation of water, the air can be effectively cooled with water consumption. Concurrently, by eliminating both compressor and refrigeration cycle, evaporative cooling has the potential to achieve superior energy efficiency, normally with a coefficient of performance (COP) of 10–20 [9].

The idea of evaporative cooling dates back to thousands of years ago, when ancient Egyptians hung wetted reeds on their windows to cool the wind that flowed in. To innovate based on this traditional cooling method, different types of evaporative coolers have been developed over the past several decades.

Direct evaporative cooler (DEC) was first invented as a mature cooler design, and it has been commercialized by many companies. Dowdy et al. [10] conducted an experimental study on a DEC, which contained an impregnated cellulose cooling media with water sprayed on it to cover all the contact surface. In DECs, the supply air flow contacts directly with water, so the air is cooled as it gradually absorbs moisture from the water evaporation process with increased air humidity. However, excessive indoor air humidity can lead to a host of issues including thermal discomfort, mould formation, bacteria and virus growth that leads to the development of diseases. Technically speaking, DEC reduces the air temperature by converting the sensible heat into the latent heat of air, which is indeed an energy conservation process.

To address the humidity problem of DEC, indirect evaporative coolers (IECs) were proposed [8], whereby the air channels were separated into dry and wet channels.

The dry channel formed the primary channel for air cooling, while the wet channel formed the secondary channel covered by water. The air stream in the wet channel acted as the working air to stimulate water evaporation, and heat was transferred from the dry to the wet channel. Unfortunately, the temperature drop achieved by the IEC was far from satisfactory with the cooling effectiveness as low as 50% when compared with the DEC. The insufficient cooling performance of IEC limits its wide-scale deployment, and it is confined to applications incertain industries with that are able to accommodatelow-efficient cooling processes.

Until recently, a breakthrough IEC was formulated, designed, engineered, and tested. A novel IEC design was suggested by Maisotsenko et al. [9, 11, 12] to improve the cooling effectiveness of the former design, known as the 'Maisotsenko cycle', or M-cycle. The concept is to pre-cool, the working air before it enters the wet channels to pick up the moisture during water evaporation. With this configuration, the theoretical limit of wet bulb (WB) temperature in evaporative cooling is overcome, and ultimately the supply air temperature is able to approach its dew-point (DP) temperature. Thus, this innovative design is often adeptly coined as the dew-point evaporative cooler (DPEC) [13, 14, 15].

The DPEC is an ideal substitute for the mechanical chiller under dry and moderate climatic conditions, as it provides efficient air sensible cooling. However, when it comes to the humid climate where the ambient humidity itself exceeds the comfortable level, a standalone DPEC is often insufficient to provide adequate cooling. Incorporating an air dehumidification process to the DPEC becomes essential for decoupling latent and sensible cooling in order to achieve the desired thermal comfort conditions.

3.2 Principle and Features of the Dew-Point Evaporative Cooling

The DPEC is principally based on the M (Maisotsenko)-cycle thermodynamic process. The cycle captures energy from the air by utilizing the psychrometric renewable energy available from the latent heat of water evaporating into the air [7, 16, [17], 18]. It combines the thermodynamic processes of heat transfer and evaporative cooling to facilitate the supply air temperature to approach the dew-point temperature of the ambient air. In otherwords, it is also a marked advancement made in the direction of indirect evaporative cooling (IEC) by which the air can be cooled to near the dew-point temperature rather than limited by the wet-bulb temperature [19]. Apart from the cooled air, the M-Cycle produces saturated hot air as its working stream, which is required by many applications. Hence, the M-Cycle is able to function for the purpose of heat recovery [9, 20, 21, 22, 23]; thereby markedly promoting its system efficiency for various industrial applications.

The direct/indirect dew-point HMX is the state-of-the-art heat exchanger that sensibly cools the supply air to below the wet-bulb temperature while approaching dew-point conditions. Its innovative approach lies in hybridizing direct and indirect evaporative cooling processes to provide thermal comfort air temperatures. In this type of exchanger, a part of the surface on the dry side is designed for the working air to pass through while the rest is allocated to the product air. Both the product and working air are guided to flow over the dry side along parallel flow channels. There are numerous holes distributed spatially on the area where the working air is retained and each of these holes allows a certain percentage of air to pass through and enter the wet-side of the sheet. The air is gradually delivered to the-wet side as it flows along the dry side, thus forming an even distribution of airstreams over the wet surface. This arrangement allows the working air to be pre-cooled before entering the wet-side of the sheet by losing heat to the opposite wet surface. The pre-cooled air delivered to the wet-side flows over the wet surface along channels arranged at right angles to the dry side channels, absorbing heat from the working and product air. Owing to effect of pre-cooling, the working air in the wet-side (working air wet channel) has a much lower temperature and therefore, is able to absorb more heat from its two adjacent sides. As a result, the cooling (wet-bulb) effectiveness of the new structure would be higher than that in the traditional cross-flow exchanger (Fig. 3.1a). The cooling process is shown on a psychometric chart in Fig. 3.1b. The structure and operating principle of the direct/indirect dew-point evaporative cooling heat and mass exchange coupled with its internal flow distributions are portrayed in Fig. 3.2.

Computational processes are often developed to optimize the geometrical sizes, evaluate several fundamental ideas, and test operating conditions of the HMX exchanger in order to enhance the cooling effectiveness (dew point and wet bulb) of the exchanger and maximize the energy efficiency of the dew-point cooling system. The results of the simulations serve to elucidate how the cooling (dew point and wet bulb) effectiveness and energy efficiency are dependent on key parameters, such as the dimensions of the airflow passages, air velocity and working-to-intake-air ratio, and the temperature of the feed water, etc.

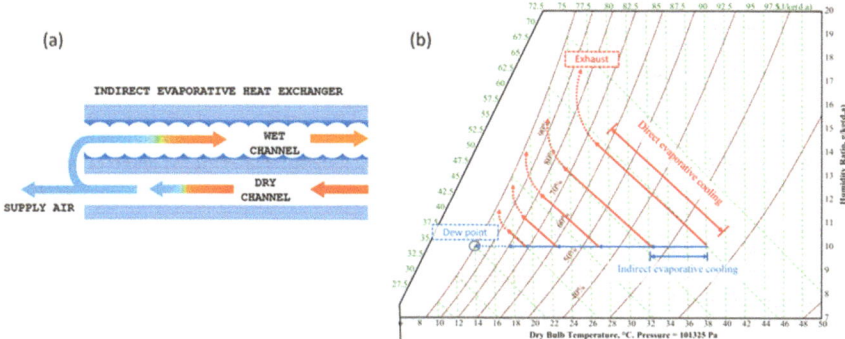

Fig. 3.1 **a** The thermal process of the direct/indirect evaporative cooling, and **b** psychrometric illustration of the process

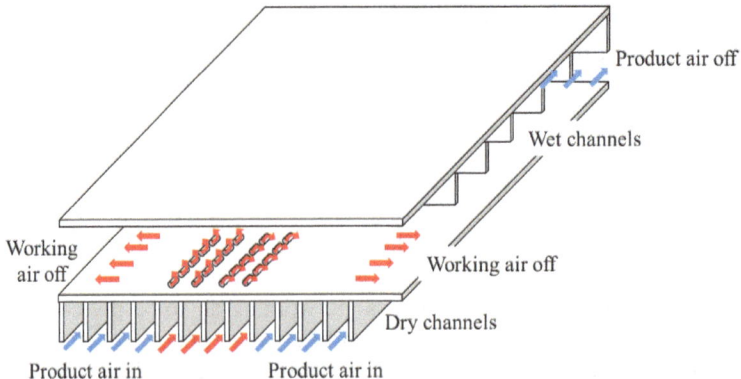

Fig. 3.2 Structure and operating principle of the direct/indirect dew point evaporative cooling HMX

3.3 Types of Dew-Point Evaporative Coolers

In practice, there are two generic types of dew-point evaporative coolers, namely, counter-flow and cross-flow, as shown in Fig. 3.3. Many initial dew-point evaporative coolers were designed based on the cross-flow configuration as portrayed in Fig. 3.3a. Later improved design was achieved by modifying the structure of the two heat and mass exchangers (HMXs) or their flow patterns. The original 'Maisotsenko cycle' was proposed by re-designing the counter-flow configuration from Fig. 3.3b into a new architecture as depicted in Fig. 3.4 [12]. As can be seen, the counter-flow M-cycle has separated the supply and the working air streams. The working air channels include both dry and wet channels and several perforations are designed

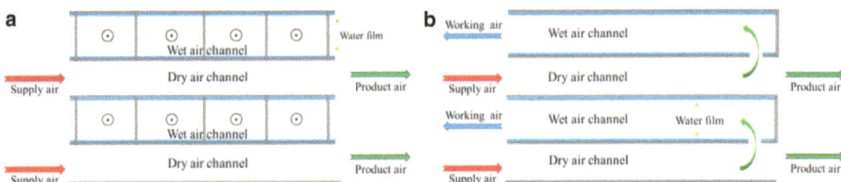

Fig. 3.3 System configuration of dew point evaporative cooling: **a** cross-flow; and **b** counter-flow

Fig. 3.4 The original M-cycle for dew point evaporative cooling [12]

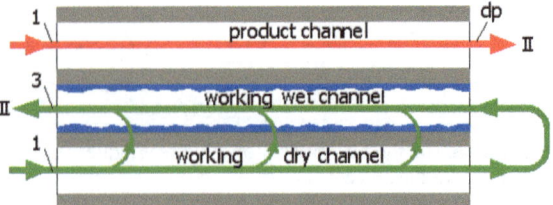

between them to reduce the pressure drop. However, early attempts to fabricate and commercialize this concept have failed due to several problems encountered [13, 15, 24] including (1) difficulty in realizing pure counter-flow regime; (2) excessive pressure drop of the air channels; (3) limited wetting method to vertical wicking; and (4) immature material and manufacturing technologies. As a result, instead of the counter-flow, the cross-flow configuration was adopted for the M-cycle, as shown in Fig. 3.5. More than two hundred different designs were mooted and tested, before it was finally commercialized by CooleradoCorporation™ [25]. In the Coolerado cooler, the dry and the wet air channels are perpendicular to each other at different channel layers. The working air first flows to the dry channel parallel to the supply air where it is precooled by the working air in the wet channel. Each of the wet channels is linked to the pre-cool dry channel via a perforated sheet so that the working air is uniformly distributed. The supply air is then gradually cooled by the adjacent wet channel layer, stage by stage. As the product air and the working air are exhausted from different sides of the cooler, the flow regime is easy to realize with simplified fabrication. Compared with the counter-flow cooler, the cross-flow M-cycle has a relatively lower pressure drop with greater potential to achieve dew-point evaporative cooling process.

Several studies on the Coolerado cooler have been carried out in the literature. The performance of an early cooler product from Coolerado was evaluated by Elberling [27]. The test facility of the cooler is presented in Fig. 3.6. The cooler was able to deliver 2549 m³/h of product air and 2243 m³/h of working air. Different supply air

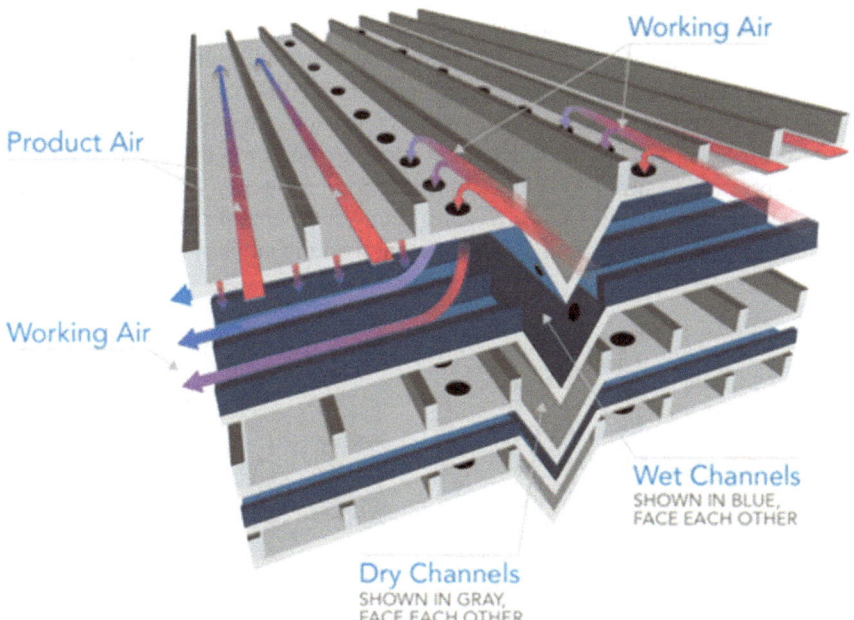

Fig. 3.5 Schematic diagram of the cross-flow M-cycle cooler [26]

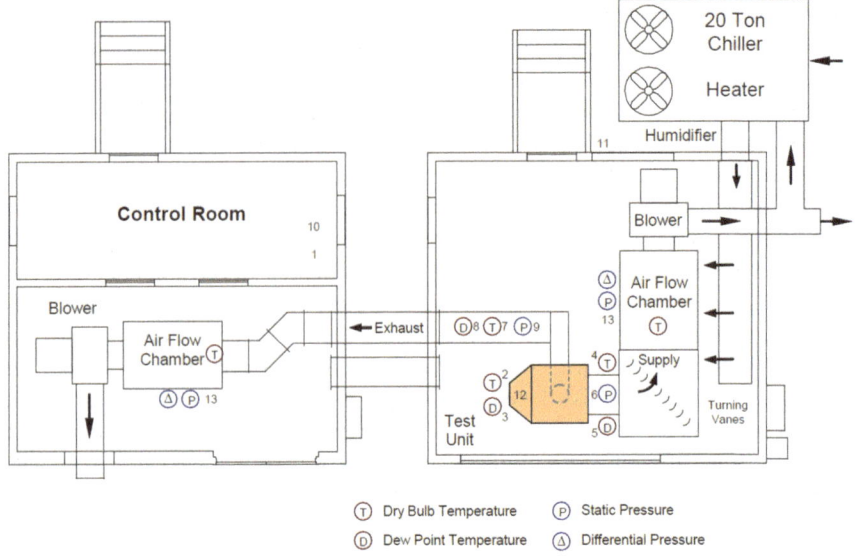

Fig. 3.6 A test system of the Coolerado cooler [27]

conditions were created for the experiments according to the ASHRAE test standard. Test results showed that the WB effectiveness spanned 0.81–0.91, while averaging at 0.86 with a COP of 9.6.

Zube and Gillan [28, 29] examined the performance of a cross-flow M-cycle HMX from Coolerado, as shown in Fig. 3.7. Temperature, humidity, and static pressure of both product and working air streams were measured at different nodal locations. The heat transfer rate, mass flow rate, and water evaporation rate were analysed based on the test results. It was observed that a large temperature drop from 38.0 to 20.8 °C of the supply air occurred when the ambient condition was dry. Higher heat transfer and water evaporation rates appeared consistently at the entrance of the HMX, while the mass flow was not uniform across different working air channels.

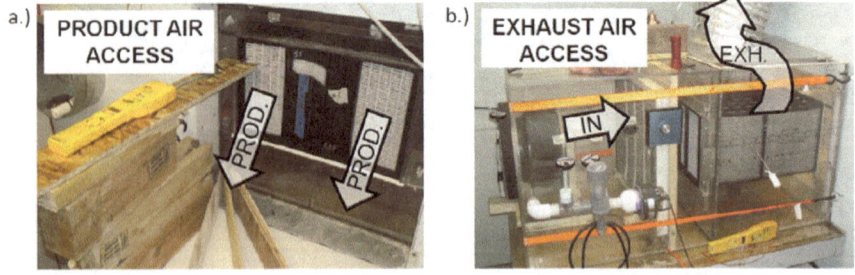

Fig. 3.7 A test facility of the cross-flow M-cycle HMX from Coolerado [28]

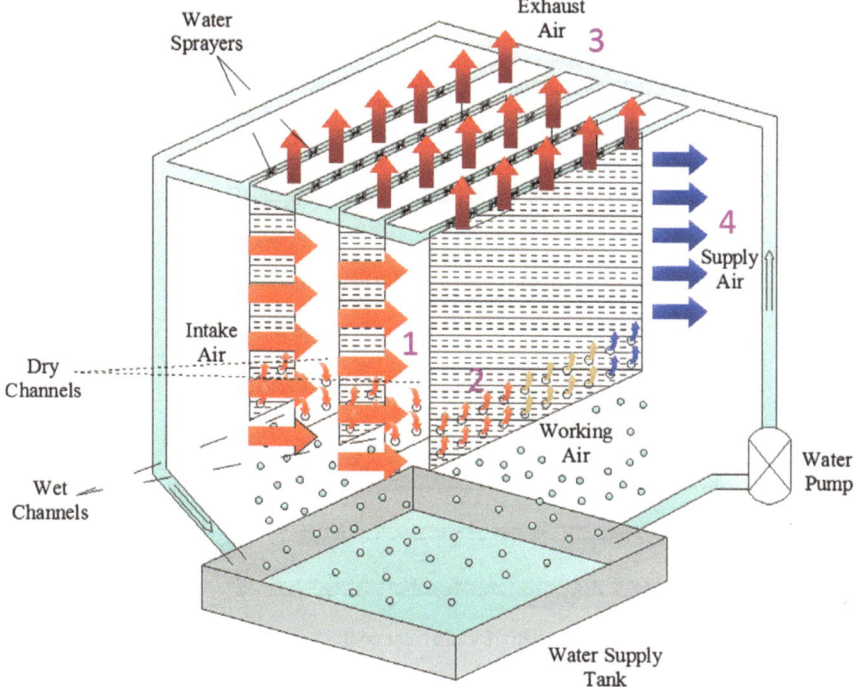

Fig. 3.8 Schematic diagram of the cross-flow dew point evaporative cooler developed by Jradi et al. [30]

Jradi et al. [30] carried out an experimental study on a similar cross-flow dew-point evaporative cooler with 66 dry channels and 65 wet channels. The concept of the cooler is illustrated in Fig. 3.8. The cooling performance was evaluated under different supply air flow rates, temperatures, and relative humidity (RH). For the supply air at 41.1 °C with RH of 14.5%, the cooler could produce product air at 17.3 °C, with cooling capacity and COP of 1054 W and 14.2, respectively. The respective ranges of WB effectiveness, cooling capacity, and COP spanned 0.70–1.16, 1054–1247 W, and 5.9–14.2.

Although the cross-flow dew-point evaporative cooler has been well engineered, it was later found that the performance of the cross-flow dew-point evaporative cooler degraded significantly under hot and humid climates. More specifically, the cooling effectiveness and energy efficiency of the cross-flow cooler were not even comparable to a conventional air conditioner. Driven by the motivation to further improve the cooling performance and the recent development of wick material and manufacturing technologies, researchers have started to revisit the counter-flow dew-point evaporative cooler.

In the literature, the earliest concept of the counter-flow dew-point evaporative cooler was experimentally evaluated by Hsu et al. [31]. As shown in Fig. 3.9, they investigated three different configurations employing wet-surface heat exchangers for

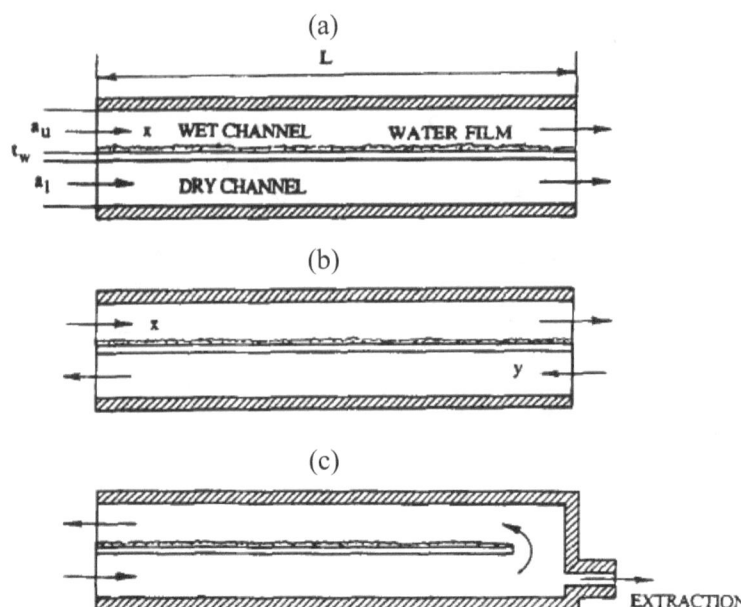

Fig. 3.9 Sketches of the cooler geometry for different configurations: **a** unidirectional; **b** counter-flow; and **c** counter-flow closed-loop [31]

indirect evaporative cooling, including unidirectional flow, counter-flow, and closed-loop flow configurations. They reported that the maximum wet-bulb effectiveness for counter-flow, cross-flow, and closed loop configurations is 1.3, higher than the unidirectional or the counter-flow configuration.

Riangvilaikul et al. [16] proposed a counter-flow dew-point evaporative cooling system with vertical flow direction. As presented in Fig. 3.10, the cooler was designed with four dry channels and five wet channels. The cooling system was tested under different inlet air temperature, humidity, and velocity. It was found that the product air temperature varied from 15.6 to 32.1 °C. The WB and DP effectiveness spanned 0.92–1.14 and 0.58–0.84, respectively. It is noteworthy that their proposed design performed well under dynamic conditions.

Zhao et al. [14] proposed a novel counter-flow heat and mass exchanger with triangular channels, as illustrated in Fig. 3.11a. They concluded that, under a typical UK summer weather condition, the system could achieve the WB effectiveness of up to 1.3. The effects of different parameters, such as air velocity, working-to-intake-air ratio, and channel size on the cooling effectiveness were also studied. Experiments were later conducted on their design by Duan [15] in Fig. 3.11b. The cooler was simulated and tested under different controlled parameters. Her results showed that the product air temperature from the cooler spanned 19.0–29.0 °C. The consequent WB and DP effectiveness varied from 0.55 to 1.10 and 0.40 to 0.85, respectively, with COP ranging from 3.0 to 12.0.

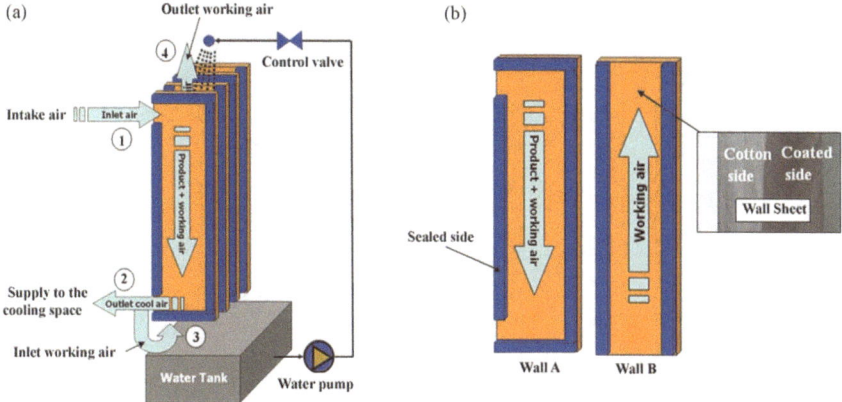

Fig. 3.10 Schematic diagram of: **a** dew point evaporative cooler; and **b** the dry and wet channels [16]

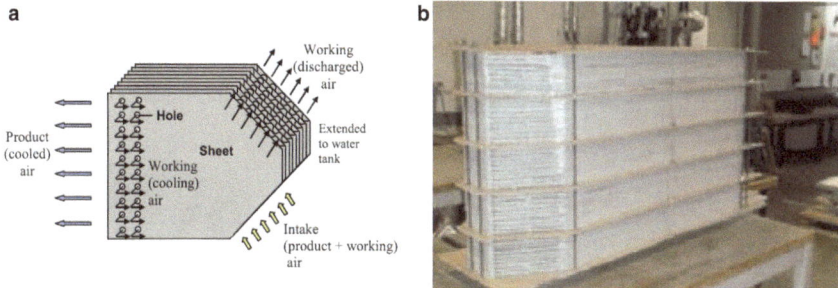

Fig. 3.11 The counter-flow dew point evaporative cooler with triangular channels: **a** model geometry; and **b** cooler fabrication [14]

Lee and Lee [24] investigated a counter-flow regenerative evaporative cooler with finned channels, which is illustrated in Fig. 3.12. The effects of inlet air conditions, water and air flow rates, and extraction ratio on product air temperature and effectiveness were studied experimentally. Their experiment showed that under the inlet condition of 32.0 °C and 50% RH, the outlet temperature was 22.0 °C, below the inlet WB temperature of 23.7 °C.

Bruno [17] carried out an experimental study of a dew-point evaporative cooler module, with a desired cooling capacity of above 10 kW. The cooler was tested for both residential and commercial buildings over a summer period, as shown in Fig. 3.13. When the daily ambient condition was 22.5–40.3 °C and 10–55% RH, the cooler in commercial application could deliver the product air to temperatures below 18.0 °C with WB effectiveness ranging from 0.93 to 1.06. For residential application, the average achievable product air temperature was below 15.0 °C with

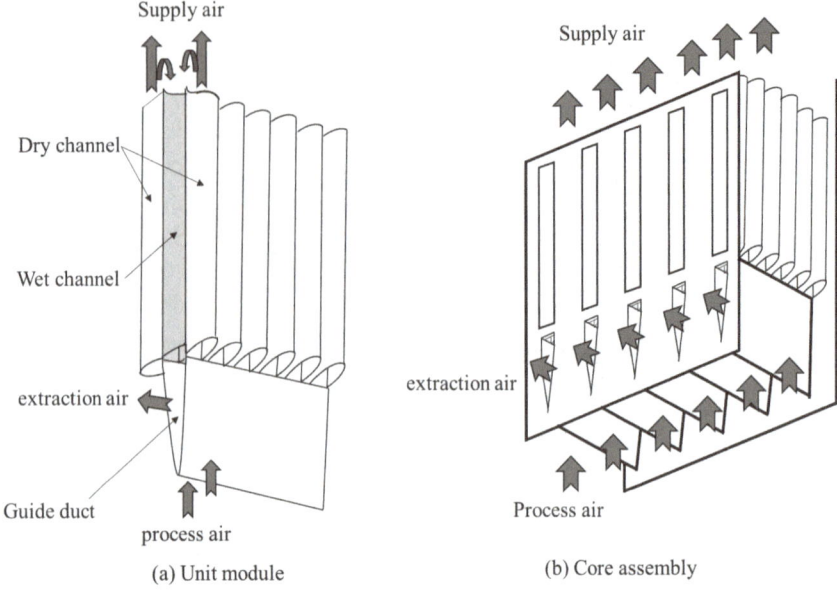

Fig. 3.12 Counter-flow regenerative evaporative cooler with finned channels [24]

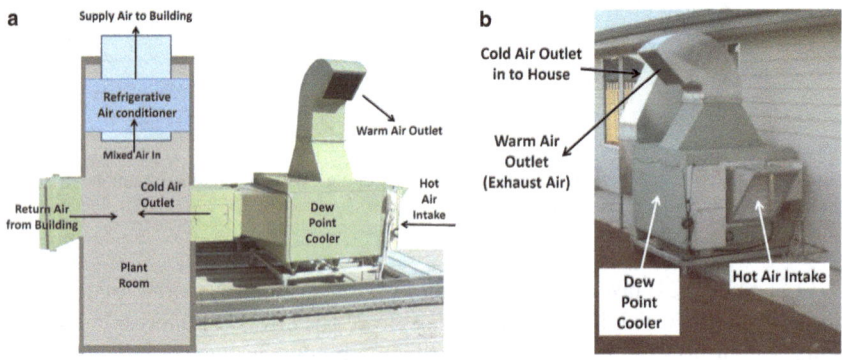

Fig. 3.13 Installation of the dew point cooler in: **a** commercial building; and **b** residential building [17]

WB effectiveness ranging from 1.18 to 1.29, when ambient air was 27.5–40.4 °C and 12–33% RH. The average COP was reported to be in the range of 4.9–11.8.

In addition, Xu et al. [32] proposed a novel irregular heat and mass exchanger with corrugated surface. The channel supporting guides were removed so that the heat transfer area between the dry and wet channels was doubled. They reported that the irregular exchanger was able to achieve more than 29.7% increment in cooling effectiveness and COP, compared with the flat-plate triangular exchanger. Later, the corrugated guideless HMX was fabricated as pictured in Fig. 3.14. The cooler

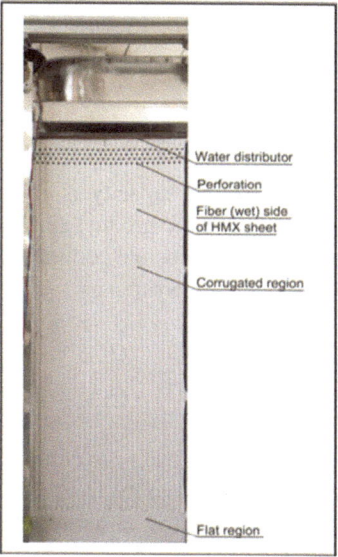

Water distributor

Perforation

Fiber (wet) side
of HMX sheet

Corrugated region

Flat region

(a) HMX

(b) Heat exchanging sheet
in position

Fig. 3.14 The general structure of the irregular corrugated HMX [32]

achieved a superior COP of 52.5 under the supply air conditions of 37.8 °C DB and 21.1 °C WB, with the optimal working air ratio of 0.364.

The overall dew-point evaporative cooling process incorporates physical mechanisms in fluid mechanics, heat and mass transfer, and thermodynamics. Therefore, in the literature, the parameters revealing the flow resistance, cooling, and energy performance of the dew-point evaporative cooler are normally investigated.

The flow resistance of the air channels is represented by the pressure drop between the air streams, is expressed as

$$\Delta P_{\mathrm{sp}} = P_{\mathrm{s}} - P_{\mathrm{p}} \tag{3.1}$$

$$\Delta P_{\mathrm{sw}} = P_{\mathrm{s}} - P_{\mathrm{w}} \tag{3.2}$$

The cooling performance of the cooler is dependent on product air temperature, cooling effectiveness, cooling capacity, and room capacity, etc. The cooling effectiveness includes WB effectiveness and DP effectiveness [8, 14]. They are defined by the ratio of temperature drop between the supply and product air temperatures to the temperature difference between the dry bulb (DB) and WB/DP of the supply air. They reflect the extent of the product air temperature approaching the WB or DP of the supply air, written as follows

$$\varepsilon_{wb} = \frac{T_s - T_p}{T_s - T_{s,wb}} \tag{3.3}$$

$$\varepsilon_{dp} = \frac{T_s - T_p}{T_s - T_{s,dp}} \tag{3.4}$$

The cooling capacity is calculated using the enthalpy difference between the supply air and the product air. As the air humidity is deemed constant along the dry channel, only sensible cooling is involved and the air enthalpy is merely a function of temperature, written as

$$Q_{sen} = \rho_a V_p (i_s - i_p) = \rho_a V_p C_a (T_s - T_p)$$
$$Q_{sen} = \rho_a V_p (i_s - i_p) = \rho_a V_p C_a (T_s - T_p) \tag{3.5}$$

The room capacity is commonly used to evaluate the amount of heat load that the cooler needs to overcome in order to maintain the thermal comfort conditions. It is defined as the enthalpy difference between the room air and the product air, expressed as

$$Q_{sen} = \rho_a V_p (i_r - i_p) \tag{3.6}$$

In general, the dew-point evaporative cooler is known to consume less electrical power than the conventional mechanical vapour compression air conditioner. The energy efficiency of the cooler, also known as COP, is evaluated to analyse its energy performance. It is defined as the ratio of the cooling capacity to the electrical power consumption

$$COP = \frac{Q_{sen}}{W} \tag{3.7}$$

3.4 Analytical Models

While different cooler prototypes are designed, fabricated, and tested, the mathematical modelling and simulation of the dew-point evaporative cooler are currently carried out. The objectives of these theoretical studies are: (1) to understand the related physical mechanisms in the cooling process; (2) to predict the cooling performance under specific test conditions and geometric parameters; (3) to improve the cooler design based on existing test results; and (4) to optimize the operating conditions and geometric parameters for a specific application.

Hsu et al. [31] first derived a lumped parameter model for different types of wet-surface heat exchangers. The convective heat and mass transfer were deemed as the major physical process between the channel surface and air streams. The longitudinal heat conduction was considered in both channel plate and water film. Based on this approach, Riangvilaikul et al. [33] performed a numerical study on their previous counter-flow dew-point evaporative cooler. It was found that their model predicted the product air temperature and cooling effectiveness within ±5.0% and ±10.0% discrepancies, respectively.

Kabeel et al. [34] improved the dew-point evaporative cooler configuration from by Riangvilaikul et al. [13] by introducing internal baffles into the dry channel. Four cooler configurations with 9, 11, 13 and 15 internal baffles were simulated and compared with the original rectangular flat-plate dew-point evaporative cooler. They claimed that the cooler's WB effectiveness increased with the number of internal baffles and at least 33.3% improvement could be achieved. Hasan [35] proposed an analytical model using modified ε-NTU method as well as the numerical method for the evaporative cooler. It was found that the results from the two models were similar and agreed well with the experimental data. The performance of four different cooler configurations was studied, i.e. single-stage and two-stage counter-flow, two-stage parallel-flow, and combined parallel-regenerative evaporative coolers. It showed that their wet-bulb effectiveness was 1.16, 1.26, 1.09, and 1.31, respectively, under 30.0 °C and 34% RH supply air conditions.

Cui et al. [36, 37, 38] numerically studied a counter-flow closed-loop dew-point evaporative cooler. A CFD model was established using the ANSYS FLUENT platform. The limits of the channel height, supply air velocity and working air ratio were examined to ensure the WB effectiveness of greater than 1. Cui and co-workers [39] also developed an analytical model using modified log mean temperature difference (LMTD) method. The need to conduct numerical simulation of the differential governing equations was eliminated by converting them into a system of algebraic equations. The air temperatures and heat transfer rate were then solved via successive substitution. The analytical model was able to predict the experimental data with a maximum discrepancy of ±8.0%.

Recently, Zhan et al. [40, 41] applied the lumped parameter model to the cross-flow dew-point evaporative cooler. The counter-flow and cross-flow dew-point cooling configurations were compared, according to their cooling effectiveness, COP, and cooling capacity. From their simulation results, the counter-flow configuration provided better effectiveness and larger cooling capacity without significant increase in energy consumption. Jradi et al. [30] also presented a two-dimensional numerical model for a cross-flow dew-point evaporative cooler, and results were validated by the experimental data from the cooler. They stated that the wet-bulb effectiveness of the cooler was 1.12 with 2017 W of cooling capacity under the supply air conditions of 30.0 °C and 50% RH.

To improve the accuracy of the earlier developed lumped parameter model, the influence of longitudinal heat conduction was partially investigated by Hettiarachchi et al. [42] and Heidarinejad et al. [43]. They developed mathematical models for cross-flow HMX concerning the longitudinal heat conduction along the plate. It was found that performance deterioration caused by conduction can be as high as 10% at some conditions.

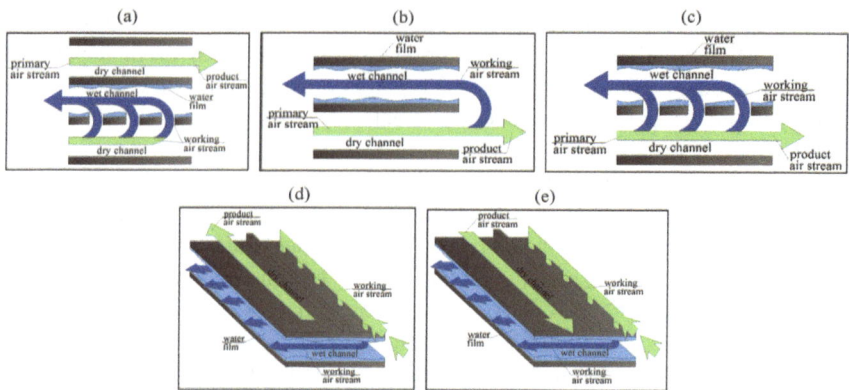

Fig. 3.15 Different versions of air flow arrangement in the dew point evaporative coolers: **a** modified counter-flow; **b** regenerative counter-flow; **c** regenerative with perforation; **d** cross-flow; and **e** modified cross-flow [44]

Subsequently, Anisimov et al. [44, 45, 46] and Pandelidis et al. [47, 48, 49] proposed a modified ε-NTU method to study the M-cycle heat and mass exchanger (HMX). Their model took into account the fin surface and air mixing process in the channels, and addressed the detailed simulation schemes and heat and mass transfer processes. They also presented a numerical study on different configurations of dew-point evaporative coolers, including three kinds of counter-flow and two kinds of cross-flow exchangers, as shown in Fig. 3.15. Their results found that the counter-flow HMX could achieve higher effectiveness, while the specific cooling capacity of cross-flow HMX was higher. A summary of the mathematical models developed by previous researchers and their respective features is depicted in Table 3.1.

In addition to the physics-based models, the emerging artificial intelligence has popularized the application of data-driven models. Recent progress in machine learning using neural network has inspired researchers to develop novel optimization algorithms for dew-point evaporative cooling. Jafarian et al. [52] presented a multi-objective optimization of the counter-flow dew-point evaporative cooler, based on the group method of data handling (GMDH) neural network model. The model was employed to design the operating and geometric conditions that simultaneously maximized the average COP and minimized the specific area of the cooler, according to different climatic conditions. Zhu et al. [53] developed a data-driven model for the counter-flow dew-point evaporative cooler using an artificial neural network (ANN) algorithm. The model was able to effectively predict the experimental data with the maximum discrepancy of ±4.0%. Their simulations found that the optimal working air ratio varied from 0.30 to 0.36 under different supply air conditions, in order to achieve the maximum cooling capacity.

Table 3.1 Summary of existing studies on mathematical modeling of the dew point evaporative cooler

Study	Cooler configuration	Modeling features
Hsu et al. [31]	Counter-flow	1. Lumped parameter model 2. Convective heat and mass transfer 3. Longitudinal heat conduction along the plate 4. Modeling of the supply air, working air and channel plate
Riangvilaikul et al. [13, 33]	Counter-flow	1. Lumped parameter model 2. Convective heat and mass transfer 3. Modeling of the supply air and working air
Duan [15]	Counter-flow with triangular air channels	1. Lumped parameter model 2. Convective heat and mass transfer 3. Latent heat transfer due to water evaporation 4. Modeling of the supply air, working air and water film
Zhan et al. [40]	Cross-flow; counter-flow	
Lee et al. [24]	Counter-flow with finned air channels	
Hasan [35]	Counter-flow	1. Both analytical ε-NTU and numerical models 2. Convective heat and mass transfer 3. Latent heat transfer due to water evaporation 4. Modeling of the supply air and working air
Kabeel et al. [34]	Counter-flow with internal baffles	1. 2-D ε-NTU model 2. Convective heat and mass transfer 3. Latent heat transfer due to water evaporation 4. Modeling of the supply air, working air and water film
Cui et al. [36, 37, 38]	Improved counter-flow	1. 2-D CFD model 2. Convective heat and mass transfer 3. Conservation of mass, momentum, thermal energy and concentration
Cui et al. [39]	Cross-flow; counter-flow	1. Modified LMTD model 2. Convective heat and mass transfer 3. Latent heat transfer due to water evaporation 4. Modeling of the supply air, working air and water film

(continued)

Table 3.1 (continued)

Study	Cooler configuration	Modeling features
Jradi et al. [30]	Cross-flow	1. Lumped parameter model 2. Convective heat and mass transfer 3. Latent heat transfer due to water evaporation 4. Modeling of the supply air, working air and water film
Anisimov et al. [44, 45, 46] Pandelidis et al. [47, 48, 49, 50, 51]	Cross-flow; counter-flow	1. Modified 2-D ε-NTU model 2. Convective heat and mass transfer 3. Latent heat transfer due to water evaporation 4. Concerning fin surface performance 5. Modeling of the supply air, working air, water film, channel plate and fins
Heidarinejad et al. [43]	Cross-flow	1. ε-NTU model 2. Convective heat and mass transfer 3. Latent heat transfer due to water evaporation 4. Longitudinal heat conduction along the plate 5. Concerning spray water temperature variation 6. Modeling of the supply air, working air, water film and channel plate
Jafarian et al. [52]	Counter-flow	1. GMDH neural network model 2. A 2-D CFD numerical model as the input data for training 3. Algebraic equations to predict results using soft-computing tools 4. Multi-objective optimization of average specific area and COP
Zhu et al. [53]	Counter-flow	1. ANN data-driven model 2. A 1-D lumped parameter model as the input data for training 3. Optimization of working air ratio

3.4.1 Cross-Flow Dew-Point Evaporative Cooler

The cross-flow dew-point evaporative cooler is designed to minimize the pressure drop of the air flow. To distribute the working air evenly, the flow arrangement of the cooler is different from a traditional cross-flow heat exchanger [40]. The internal structure of the cooler has been highlighted by Anisimov et al. [46, 51]. Figure 3.16a and b, respectively, shows the dry and wet channel arrangements for a dew-point cooler prototype. The dry channel is aligned along the y-direction, and the wet channel follows the x-direction. For the dry channels, the first three rows are the pre-cooling channels of the working air. The airflow in these three channels is divided into the 18 columns of wet channels in the adjacent layer. In the wet channels, the first nine columns are linked to the first row of dry channels through holes, while the rest are connected with the second row. The air in the third row of dry channels gradually flows to the second row via three openings on the side of the channel. The air mixing and separation process take place near the holes and openings. The last three rows of dry air are the supply air for evaporative cooling and are eventually delivered as the product air.

Based on the flow arrangement and the heat and mass transfer process for the designed evaporative cooling, a mathematical model is formulated. The energy and mass balance equations are expressed for the following control volume shown in Fig. 3.17a–d, i.e. dry channel air, wet channel air, channel plate, and water film. Several assumptions are made and are presented as follows:

(1) The supply air and working air are evenly distributed into each channel or hole;
(2) The physical properties inside each control volume are assumed to be uniform;
(3) There is negligible heat transfer between the cooler and the environment; and
(4) The wet channel surface is fully covered by a layer of water film, and the water surface is saturated.

The governing equations of the cross-flow dew-point evaporative cooler are expressed as follows:

(1) Dry air channel.

In dry air channel, the energy and heat entering and leaving the control volume are denoted as E_{in}, q_{in} and E_{out}, q_{out}, respectively. The general energy balance equation is expressed as

$$\Delta E_{CV} = E_i - E_o + q_i - q_o \tag{3.8}$$

According to the control volume shown in Fig. 3.17a, Eq. (3.8) can be written as

$$\Delta U_{CV} = q_{d,y} - q_{d,y+dy} - q_{d,z} + m_d i_{d,y} - m_d i_{d,y+dy} \tag{3.9}$$

where $q_{d,y}$ and $q_{d,y+dy}$ represent the heat conduction along longitudinal direction, while $q_{d,z}$ is the convective heat transfer between the air stream and the water film.

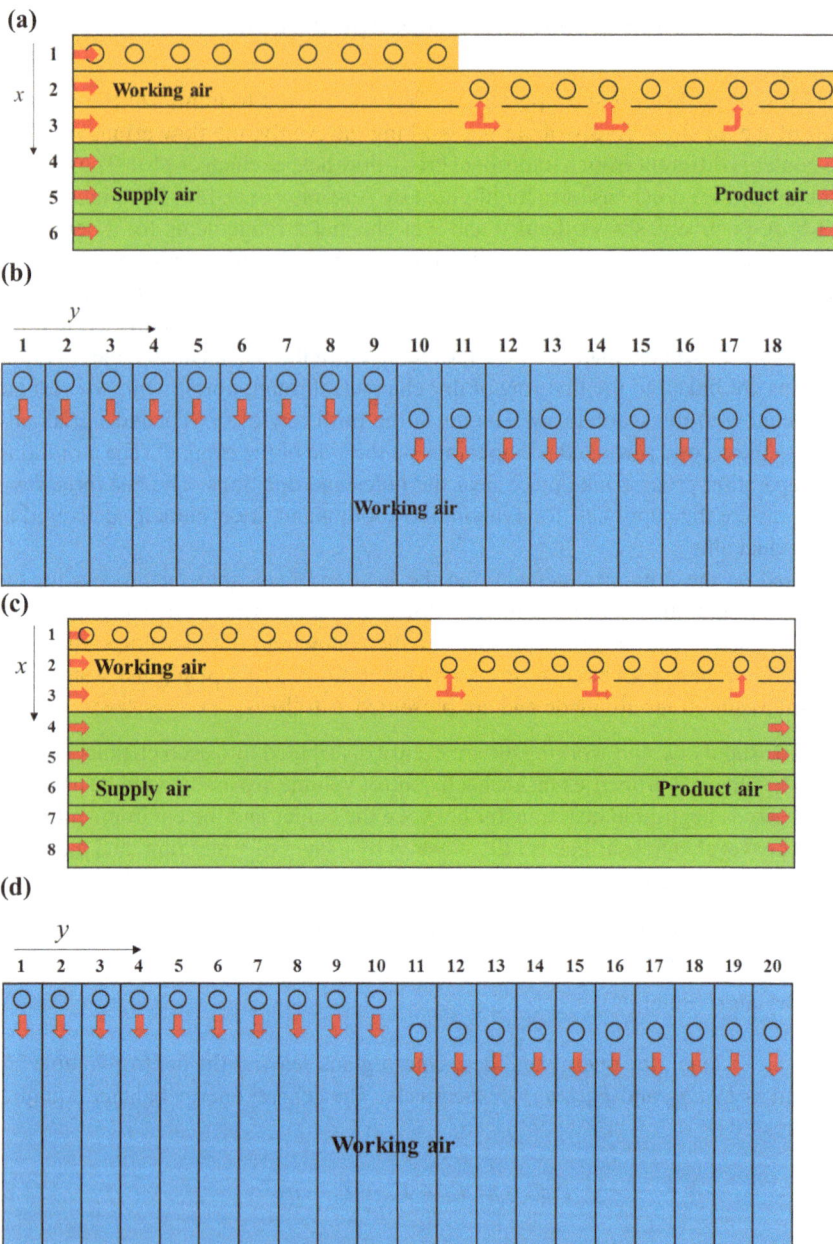

Fig. 3.16 Flow arrangement of the cross-flow dew point evaporative cooler prototype: **a** dry channel layer of the cooler prototype; **b** wet channel layer of the cooler prototype; **c** dry channel layer of the cooling system; and **d** wet channel layer of the cooling system

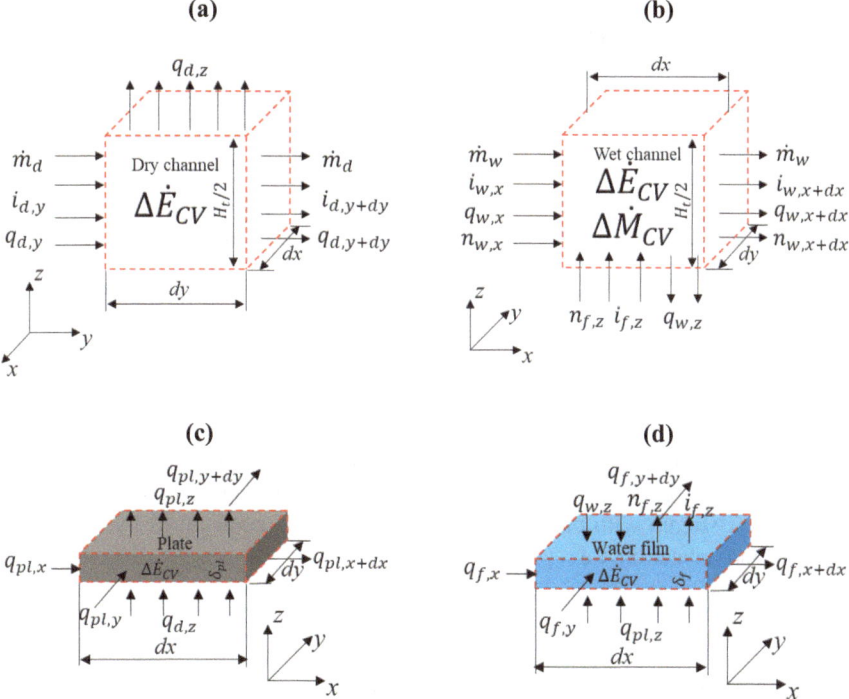

Fig. 3.17 Control volume for the mathematical model: **a** dry channel air; **b** wet channel air; **c** channel plate; and **d** water film

The enthalpy and the internal energy of moist air in the model are contributed by both the dry air and the water vapour, namely

$$i = i_a + \omega i_v = c_{pa}T + \omega(i_0 + c_{pv}T) \tag{3.10}$$

$$u = u_a + \omega u_v = c_{va}T + \omega(u_0 + c_{vv}T) \tag{3.11}$$

Similarly, the specific heats of the moist air can be expressed in the following equations

$$c_p = c_{pa} + \omega c_{pv} \tag{3.12}$$

$$c_v = c_{va} + \omega c_{vv} \tag{3.13}$$

At steady state, the left side of Eq. (3.8) is set to zero. Therefore, rearranging this equation, the final expression is obtained as below

$$\rho_a c_p v_{dm} \frac{dT_d}{dy} = k_a \frac{d^2 T_d}{dy^2} - \frac{2\overline{h}_d}{H_t}(T_d - T_{pl}) \tag{3.14}$$

(2) Wet air channel.

In the wet air channel, the heat and mass transfer are coupled as the existence of water evaporation causes the air humidity to change. According to the control volume in Fig. 3.17b, the energy balance and mass balance equations are derived as follows

$$\Delta U_{CV} = q_{w,x} - q_{w,x+dx} - q_{w,z} + m_w i_{w,x} - m_w i_{w,x+dx} + n_{f,z} i_{f,z} \tag{3.15}$$

$$\Delta M_{CV} = n_{w,x} - n_{w,x+dx} + n_{f,z} \tag{3.16}$$

where $n_{w,x}$ and $n_{w,x+dx}$ denote the mass fluxes entering and leaving the control volume, and $n_{f,z}$ represents the amount of the water vapour entering the boundary.

For an air stream with bulk motion, the mass flux is contributed by both mass diffusion and advection, i.e.

$$n''_{w,x} = j_v - \rho_v v_{wm} \tag{3.17}$$

The water evaporation rate is calculated by the water vapour density difference

$$n_{f,z} = \overline{h}_m (\rho_{v,sa} - \rho_v)(W dx) \tag{3.18}$$

Rearranging Eqs. (3.15) and (3.16) leads to the final set of equations

$$\rho_a c_p v_{wm} \frac{dT_w}{dx} = k_a \frac{d^2 T_w}{dx^2} - \frac{2\overline{h}_w}{H}(T_w - T_f)$$
$$\quad - i_v(T_w) \cdot v_{wm} \frac{d\rho_v}{dx} + \frac{2\overline{h}_m}{H_t}(\rho_{v,sa} - \rho_v) \cdot i_v(T_f) \tag{3.19}$$

$$v_{wm} \frac{d\rho_v}{dx} = D_{va} \frac{d^2 \rho_v}{dx^2} + \frac{2\overline{h}_m}{H_t}(\rho_{v,sa} - \rho_v) \tag{3.20}$$

where $i_v(T_w)$ and $i_v(T_f)$ are the enthalpy of water vapour calculated at the temperature of working air and water film, respectively. $\rho_{v,sa}$ is the saturation water vapour density on the water film.

The water vapour content of moist air varies along the wet channel. Hence, the enthalpy of the air changes with both temperature and humidity. The last two terms on the right side of Eq. (3.19) describe the enthalpy change due to humidity change and water evaporation, and the vapour density gradient along the flow direction gives the longitudinal mass diffusion and water evaporation rate.

(3) Channel plate.

The channel plate temperature is modelled by considering the lateral temperature difference between the plate in dry channel and the water film. The description of channel plate in the differential control volume is shown in Fig. 3.17c. The expression for the channel plate using energy balance is given as

$$k_{pl}\delta_{pl}\frac{\partial^2 T_{pl}}{\partial x^2} + k_{pl}\delta_{pl}\frac{\partial^2 T_{pl}}{\partial y^2} + \bar{h}_d(T_d - T_{pl}) - k_{pl}\frac{(T_{pl} - T_f)}{\delta_{pl}} = 0 \qquad (3.21)$$

The first and second terms in Eq. (3.21) represent the longitudinal heat conduction along x and y directions. The third and fourth terms denote the heat transfer from the dry channel and to the water film, respectively.

(4) Water film.

Similar to the channel plate, the energy balance equation for the control volume of the water film presented in Fig. 3.17d is expressed as

$$k_f\frac{\partial^2 T_f}{\partial x^2} + k_f\frac{\partial^2 T_f}{\partial y^2} + \frac{k_{pl}}{\delta_f\delta_{pl}}(T_{pl} - T_f) + \frac{\bar{h}_w}{\delta_f}(T_w - T_f)$$
$$+ \frac{\bar{h}_m}{\delta_f}(\rho_{v,sa} - \rho_v)(c_f T_f - i_v(T_f)) = 0 \qquad (3.22)$$

In Eq. (3.22), the third term is related to the heat conduction from the plate, while the fourth and fifth terms quantify the convective heat transfer from the wet channel and the latent heat transfer by water evaporation, respectively.

The saturation water vapour density or humidity is calculated via the relationship between air humidity ratio and the partial pressure of water vapour accordingly

$$\omega = 0.62198\frac{\phi P_{sa}}{P - \phi P_{sa}} \qquad (3.23)$$

The saturation pressure is calculated using the expression provided by ASHRAE Handbook-Fundamentals [54, 55].

$$\ln P_{sa} = C_1/T + C_2 + C_3 T + C_4 T^2 + C_5 T^3 + C_6 \ln T \qquad (3.24)$$

where T is in absolute temperature of K, and the coefficients are as follows

$$C_1 = -5800.2206, \quad C_2 = 1.3914993, \quad C_3 = -4.8640239 \times 10^{-2},$$
$$C_4 = 4.1764768 \times 10^{-5}, \quad C_5 = -1.4452093 \times 10^{-8}, \quad C_6 = 6.5459673$$

The convective heat and mass transfer coefficients are calculated using Nusselt number and Lewis number. Due to the small channel size of the cooler, the flow regime inside the air channel is assumed to be fully developed laminar flow ($Re_D < 2300$).

Therefore, for the dry air channel, the Nusselt number is a constant along the channel [56]

$$Nu_D = 5.60 \qquad (3.25)$$

In this section, a particular Nusselt number correlation proposed by Dowdy and Karabash [10] is employed for evaporative cooling. It is presented as

$$Nu_D = 0.10\left(\frac{l_e}{\delta}\right)^{0.12} Re_l^{0.8} Pr^{1/3} \qquad (3.26)$$

where l_e is the thickness of the water film, and δ is the total thickness of the channel plate including the water film. The Reynolds number is defined as

$$Re_l = \frac{\rho_a v l_e}{\mu} \qquad (3.27)$$

Finally, the mass transfer coefficient can be derived via the Chilton-Colburn analogy [56]

$$\frac{\overline{h}_w}{\overline{h}_m} = \rho_a c_p Le^{2/3} \qquad (3.28)$$

where $Le = \frac{\alpha_a}{D_{va}}$.

The above equations are numerically simulated using the MATLAB environment. The performance of the dew-point evaporative cooler is evaluated through several performance indices, namely, cooling effectiveness, cooling capacity, energy efficiency, and room capacity.

The power consumption of the system comprises the required electrical power for operation, and the thermal energy embedded in the fresh water supply. However, the necessary water supply rate for the dew-point evaporative cooling system is found to be less than 3.2 g/s under all operating conditions in this study. In contrast, the thermal energy of the liquid water is not comparable to electricity and should be converted in the calculation of COP. The electrical power contributed by the fresh water input can be estimated by the amount of electricity required to produce the water. Currently, the major electricity-driven technologies such as seawater desalination is able to produce fresh water at the energy efficiency of 2.0–4.0 kWh/m^3 [57, 58]. The corresponding electrical power to meet the water demand of the cooler is within 10% of the total power consumption. Therefore, the energy consumption imbedded in the water has marginal impact on the energy efficiency of the evaporative cooler.

3.4.1.1 Model Validation

Simulated results from the model are first validated with test results from the cross-flow cooler prototype under three specific test conditions. The outlet product and working air temperatures of different channels are plotted in Fig. 3.18a, b. It is observed that the proposed model is able to predict the cooler performance with good accuracy. The maximum discrepancies for the product and working air temperatures are within ±1.0% and ±3.0%, respectively. For the three test conditions, the average product air temperatures are 20.6, 20.3, and 19.4 °C, and the average working air temperatures are 25.1, 24.1, and 21.1 °C. The model is capable of simulating these average values to within ±0.8% discrepancy.

Simulation results from the mathematical model are also validated with data acquired from the cooling system. The system was tested under various ambient and room conditions, and the average product and working air temperatures of different air channels were measured and displayed in Fig. 3.18c, d. It is found that the model

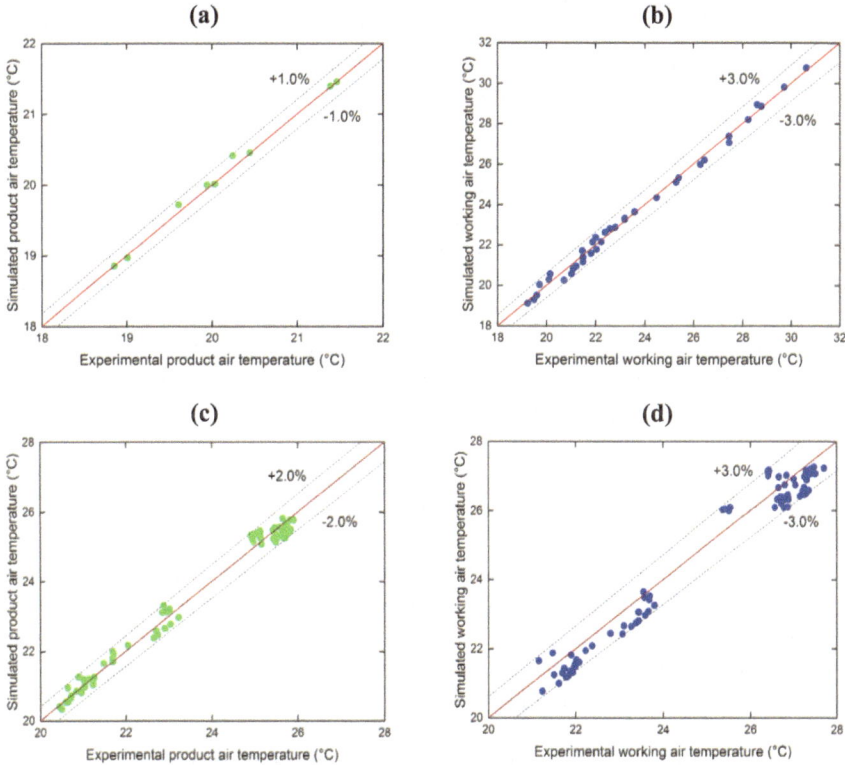

Fig. 3.18 Model validation with the experimental data: **a** product air temperatures of the cooler prototype; **b** working air temperatures of the cooler prototype; **c** Average product air temperatures of the cooling system; and **d** Average working air temperatures of the cooling system

agrees well with the experimental values. The discrepancy between simulated results and experimental data are within ±2.0% for the product air temperature, and ±3.0% for the working air temperature. Therefore, upon conducting the validation exercise, the model is now capable of simulating precisely experimental results to ±3.0% accuracy.

3.4.1.2 Simulated Results

The cooler prototype was tested under three different supply air conditions, namely, two conditions generated through a climatic chamber and one using the room condition. The average air velocity to each channel was maintained at 1.50 m/s. Figure 3.19a shows the working and product air temperature profiles measured at the channel outlet. For product air, the channel number 1 to 3 represents the fourth to sixth row of dry channels. It is apparent that the working air temperature gradually decreases along the y-direction, as it gets farther from the inlet of the supply air. However, the outlet temperature gradient in the first nine column is larger than the second part, and a sudden transition appears between the channels $y = 9$ and $y = 10$. This observation is attributed to the flow arrangement of the cooler, where the mass flow rate is designed to be different for the two parts of the wet channels. It is also observed that the product air temperature increases along the x-direction, and the minimum product air temperature appears in the first supply air channel.

The wet-bulb and dew-point effectiveness of the cooler for each channel are further presented in Fig. 3.19b, c. There is marginal difference between the first two channels, while the third channel has relatively poor effectiveness. In general, the cooler is able to achieve the wet-bulb and dew-point effectiveness above 1.20 and 0.80, respectively. The average wet-bulb effectiveness is approximately 1.25 with the dew-point effectiveness at 0.85. These results have demonstrated that the cross-flow dew-point evaporative cooler is able to achieve similar cooling potential, when compared with its counter-flow counterpart.

3.4.1.3 Temperature Profiles Along Channels

In this section, the cross-flow dew-point evaporative cooler prototype is further studied to understand the temperature distribution along each air channel. The cooler is simulated with the supply air of 30.0 °C temperature and 12.0 g/kg humidity. Figure 3.20a, b illustrates the temperature profiles for dry and wet layers of the air channels. The lowest temperature of the dry channel appears at the end of the pre-cooling working air channel, i.e. $x = 2$. As the channel number increases, the exit air temperature gradually rises. Further, through the pre-cooling process, the working air is cooled from 30.0 to 18.0 °C and gradually directed to the wet channels. This demonstrates the importance of the pre-cooling channels, which widen the tempera-ture difference between the supply and working air. Consequently, the heat transfer rate and cooling effectiveness are improved. The working air temperature is lower

Figs. 3.19: Test results of the cross-flow cooler prototype: **a** working and product air temperatures; and **b** wet bulb effectiveness; and **c** dew point effectiveness

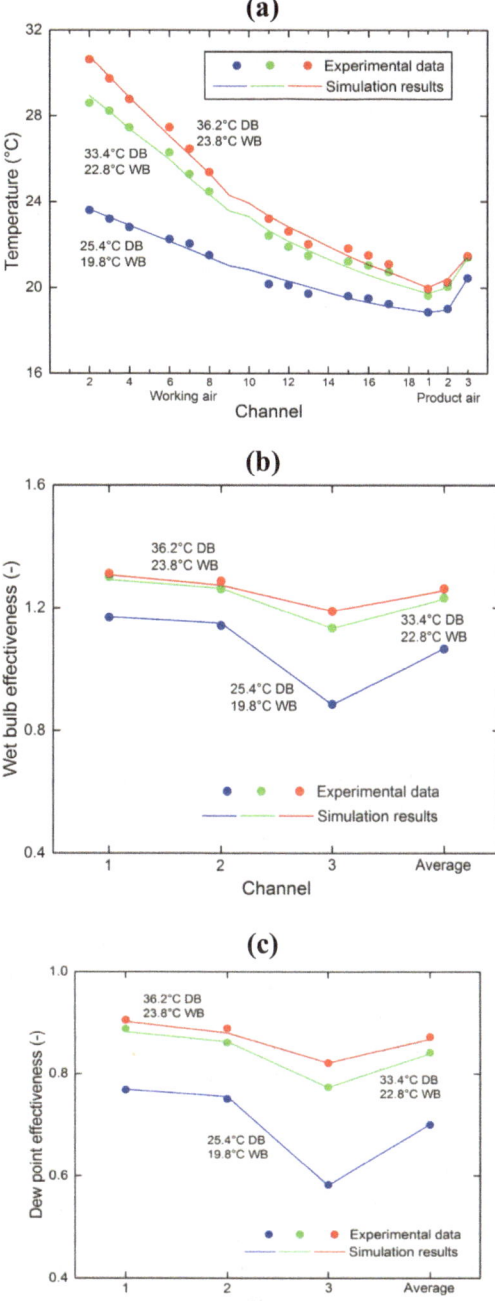

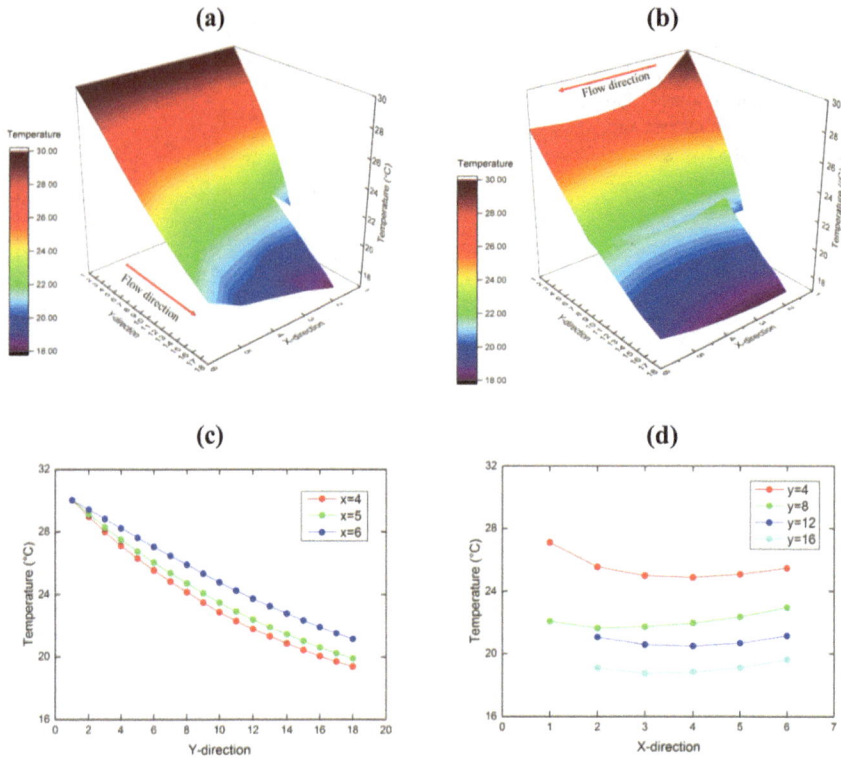

Fig. 3.20 Simulation results of the cooler prototype with the supply air condition of 30.0 °C and 12.0 g/kg: **a** temperature distribution of the dry channel; **b** temperature distribution of the wet channel; **c** temperature profile of the supply air; and **d** temperature profile of the working air

when the wet channel is closer to the dry channel exit. However, there is a disparity between channels at $y = 9$ and $y = 10$. This observation is attributed to the shift of the channel inlet and the change in air flow rate, as the second phase of the wet channels ($y = 10$–18) is linked to the second and third dry channels ($x = 2$ and 3).

The respective supply air and four working air temperature profiles are plotted in Fig. 3.20c, d. It is apparent that the supply air temperature in the first channel ($x = 4$) drops faster and the gradient of the temperature profiles reduces along the flow direction. This indicates that the cooling effect is more prominent near the channel entrance even though the working air temperature in the adjacent wet channel is higher. For the wet channels, the working air temperature first decreases along the flow direction as its sensible heat is absorbed by the water evaporation on the channel surface. At a later stage, the working air temperature increases again when heat is transferred from the supply air.

The performance of the cross-flow dew-point evaporative cooler is also investigated under normal ambient conditions for practical reasons. Figure 3.21 presents the test results and performance of the cooling system. During testing, the average

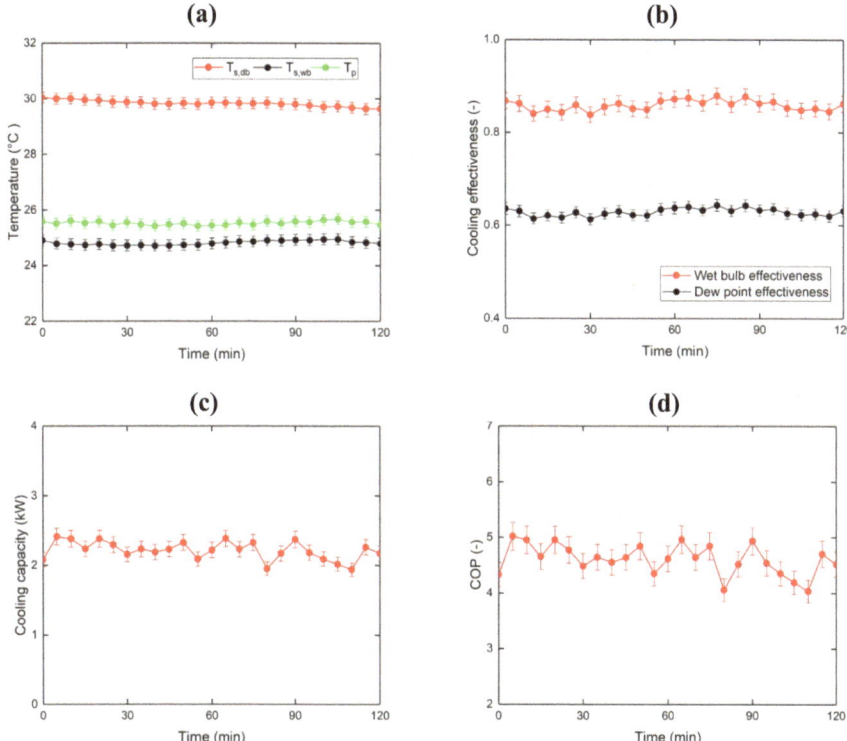

Fig. 3.21 Test results of the cross-flow dew point evaporative cooling system: **a** supply and product air temperatures; **b** cooling effectiveness; **c** cooling capacity; and **d** COP

ambient temperature was 29.8 °C dry bulb and 24.8 °C wet bulb ($\omega = 17.6$ g/kg). The power consumption of the system was measured to be approximately 480 W.

Data plots in Fig. 3.21a show that the average product air temperature is 25.5 °C with a humidity ratio of 17.8 g/kg. Due to the high ambient humidity, the cooling of the supply air is not sufficient. It is worthy to note that the wet-bulb effectiveness in Fig. 3.21b ranges only from 0.83 to 0.86. Therefore, the evaporative cooler is not able to cool the air below its inlet wet-bulb temperature. Further, the average cooling capacity is confined to 2.2 kW, with a COP of 4.6 as depicted in Fig. 3.21c, d. Although the energy efficiency of the cooling system is still higher than a conventional air conditioner with a typical COP of 2.0–4.0 [59, 60], it is far from its optimal performance where the COP can be above 10.0 [9, 26].

The performance of the dew-point evaporative cooling system reveals that both temperature reduction and cooling capacity depreciate when the air humidity rises above the comfortable level. Therefore, the product air condition needs to be further improved. To lower the ambient humidity to an appropriate level, an air dehumidification process is required before the evaporative cooling.

In the previous few sections, the cross-flow dew-point evaporative cooler is investigated numerically and experimentally via a cooler prototype. This study has demonstrated that the cross-flow cooler is able to achieve good cooling potential when the supply air humidity is suitable for indoor conditions. However, its performance degrades when it operates under the humid weather conditions due to the limited evaporation rate. An air dehumidification process is proposed prior to the evaporative cooler. Key findings that emerged from this part of the study are summarized as.

(1) A cross-flow cooler prototype has been tested for the supply air temperature and humidity of 25.0–37.0 °C and 12.0–13.0 g/kg, respectively. It is observed that the overall wet-bulb and dew-point effectiveness of the cooler approach 1.25 and 0.85, respectively.
(2) When the cross-flow dew-point evaporative cooling system is tested under the humid air condition of 29.8 °C temperature and 17.6 g/kg humidity, it is only capable of achieving a maximum wet-bulb effectiveness of 0.86. The cooling capacity and the system COP are confined to 2.2 kW and 4.6, respectively.

3.4.2 Counter-Flow Dew-Point Evaporative Cooler

In the literature, both cross-flow and counter-flow dew-point evaporative coolers have been investigated using different design parameters and test conditions. Comparatively, the performance of the counter-flow configuration is deemed to be superior as it is able to achieve lower product air temperature and higher COP in practice. However, as an emerging cooling technology, there remain several critical issues yet to be addressed for the counter-flow dew-point evaporative cooler, including (1) most of the earlier studies merely focus on the steady-state performance of the cooler, while its transient characteristics when subjected to a change in the input conditions have not been experimentally studied. As the inlet and the operating conditions of the cooler may consistently vary during operations, a comprehensive understanding of the cooler's transient response is crucial [61]; (2) the existing coolers usually employ a water distribution system to spray the water into the wet channels in a vertical orientation, while the effect of water spray on its dynamic performance remains unclear. The water supply at a constant temperature may disrupt the surface temperature distribution along the wet channel, causing disturbance to the state of the cooling process [24]; and (3) the cooling and energy performance of the dew-point evaporative cooler have been extensively studied, while the flow resistance of the air channels in the cooler is currently uncertain. The magnitude of the flow resistance is important in designing the geometric parameters and operating conditions of the cooler as well as in selecting the appropriate air blower [62].

To bridge the aforementioned research gaps, this part of the study aims to facilitate a deeper understanding of the transient and steady-state performance of the counter-flow dew-point evaporative cooler. Two counter-flow dew-point evaporative cooler prototypes with horizontal orientation are judiciously developed for experimental investigation. In dew-point evaporative coolers, water is spread slowly and evenly over the channel surface by capillary force to eliminate the need for a water distribution system. Concurrently, both lumped parameter and CFD mathematical models are formulated to simulate the transient and steady-state performance of the cooler. The content of the following sections includes: (1) the transient response of the channel plate temperature after water spray is tested in both vertical and horizontal orientations; (2) the initial-state response of the coolers is measured under a sudden change in supply air conditions; (3) the steady-state flow resistance, cooling performance and energy efficiency of the coolers are investigated under a wide range of supply air and operating conditions; and (4) the transient and steady-state characteristics of the dew-point evaporative coolers are also discussed via numerical methods.

3.4.2.1   Water Distribution on the Channel Surface

In existing dew-point evaporative coolers, a water distribution system is usually installed on top to spray the water into the wet channels. However, if the water supply rate is not well controlled, the steady-state of the cooler will be continuously disturbed whenever the spraying process takes place. It is hypothesized that a quick water flow with constant temperature will interrupt the surface temperature development along the wet channel, leading to the deterioration of cooling effectiveness.

To study and investigate the effect of water spray, the transient characteristics of the channel surface temperature are measured. Figure 3.22 shows the test facility for the channel plate. The channel plate was formed by covering one layer of impervious polymer (dry side) with another layer of wick material (wet side) for water retention and distribution. Two plate orientations, i.e. vertical and horizontal, were tested. Water was sprayed on the top section of the channel plate and was gradually allowed to distribute to the entire surface by capillary and gravity force. The channel plate was placed in a forced air flow with an air velocity of approximately 1.50 m/s at the plate surface. Four surface RTD temperature sensors (diameter: 0.8 mm, accuracy: class A) were placed on both the dry and wet sides of the channel plate along the centre-line, denoted as P1 to P4. The transient behaviour of the plate surface temperature at each measurement point after water spraying was recorded via an Agilent data logger (34970A).

Based on the test results from the water spray experiments on two channel orientations, a horizontal counter-flow dew-point evaporative cooler was designed and engineered in the laboratory. As shown in Fig. 3.23a, the cooler was stacked through pairs of dry and wet air channels, and the channels were separated by similar channel plates. The following key points were taken into account during the design and fabrication process:

(1) For ease of machining and assembly, an acrylic material was used for the basic structure of the HMX. In addition, this material provided a clear view of the internal structure for testing and demonstration.

(a)

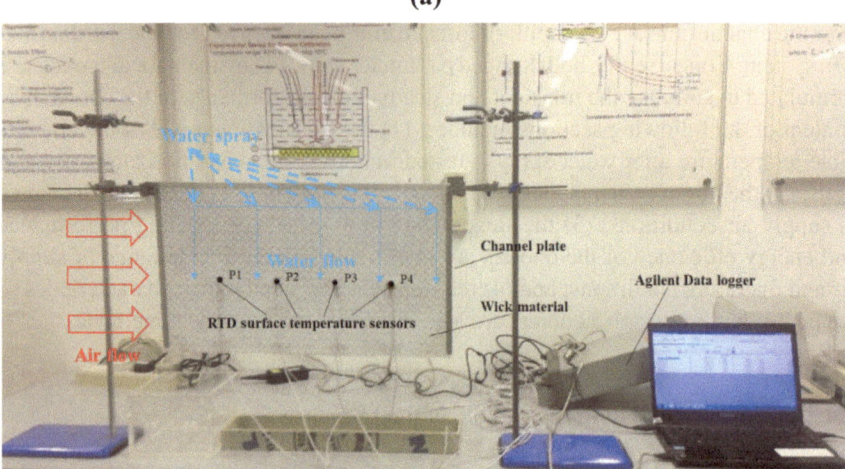

(b)

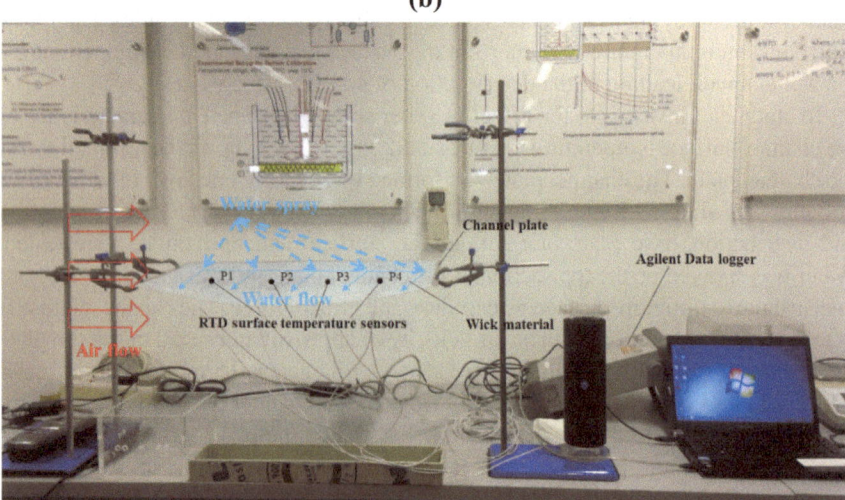

Fig. 3.22 Test facility for the channel plate in dew point evaporative cooling: **a** vertical orientation; and **b** horizontal orientation

(2) Several types of wick material (fibre, cotton, gauze, and tissue) were tested to compare their absorbency, and a thin, porous, natural fibre material was selected. One side of the wick material was dipped into the water tank to enable the absorbing and spreading of water over the channel plate. A test was conducted to demonstrate that water can be evenly spread across the entire plate.

(3) Several baffle plates were installed in each channel to prevent any bending or folding of the plates.

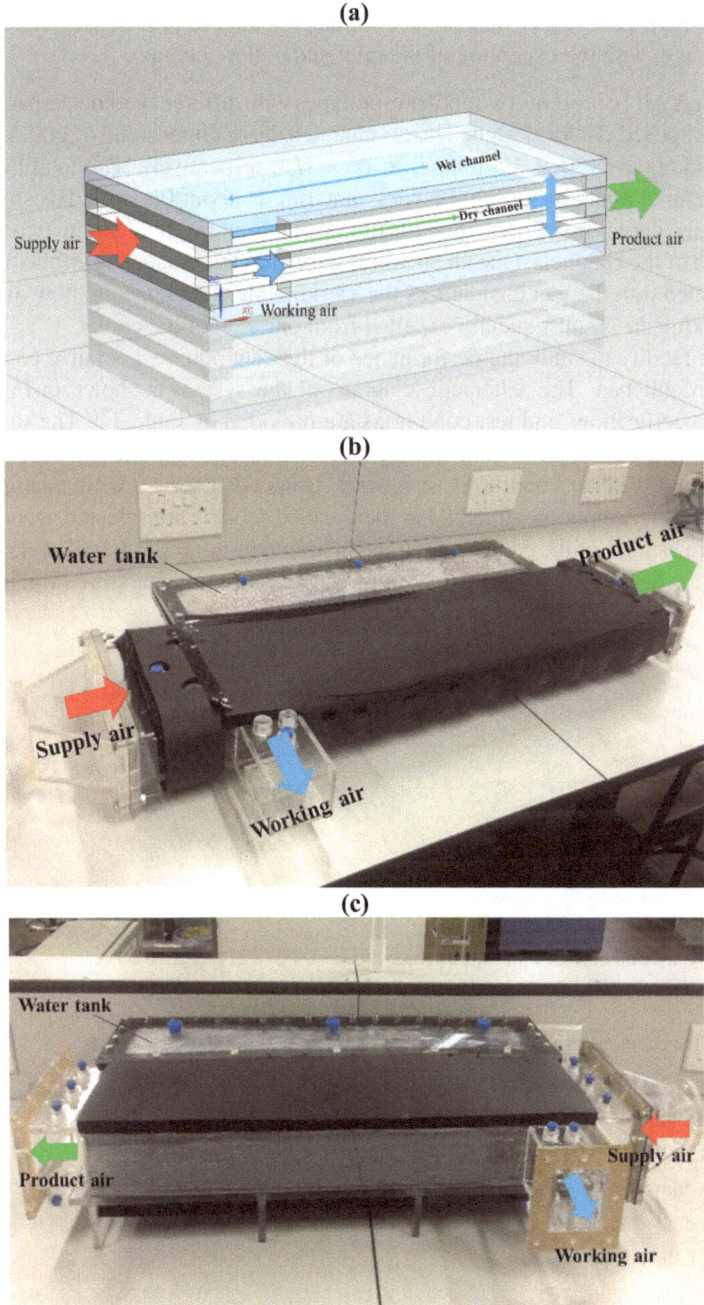

Fig. 3.23 Design and fabrication of the horizontal counter-flow dew point evaporative cooler: **a** 3D modeling; **b** cooler prototype V1; and **c** cooler prototype V2

(4) The HMX and water tank were eventually sealed to prevent any water or air leakage, with the exception of the inlet and outlet openings.

Figure 3.23b, c presents two cooler prototypes with different geometric parameters for the proposed cooler design. The air channels have dimensions of $600 \times 150 \times 3$ mm and $600 \times 150 \times 4.5$ mm ($L \times W \times H_t$), respectively. The total thickness of the channel plate, including the wick material, is about 0.4 mm. The supply air entering the cooler first flows along the dry channels and is divided into two streams at the end of the channels. One air stream is purged as the product air, while the other is redirected into the wet channels as the working air in an opposite flow direction. The working air is subsequently expelled from one side of the cooler

A test facility to study the performance of the dew-point evaporative cooler was further established. The schematic diagram of the system is shown in Fig. 3.24, and the specifications and test conditions are provided in Table 3.2. The supply air conditions were controlled using a temperature–humidity climatic chamber to simulate different weather conditions. A blower (Delta AFB series) with a damper was installed at the supply air stream, and the detailed models are selected according to the flow resistance of the cooler prototypes at nominal supply air flow. The flow resistance of the dry/wet channel was obtained by measuring the pressure drop between

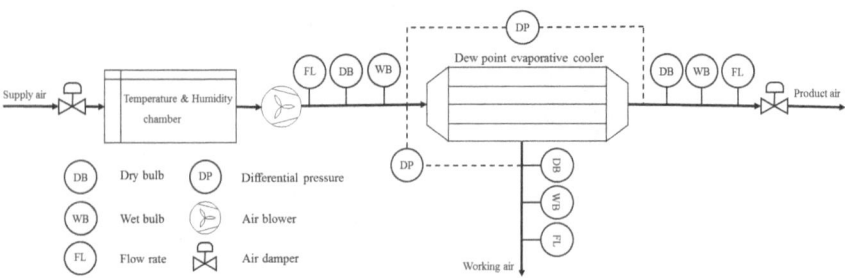

Fig. 3.24 Experimental set-up of the counter-flow dew point evaporative cooler

Table 3.2 Specifications and test conditions of the dew point evaporative cooler

Parameter	Symbol	Nominal value		Range	
Cooler prototypes		V1	V2	V1	V2
Channel length (m)	L	0.6	0.6	–	–
Channel width (mm)	W	150	150	–	–
Channel height (mm)	H_t	4.5	3.0	–	–
Number of channel pairs (–)	N	5	10	–	–
Supply air temperature (°C)	T_s	30.0	35.0	28.0–35.0	30.0–40.0
Supply air humidity (g/kg)	ω_s	17.0	11.0	11.5–21.5	10.5–12.0
Supply air velocity (m/s)	v_s	1.50	2.00	0.98–2.27	1.11–2.00
Working air ratio (–)	r	0.50	0.33	0.21–0.69	0.22–0.79

the supply and product/working air streams. The working air ratio of the cooler was regulated using an air damper at the product air stream outlet. The DB temperature, WB temperature and flow rate of the supply, working and product air streams were simultaneously measured during each experiment.

In this system, RTD temperature sensors (range: 100–250 °C, accuracy: 1/10 DIN), differential pressure sensors (range: ±200 Pa, accuracy: ±1.0% FS) and air velocity meters (range: 0–5.08 m/s, accuracy: ±2.0% FS) were employed to measure temperatures, differential pressures, and flow rates. The Agilent data logger was employed to record all the experimental data during testing. The scan time for data acquisition was designated to be every 2 s. To measure the transient state response of the cooler, experiments were carried out after the supply air conditions in the temperature–humidity climatic chamber had reached the desired set points. As for the steady-state performance of the evaporative cooler, the experimental data were evaluated when the product air conditions remained highly stable, i.e. temperature variations were less than 0.2 °C.

3.4.2.2 Model Development

The lumped parameter mathematical model of the counter-flow dew-point evaporative cooler is developed based on a single pair of dry and wet air channels, as shown in Fig. 3.25. Employing the grid meshing technique, a pair of dry and wet channels is divided into many identical cell elements. Each cell contains half the height of the dry and wet channels, the separating plate and water film in between. A transient model is first derived to obtain the spatial and temporal temperature distributions for the supply air, the working air, the water film, and the humidity distribution of the working air. Subsequently, an improved steady-state model is formulated from the transient model to capture the nominal cooling performance under different simulation conditions. The assumptions installed for the lumped-parameter model include:

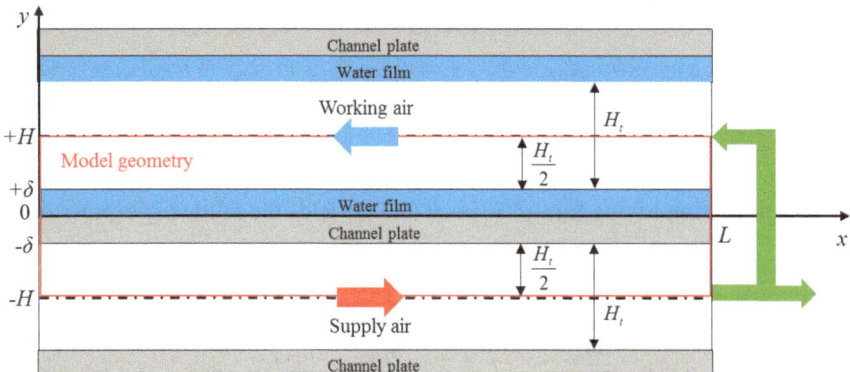

Fig. 3.25 Model geometry of the counter-flow dew point evaporative cooler

(1) The cooling system is well insulated and heat interaction with the surrounding
    is negligible;
(2) The air flow is steady, and the influence of pressure difference along the air flow
    is neglected;
(3) The entire wet channel surface is covered by a layer of stagnant and saturated
    water film, and the influence of wick material is ignored;
(4) The air and water properties are uniform in each control volume, and their bulk
    average values are used; and
(5) The channel plate is integrated with the water film in the transient modelling
    and their temperature difference is neglected.

Based on the energy and mass balances conducted for an infinitesimal time interval
in each of the control volume shown in Fig. 3.26, the modelling exercise of the
counter-flow cooler follows a similar process to that of the earlier cross-flow cooler.
The governing differential equations are derived below:

(1) Supply air

According to the control volume shown in Fig. 3.26a, the energy balance equation
of the supply air can be expressed as

$$\Delta U_{CV} = q_{d,x} - q_{d,x+dx} - q_{d,y} + m_d i_{d,x} - m_d i_{d,x+dx} \qquad (3.29)$$

Expanding and rearranging each term in Eq. (4.1), the equation is simplified into

$$\rho_a c_v \frac{\partial T_d}{\partial t} = k_a \frac{\partial^2 T_d}{\partial x^2} - \frac{2\overline{h}_d}{H_t}(T_d - T_f) - \rho_a c_p v_{dm} \frac{\partial T_d}{\partial x} \qquad (3.30)$$

(2) Working air

Refer to the control volume in Fig. 3.26b, the energy and mass balance equations of
the working air are written as

$$\Delta U_{CV} = q_{w,x} - q_{w,x+dx} - q_{w,y} + m_w i_{w,x+dx} - m_w i_{w,x} + n_{f,y} i_{f,y} \qquad (3.31)$$

$$\Delta M_{CV} = n_{w,x} - n_{w,x+dx} + n_{f,y} \qquad (3.32)$$

Therefore, simplifying Eqs. (3.31) and (3.32) yields the following expressions

$$\rho_a c_v \frac{\partial T_w}{\partial t} + u_v(T_w) \cdot \frac{\partial \rho_v}{\partial t} = k_a \frac{\partial^2 T_w}{\partial x^2} - \frac{2\overline{h}_w}{H_t}(T_w - T_f)$$

$$+ \rho_a c_p v_{wm} \frac{\partial T_w}{\partial x} + i_v(T_w) \cdot v_{wm} \frac{\partial \rho_v}{\partial x} + \frac{2\overline{h}_m}{H_t}(\rho_{v,sa} - \rho_v) \cdot i_v(T_f) \qquad (3.33)$$

Fig. 3.26 Differential control volume in the transient mathematical model for: **a** dry channel; **b** wet channel; and **c** water film

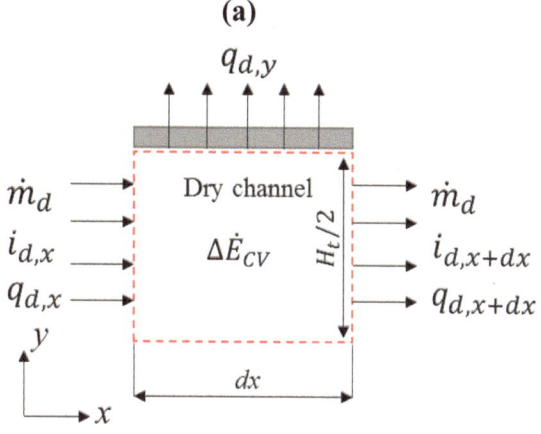

(a)

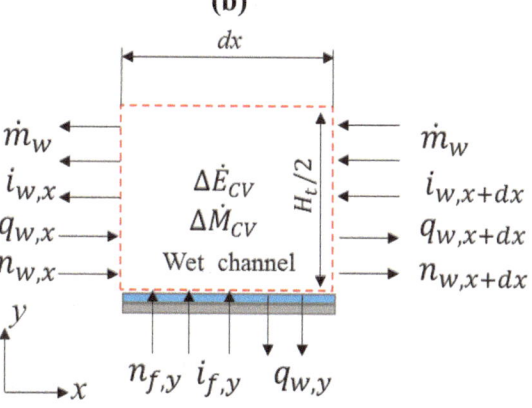

(b)

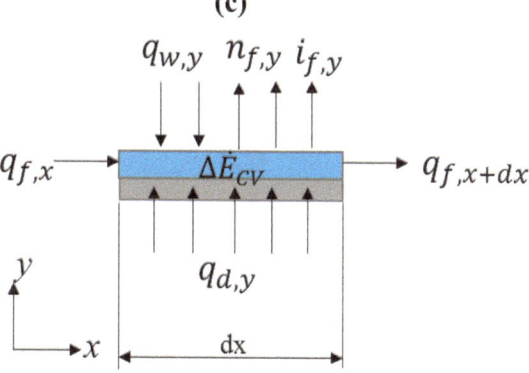

(c)

$$\frac{\partial \rho_v}{\partial t} = D_{va}\frac{\partial^2 \rho_v}{\partial x^2} + v_{wm}\frac{\partial \rho_v}{\partial x} + \frac{2\overline{h}_m}{H_t}(\rho_{v,sa} - \rho_v) \tag{3.34}$$

(3) Water film

On the water film, the combined heat and mass transfer process between the water and the air streams is considered in Fig. 3.26c. Similar approaches are adopted to enable appropriate energy and mass balances.

$$\Delta U_{CV} = c_f m_f \frac{\partial T_f}{\partial t} + c_f T_f \frac{\partial m_f}{\partial t} = q_{f,x} - q_{f,x+dx} + q_{d,y} + q_{w,y} - n_{f,y}i_{f,y} \tag{3.35}$$

$$\Delta M_{CV} = \frac{\partial m_f}{\partial t} = -n_{f,y} = -\overline{h}_m(\rho_{v,sa} - \rho_v)(Wdx) \tag{3.36}$$

As the concentration of the water film remains constant, the water film temperature is, therefore, a key concern in the model. Rearranging Eqs. (3.35) and (3.36) yields the governing equation for the water film as

$$\rho_f c_f \frac{\partial T_f}{\partial t} = k_f \frac{\partial^2 T_f}{\partial x^2} + \frac{\overline{h}_d}{\delta_f}(T_d - T_f) + \frac{\overline{h}_w}{\delta_f}(T_w - T_f)$$

$$+ \frac{\overline{h}_f}{\delta_f}(\rho_{v,sa} - \rho_v)(c_f T_f - i_v(T_f)) \tag{3.37}$$

The initial and boundary conditions are established in Table 3.3 for temperature and humidity of the air streams and water film. The initial temperature inside the cooler is assumed to be uniform everywhere at the ambient temperature.

The steady-state model is obtained by taking into account the temperature difference between the channel plate and water film, while setting the time derivative terms

Table 3.3 Initial and boundary conditions for the model installation

Supply air	Working air
$t \le 0, 0 \le x \le L: \quad T_d = T_0$	$t \le 0, 0 \le x \le L: \quad T_w = T_0, \rho_v = \rho_{v,sa}(T_0)$
$t > 0, x = 0: \quad T_d = T_s, \frac{\partial T_d}{\partial x} = 0$	$t > 0, x = L: \quad T_w = T_d, \rho_v = \rho_{v,s}$
	$t > 0, x = 0: \quad \frac{\partial T_w}{\partial x} = 0, \frac{\partial \rho_v}{\partial x} = 0$

Water film
$t \le 0, 0 \le x \le L: \quad T_f = T_0$
$t > 0, x = 0: \quad \frac{\partial T_f}{\partial x} = 0$
$t > 0, x = L: \quad \frac{\partial T_f}{\partial x} = 0$

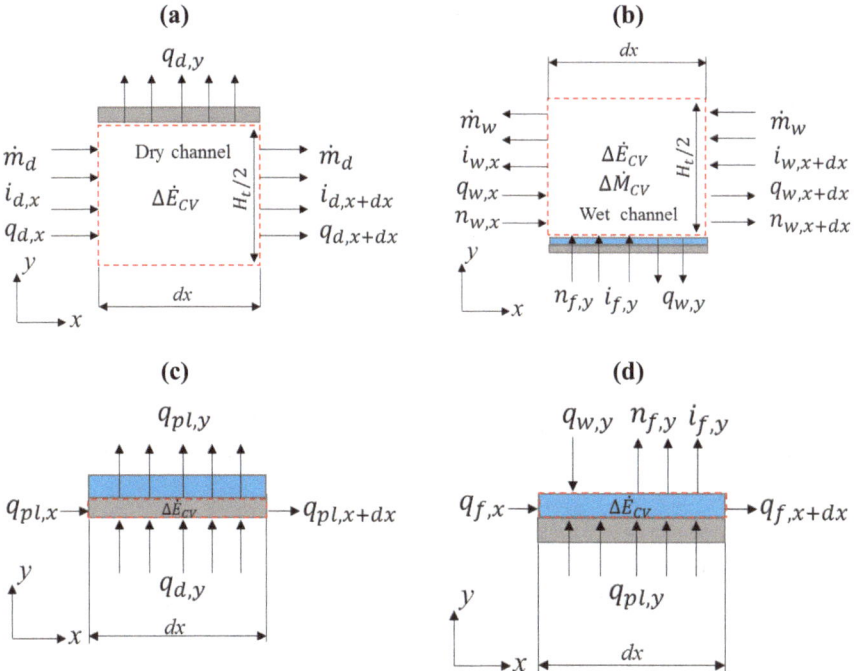

Fig. 3.27 Differential control volume in the steady-state mathematical model for: **a** dry channel; **b** wet channel; **c** plate surface; and **d** water film

in the governing equations to zero. As shown in Fig. 3.27, the control volume of the channel plate and water film are separated to improve the accuracy of the model.

Details of the governing differential equations are outlined below:

(1)   Supply air

$$\rho_a c_p v_{dm} \frac{dT_d}{dx} = k_a \frac{d^2 T_d}{dx^2} - \frac{2\bar{h}_d}{H_t}(T_d - T_{pl}) \tag{3.38}$$

(2)   Working air

$$\rho_a c_p v_{wm} \frac{dT_w}{dx} = -k_a \frac{d^2 T_w}{dx^2} + \frac{2\bar{h}_w}{H_t}(T_w - T_f)$$
$$- i_v(T_w) \cdot v_{wm} \frac{d\rho_v}{dx} - \frac{2\bar{h}_m}{H_t}(\rho_{v,sa} - \rho_v) \cdot i_v(T_f) \tag{3.39}$$

$$v_{wm}\frac{d\rho_v}{dx} = -D_{va}\frac{d^2\rho_v}{dx^2} - \frac{2\overline{h}_m}{H_t}(\rho_{v,sa} - \rho_v) \tag{3.40}$$

(3) Channel plate

$$k_{pl}\delta_{pl}\frac{d^2 T_{pl}}{dx^2} + \overline{h}_d(T_d - T_{pl}) - k_{pl}\frac{(T_{pl} - T_f)}{\delta_{pl}} = 0 \tag{3.41}$$

(4) Water film

$$k_f\frac{d^2 T_f}{dx^2} + \frac{k_{pl}}{\delta_f\delta_{pl}}(T_{pl} - T_f) + \frac{\overline{h}_w}{\delta_f}(T_w - T_f) + \frac{\overline{h}_m}{\delta_f}(\rho_{v,sa} - \rho_v)(c_f T_f - i_v(T_f)) = 0 \tag{3.42}$$

The convective heat transfer coefficient is calculated using Nusselt number. Due to the fact that the channel size and the air velocity are relatively small, the air stream in the fully developed region is considered to be laminar flow.

In the dry channel, the entry length is calculated from [56]

$$\frac{l_{th}}{D_h} = 0.05 Re_D Pr \tag{3.43}$$

The corresponding Nusselt number in the entry region for a rectangular channel can be calculated using the following correlation [56, 63]

$$Nu_D = \frac{\dfrac{7.54}{\tanh\left(2.264 Gz_D^{-\frac{1}{3}} + 1.7 Gz_D^{-\frac{2}{3}}\right)} + 0.0499 Gz_D \tanh\left(Gz_D^{-1}\right)}{\tanh\left(2.432 Pr^{\frac{1}{6}} Gz_D^{-\frac{1}{6}}\right)} \tag{3.44}$$

where $Gz_D = \left(\frac{D_h}{x}\right) Re_D Pr$.

For a fully developed laminar flow, the Nusselt number is constant either at constant surface temperature or constant heat flux. But in the real situation, the plate surface is neither of the two cases. Both types of the Nusselt number have been employed to approximate the heat transfer coefficient in the literature [15, 30, 33]. In this paper, Nusselt number at constant surface temperature is used for the dry channel [56]

$$Nu_D = 7.54 \tag{3.45}$$

The Nusselt number for the wet channel is calculated using the Dowdy and Karabash correlation and the mass transfer coefficient is determined using the Chilton-Colburn analogy, as presented as $\frac{\overline{h}_w}{\overline{h}_m} = \rho_a c_p Le^{2/3}$ where $Le = \frac{\alpha_a}{D_{va}}$.

Furthermore, in the wet channel, the heat transfer incorporates the overall effect of sensible and latent heat transfer. Although the latent heat transfer is maintained by the mass transfer through water evaporation, it can be approximated using the sensible heat transfer approach. Therefore, an overall heat transfer coefficient for the working air stream is defined by combining the sensible and latent heat transfer in the wet channel [42]

$$\left|\overline{h}_{\text{overall}}(T_f - T_w)\right| = \left|\overline{h}_w(T_f - T_w) + \overline{h}_m(\rho_{v,sa} - \rho_v) \cdot i_v(T_f)\right| \tag{3.46}$$

$$\overline{h}_{\text{overall}} = \left|\overline{h}_w + \frac{\overline{h}_m(\rho_{v,sa} - \rho_v) \cdot i_v(T_f)}{(T_f - T_w)}\right| \tag{3.47}$$

The proposed lumped parameter model is employed to predict the thermodynamic performance of the cooler prototype and to analyse the transient and steady-state behaviour of the counter-flow dew-point evaporative cooler. It is simulated using the MATLAB platform. The differential governing equations in the model are discretized into nonlinear equations using finite difference scheme. Because the inlet conditions are known for all variables, a forward difference scheme can be appropriately employed. The Newton Raphson iteration method was adopted to solve the coupled governing and heat and mass transfer equations simultaneously. In the simulation, the spatial grid size is set at 0.05 m, i.e. 20 cell elements along the channel. A grid independence test revealed that when the number of spatial cell elements increases from 20 to 100, the change of product temperature is within 0.05 °C or 0.20%. For the progressive iterations, the time step is set to be 0.01 s, while the number of temporal elements depends on the duration of simulation.

3.4.2.3 Computational Fluid Dynamics Model

Thus far, existing mathematical models are based on the lumped parameter method or bulk average values of the air flow. The flow field, temperature distribution in the control volume, and humidity distribution across the working air are typically ignored. The heat and mass transfer rates across the channels are often estimated using the transfer coefficients which are derived from the correlations for the Nusselt and Sherwood numbers [56]. Although these models are able to predict the cooler with reasonably good accuracy, they are not suitable for conducting scaling and dimensional analysis for the following reasons: (1) the relative importance of each physical mechanism cannot be easily determined, as some key terms have been neglected, such as the momentum and the continuity equations; and (2) the relevant dimensionless groups/numbers cannot be identified as some key factors have been simplified by other parameters, such as Reynolds number.

Therefore, in this section, a general 2D computational fluid dynamics (CFD) model is developed based on the basic geometry of a counter-flow dew-point evaporative cooler. The control volume of each domain in the model is further presented in Fig. 3.28. By considering the momentum, continuity, energy and species balance equations, the mathematical model involves the distribution of flow, temperature and humidity in x and y directions. Several key assumptions are adopted prior to the modelling exercise:

(1) The channel width is several orders of magnitude greater than the channel height, so that the geometry of the cooler can be viewed as the air flow between two parallel flat plates;
(2) The air flow is assumed to be an incompressible Newtonian fluid due to the small change of air density in the cooling process;
(3) The physical properties of the material are constant within the small temperature range, and the air properties are calculated based on its inlet conditions;
(4) The channel plate and the water film have almost same thickness in order to simplify the geometric expression; and

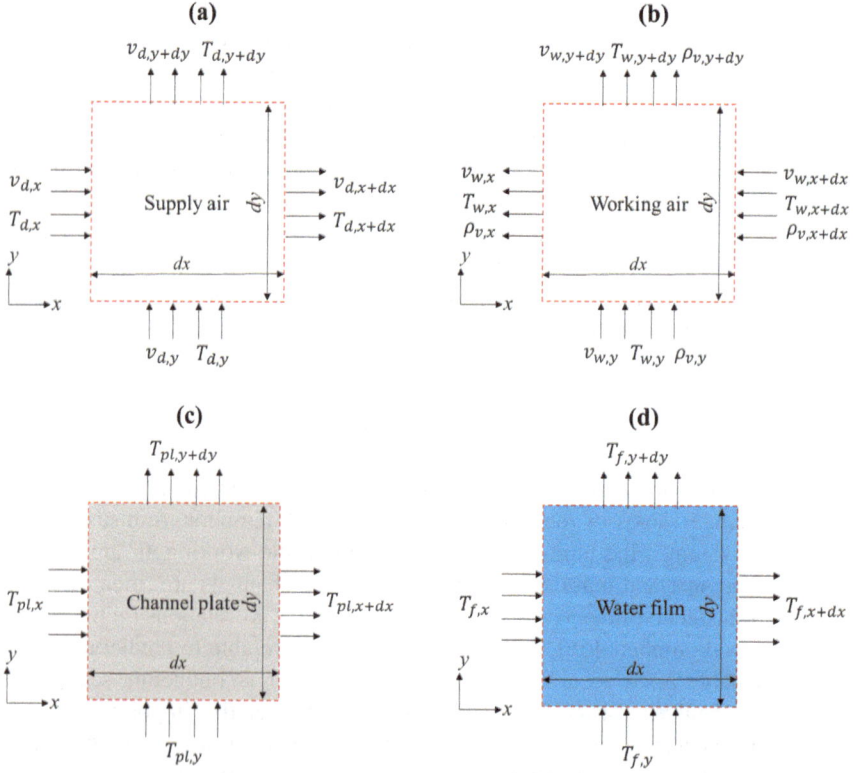

Fig. 3.28 Control volume of the model geometry: **a** supply air; **b** working air; **c** channel plate; and **d** water film

(5) The heat loss to the environment and the heat generation by viscous dissipation are taken to be negligible.

The following governing equations capture the physics of the counter-flow dew-point evaporative cooler:

(1) Supply air flow

$$X \text{ - momentum}: \rho_a \frac{\partial v_{dx}}{\partial t} + \rho_a v_{dx} \frac{\partial v_{dx}}{\partial x} + \rho_a v_{dy} \frac{\partial v_{dx}}{\partial y} =$$
$$- \frac{\partial P_d}{\partial x} + \mu_a \frac{\partial^2 v_{dx}}{\partial x^2} + \mu_a \frac{\partial^2 v_{dx}}{\partial y^2} \tag{3.48}$$

$$Y \text{- momentum}: \rho_a \frac{\partial v_{dy}}{\partial t} + \rho_a v_{dx} \frac{\partial v_{dy}}{\partial x} + \rho_a v_{dy} \frac{\partial v_{dy}}{\partial y} =$$
$$- \frac{\partial P_d}{\partial y} + \mu_a \frac{\partial^2 v_{dy}}{\partial x^2} + \mu_a \frac{\partial^2 v_{dy}}{\partial y^2} \tag{3.49}$$

$$\text{Continuity}: \frac{\partial v_{dx}}{\partial x} + \frac{\partial v_{dy}}{\partial y} = 0 \tag{3.50}$$

$$\text{Energy}: \rho_a c_p \frac{\partial T_d}{\partial t} + \rho_a c_p v_{dx} \frac{\partial T_d}{\partial x} + \rho_a c_p v_{dy} \frac{\partial T_d}{\partial y} = k_a \frac{\partial^2 T_d}{\partial x^2} + k_a \frac{\partial^2 T_d}{\partial y^2} \tag{3.51}$$

(2) Working air flow.

X-momentum:

$$\rho_a \frac{\partial v_{wx}}{\partial t} + \rho_a v_{wx} \frac{\partial v_{wx}}{\partial x} + \rho_a v_{wy} \frac{\partial v_{wx}}{\partial y} =$$
$$- \frac{\partial P_w}{\partial x} + \mu_a \frac{\partial^2 v_{wx}}{\partial x^2} + \mu_a \frac{\partial^2 v_{wx}}{\partial y^2} \tag{3.52}$$

Y-momentum:

$$\rho_a \frac{\partial v_{wy}}{\partial t} + \rho_a v_{wx} \frac{\partial v_{wy}}{\partial x} + \rho_a v_{wy} \frac{\partial v_{wy}}{\partial y} = - \frac{\partial P_w}{\partial y} + \mu_a \frac{\partial^2 v_{wy}}{\partial x^2} + \mu_a \frac{\partial^2 v_{wy}}{\partial y^2} \tag{3.53}$$

$$\text{Continuity}: \frac{\partial v_{wx}}{\partial x} + \frac{\partial v_{wy}}{\partial y} = 0 \tag{3.54}$$

$$\text{Energy}: \rho_a c_p \frac{\partial T_w}{\partial t} + \rho_a c_p v_{wx} \frac{\partial T_w}{\partial x} + \rho_a c_p v_{wy} \frac{\partial T_w}{\partial y} = k_a \frac{\partial^2 T_w}{\partial x^2} + k_a \frac{\partial^2 T_w}{\partial y^2} \tag{3.55}$$

$$\text{Species}: \frac{\partial \rho_v}{\partial t} + v_{wx} \frac{\partial \rho_v}{\partial x} + v_{wy} \frac{\partial \rho_v}{\partial y} = D_{va} \frac{\partial^2 \rho_v}{\partial x^2} + D_{va} \frac{\partial^2 \rho_v}{\partial y^2} \tag{3.56}$$

(3)   Channel plate

$$\text{Energy} : \rho_{\text{pl}} c_{\text{pl}} \frac{\partial T_{\text{pl}}}{\partial t} = k_{\text{pl}} \frac{\partial^2 T_{\text{pl}}}{\partial x^2} + k_{\text{pl}} \frac{\partial^2 T_{\text{pl}}}{\partial y^2} \tag{3.57}$$

(4)   Water film

$$\text{Energy} : \rho_{\text{f}} c_{\text{f}} \frac{\partial T_{\text{f}}}{\partial t} = k_{\text{f}} \frac{\partial^2 T_{\text{f}}}{\partial x^2} + k_{\text{f}} \frac{\partial^2 T_{\text{f}}}{\partial y^2} \tag{3.58}$$

The initial and boundary conditions of the governing equations are listed in Table 3.4. The proposed CFD model is employed to simulate the thermodynamic performance of the cooler prototype V2 and to conduct fundamental analysis of the counter-flow dew-point evaporative cooler. The governing equations, initial conditions, and boundary conditions are defined for the respective supply air, working air, channel plate, and water film domains. The entire model geometry is firstly meshed into

Table 3.4 Initial and boundary conditions of the cooler model

Supply air
$t \leq 0, 0 \leq x \leq L, -H \leq y \leq -\delta :$ $v_{\text{dx}} = 0, v_{\text{dy}} = 0, T_{\text{d}} = T_0$
$t > 0, x = 0 :$ $v_{\text{dx}} = v_{\text{s}}, v_{\text{dy}} = 0, T_{\text{d}} = T_{\text{s}}$
$t > 0, x = L :$ $P_{\text{d}} = 0, \frac{\partial T_{\text{d}}}{\partial x} = 0$
$t > 0, y = -\delta :$ $v_{\text{dx}} = 0, v_{\text{dy}} = 0, k_{\text{a}} \frac{\partial T_{\text{d}}}{\partial y} = k_{\text{pl}} \frac{\partial T_{\text{pl}}}{\partial y}$
$t > 0, y = -H :$ $\frac{\partial v_{\text{dx}}}{\partial y} = 0, \frac{\partial T_{\text{d}}}{\partial y} = 0$

Working air
$t \leq 0, 0 \leq x \leq L, \delta \leq y \leq H :$ $v_{\text{wx}} = 0, v_{\text{wy}} = 0, T_{\text{w}} = T_0, \rho_{\text{v}} = \rho_{\text{v,sa}}(T_0)$
$t > 0, x = L :$ $v_{\text{wx}} = -r v_{\text{s}}, v_{\text{wy}} = 0, T_{\text{w}} = T_{\text{d}}, \rho_{\text{v}} = \rho_{\text{v,s}}$
$t > 0, x = 0 :$ $P_{\text{w}} = 0, \frac{\partial T_{\text{w}}}{\partial x} = 0, \frac{\partial \rho_{\text{v}}}{\partial x} = 0$
$t > 0, y = \delta :$ $v_{\text{wx}} = 0, v_{\text{wy}} = 0, \rho_{\text{v}} = \rho_{\text{v,sa}}(T_{\text{f}}), k_{\text{a}} \frac{\partial T_{\text{w}}}{\partial y} + h_{\text{fg}} D_{\text{va}} \frac{\partial \rho_{\text{v}}}{\partial y} = k_{\text{f}} \frac{\partial T_{\text{f}}}{\partial y}$
$t > 0, y = H :$ $\frac{\partial v_{\text{wx}}}{\partial y} = 0, \frac{\partial T_{\text{w}}}{\partial y} = 0, \frac{\partial \rho_{\text{v}}}{\partial y} = 0$

Channel plate	Water film
$t \leq 0, 0 \leq x \leq L, -\delta \leq y \leq 0 :$ $T_{\text{pl}} = T_0$	$t \leq 0, 0 \leq x \leq L, 0 \leq y \leq \delta :$ $T_{\text{f}} = T_0$
$t > 0, x = 0 :$ $\frac{\partial T_{\text{pl}}}{\partial x} = 0$	$t > 0, x = 0 :$ $\frac{\partial T_{\text{f}}}{\partial x} = 0$
$t > 0, x = L :$ $\frac{\partial T_{\text{pl}}}{\partial x} = 0$	$t > 0, x = L :$ $\frac{\partial T_{\text{f}}}{\partial x} = 0$
$t > 0, y = 0 :$ $k_{\text{pl}} \frac{\partial T_{\text{pl}}}{\partial y} = k_{\text{f}} \frac{\partial T_{\text{f}}}{\partial y}$	

small quadrilateral finite elements so that the governing partial differential equations are discretized into algebraic equations via polynomial interpolation. The flow, temperature, and humidity fields can then be solved simultaneously.

The mesh scheme of the model geometry is shown in Fig. 3.29. A grid independence test is carried out and results are presented in Fig. 3.30. The simulation conditions are set at the nominal test conditions of cooler prototype listed in Table 3.5 ($T_s = 35.0\,°C$, $\omega_s = 11.0$ g/kg, $v_s = 2.00$ m/s and $r = 0.33$). The original mesh

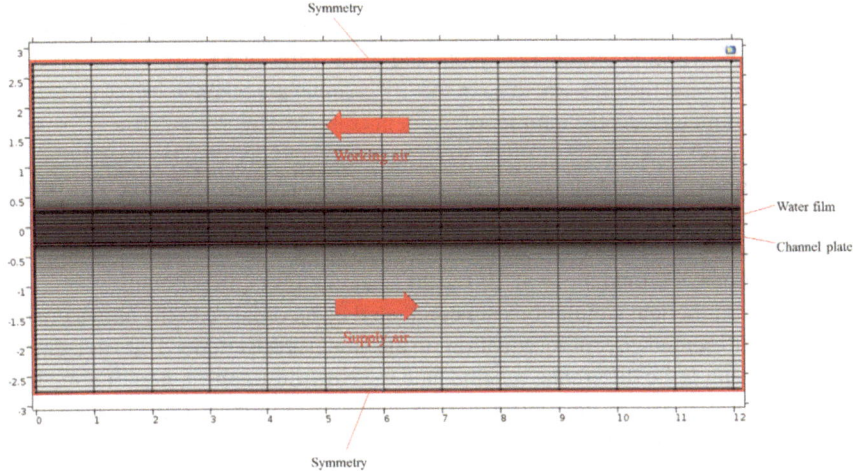

Fig. 3.29 Mesh scheme of the cooler model

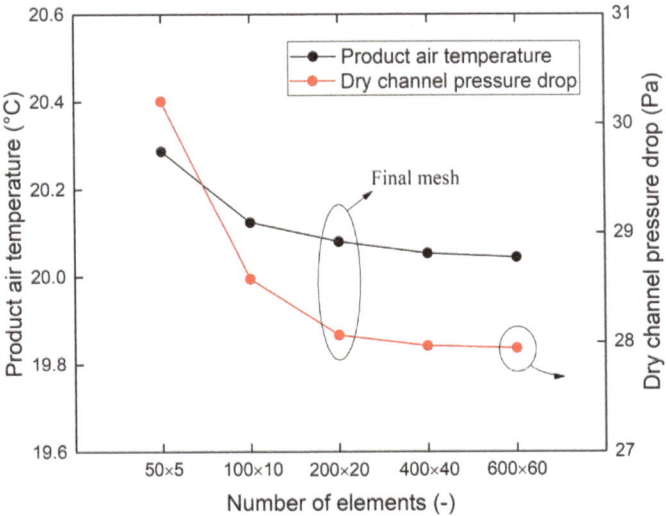

Fig. 3.30 Grid independence test of the numerical simulation

Table 3.5 Nominal cooling system specifications

Parameter	Value
Channel length (m)	1
Channel width (mm)	100
Channel height (mm)	5
Working to supply air ratio	0.3
Inlet air velocity (m/s)	2.0

is defined to be 50 elements in the x direction and 5 elements in the y direction for the supply air, working air, channel plate and water film, respectively. The product air temperature and the pressure drop in the dry channel are taken to be the reference objective parameters. As the number of elements rises to $100 \times 10, 200 \times 20, 400 \times 40$ and 600×60 ($x \times y$), the product air temperature and dry channel pressure drop gradually approach 20.04 °C and 27.94 Pa. It is observed that the variations of the simulation results are within 0.5% when the number of elements increases to 200×20. Therefore, taking into consideration the processing time for simulation and the result variations, the final mesh of 200×20 is adopted for each domain.

Dynamic numerical simulations are conducted in accordance to the above conditions. The differential governing equations in the model are discretized into non-linear equations using finite difference scheme, and the Newton Raphson iteration method was adopted to solve the coupled governing and heat and mass transfer equations simultaneously in the MATLAB environment. The simulation process is further illustrated using the flow chart in Fig. 3.31. In the simulation, the spatial grid size is set at 0.05 m, i.e. 20 cell elements along the channel. A grid independence test revealed that when the number of spatial cell elements increases from 20 to 100, the change of product temperature is within 0.05 °C or 0.20%. On the other hand, the time step is set to be 0.01 s, while the number of temporal elements depends on the duration of simulation.

3.4.2.4 Effect of Water Spray

The transient characteristics of the channel plate when subjected to the water spray are first investigated. Figure 3.32 shows the plate surface temperatures at four different measurement points in both vertical and horizontal orientations. The data points are plotted for every 10 s, while more data are provided in the transient response regions. During testing, water was sprayed on the channel plate for three times, with a sufficient time interval in between. The temperatures of the supply water and the environment were maintained at around 23.0 °C. Additionally, the plate temperature difference between the dry side and wet side was found to be within 0.3 °C, so the temperature data on the dry side are omitted in the figure.

Before testing, the channel plate was fully wetted and surrounded by a forced air flow. The channel plate temperature gradually decreased from the initial water supply temperature due to the evaporative cooling process. Finally, a temperature

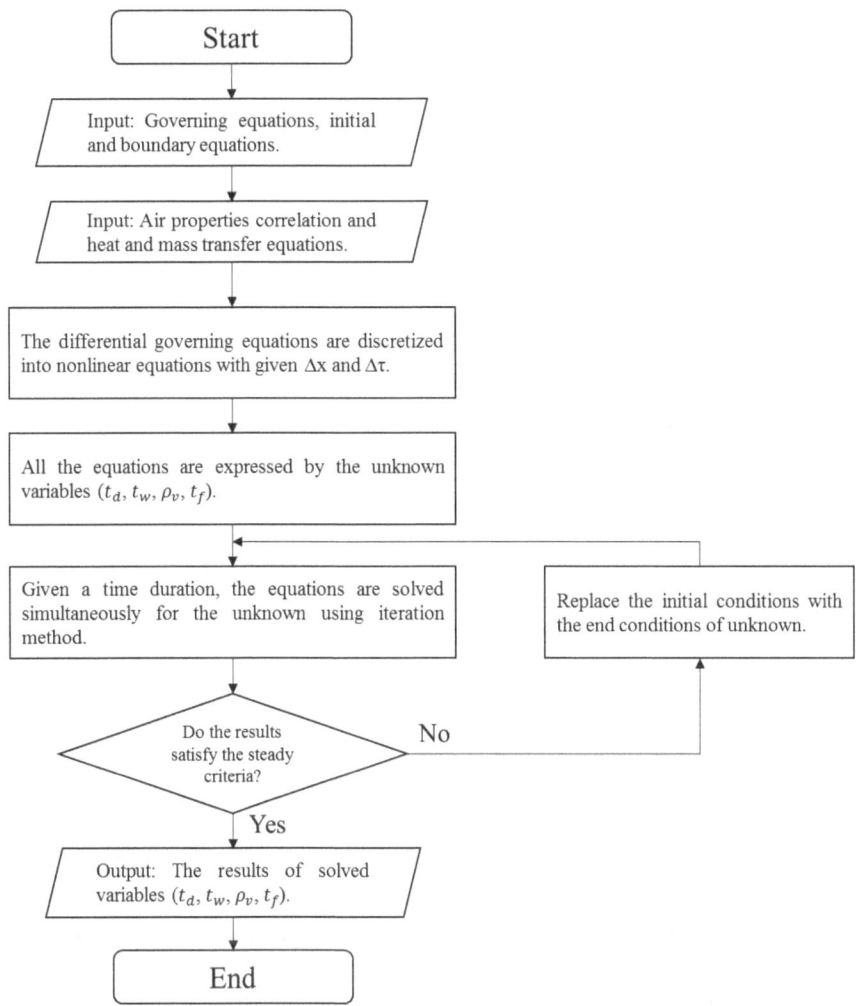

Fig. 3.31 Flow chart of simulation process

distribution is developed in the channel plate along the flow direction. As shown in Fig. 3.32a, c, the first water spray takes place immediately after the experiment starts, followed by a second and third spray at $t = 300$ and 600 s, respectively. It is apparent that the channel plate temperature at all measurement points undergoes a dramatic fluctuation after each water spray. The plate temperature increases to a peak value within a short time period of 20 s, and then gradually recovers to the steady-state after 150–200 s. This result indicates that the sprayed water has a marked effect on the state of the channel plate. When water is rapidly supplied to the channel plate, the initial fully developed temperature distribution along the channel plate is disrupted, leading to a significant transient response in the temperature profile. Consequently,

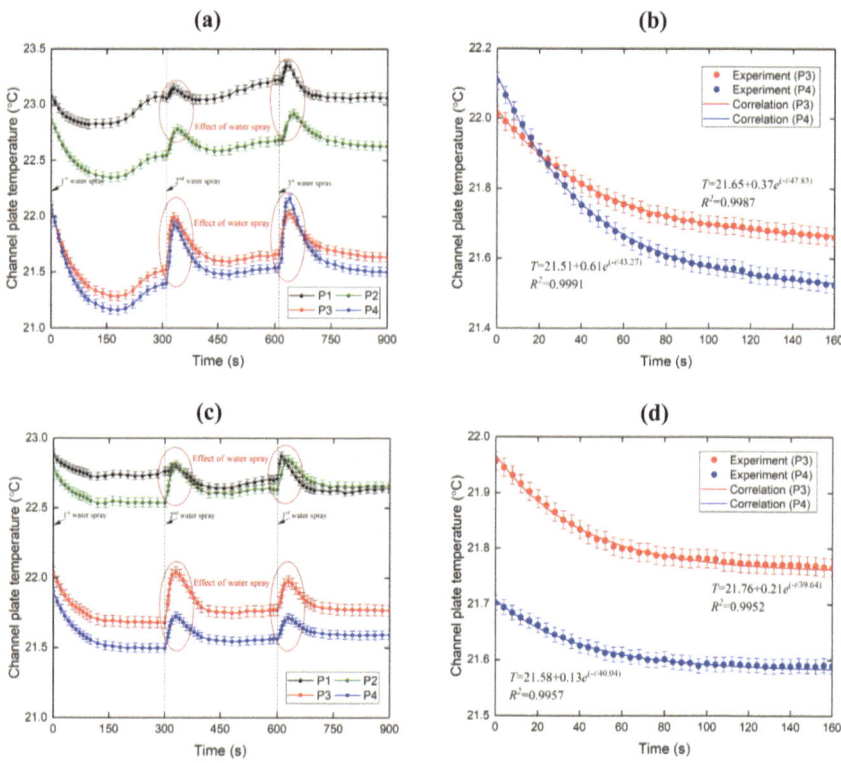

Fig. 3.32 Effect of water spray on the channel plate temperature: **a** temperature profiles in vertical plate orientation; **b** temperature decay at P3 and P4 (vertical orientation); **c** temperature profiles in horizontal plate orientation; and **d** temperature decay at P3 and P4 (horizontal orientation)

the increment in the plate temperature degrades the dynamic cooling performance of the cooler, as more time is required for the channel plate to become stable.

Furthermore, it is noted that the recovery of the channel plate temperature after a transient response takes significantly longer time than the interruption of the plate at a steady-state. This is attributed to the observation that the temperature profile after reaching the peak value follows an exponential decay function, as expressed below [64]

$$T(t) = T_{st} + Ce^{-t/t_c} \tag{3.59}$$

where $T(t)$ denotes the channel plate temperature at a certain measurement point, T_{st} is the steady-state temperature, C is the temperature difference between the peak point and the steady-state, and t_c is the time constant or mean lifetime for temperature decay.

Accordingly, the detailed temperature profiles at P3 and P4 after the third spray are selected for plotting in Fig. 3.32b, d, together with their respective decay functions. It is observed that the time constant of the temperature profiles lies in the range of 40–50 s. The corresponding settling time to reach the steady-state, usually calculated as 4τ, varies from 158.6 to 191.3 s [65]. This finding provides the guideline for the time interval between every two sprays. If the time interval is shorter than the settling time, the channel plate temperature will periodically fluctuate and its steady state cannot be reached. The amplitude of the temperature fluctuation depends on the water supply temperature that is normally larger than the plate temperature. In this case, the cooling process remains unstable and the dynamic cooling performance is not able to achieve the optimal state. In order to minimize the disturbance due to the changing channel plate temperature, the time interval between water sprays should be longer than the settling time. Alternatively, instead of using water spray, water can be distributed to the entire channel plate by capillary force at a slow and steady supply rate.

3.4.2.5 Model Validation

The proposed transient model is first simulated by varying the inlet air temperature from 25 to 40 °C and humidity ratio between 10 and 18 g/kg. The simulation was conducted for a sufficient time interval until the objective variables remained almost unchanged, which was perceived as reaching the steady-state condition. In order to evaluate the discrepancy of the dynamic simulation with reference to the steady-state simulation, the final product air temperatures are compared with the results obtained from the steady-state model presented earlier. Figure 3.33 compares the simulated product air temperature development in transient and steady-state modelling. It is observed that the product air temperatures under steady-state condition agree well with the results of steady-state model, and the errors are within ±0.02 °C.

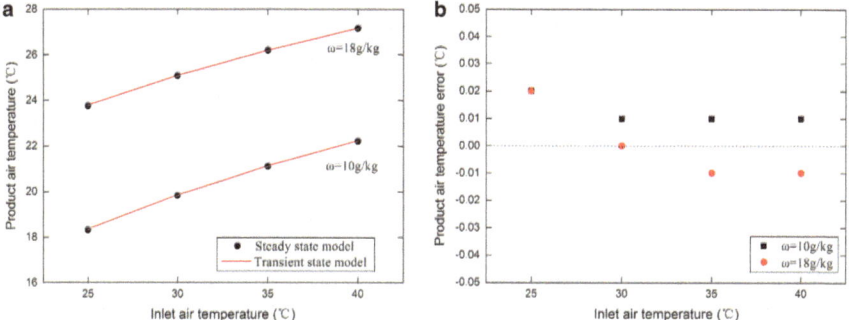

Fig. 3.33 Comparing transient and steady state models: **a** product air temperature; and **b** error in product air temperature

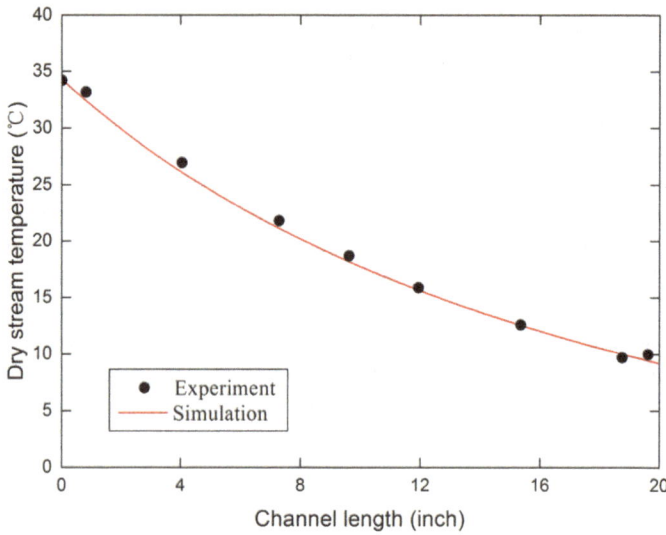

Fig. 3.34 Model validation with dry stream temperature along the channel from Hsu et al. [31]

Simulations are further carried out to validate the proposed model with acquired experimental data from literature, including both steady-state and transient responses. Hsu et al. [31] conducted an experimental study on a counter-flow, closed-loop heat exchanger with a supply air stream of 34.2 °C dry bulb and 15 °C wet bulb. The supply air stream temperatures along the channel were measured at steady state. The present model was validated using their experimental data and is displayed in Fig. 3.34. It is apparent that our proposed model predicts the dry stream temperature with close proximity to within 4.3%.

Riangvilaikul and Kumar [13] carried out a series of experiments on a dew-point evaporative cooler with four dry channels and five wet channels. The influences of inlet air temperature, humidity, and velocity on steady-state performance were investigated individually while other parameters were kept unchanged during the experiments. The present model is validated using these experimental data as can be seen in Fig. 3.35. It is observed that the maximum discrepancy of the model is 3.1% under different inlet temperature or humidity, and 2.2% under different inlet air velocity.

Bruno [59] tested the dynamic performance of a counter-flow dew-point evaporative cooling system in a commercial building. The inlet and outlet air conditions of the cooler were measured throughout the day. However, only the dry bulb air temperatures are provided by the author. To evaluate the model's transient predictions, the inlet air humidity is estimated based on the dew-point temperature range, spanning 1–18 °C, equivalent to 4.1–s13.0 g/kg in humidity, while the average air humidity is about 8.0 g/kg. We validate our transient model with the outlet temperature for the initial period of 150 min in Fig. 3.36. Apparently, the profile of simulation results follows the experimental data for most of the points, albeit some insignificant delay

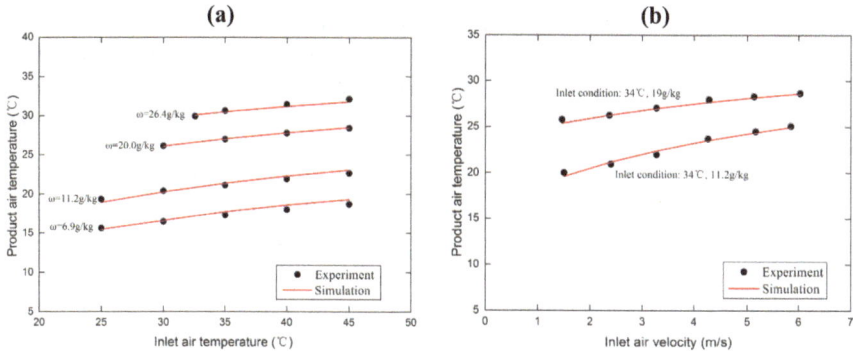

Fig. 3.35 Model validation with product air temperature data acquired from Riangvilaikul and Kumar [13] under different inlet conditions: **a** temperature and humidity ratio; and **b** velocity

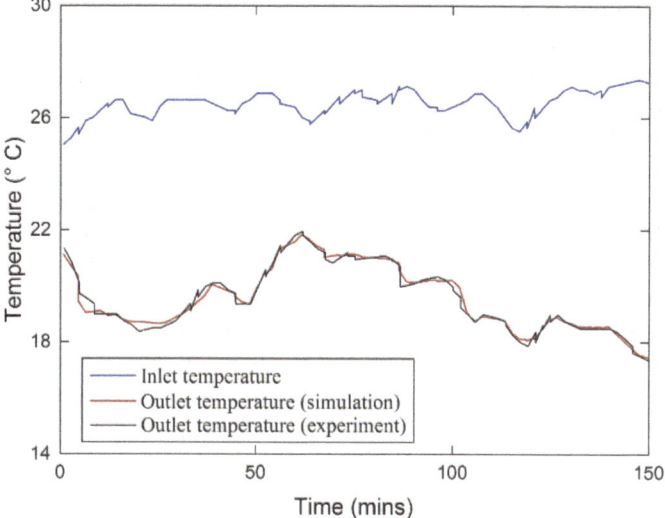

Fig. 3.36 Model transient results compared with outlet air temperature from Bruno [59]

and disagreement. The maximum discrepancy of the outlet temperature in this period is about 2.8%.

3.4.3 Modified LMTD Model

The log mean temperature difference (LMTD) method is a well-known and simple method that is capable of evaluating the thermal performance of a sensible heat exchanger. The LMTD method, when compared with the arithmetic mean temperature method, is able to predict with better accuracy, particularly when the temperature change in the heat exchanger is non-linear and the greatest temperature difference at one end of the heat exchanger is more than twice the temperature difference at the other end of the heat exchanger.

However, thus far, the LMTD method has not been effectively applied to study a heat exchanger where the latent heat is associated. To the best of our knowledge, no research work has been conducted to extend the LMTD method to investigate the indirect evaporative cooling system. The current work is proposed to address these issues via a modified LMTD analytical model.

To develop a modified LMTD method suitable for studying evaporative cooling systems, proper modifications are necessary. Appropriate assumptions for the modified LMTD method include: (1) water film is evenly distributed on wet surfaces; (2) heat and mass transfer occurs in the direction normal to the air flow; (3) Lewis factor equals to one; (4) the air flow is fully developed and laminar; and (5) no heat transfer occurs to the surrounding.

The modified LMTD method is developed by using a counter-flow plate type indirect evaporative cooler as an example. A computational element is first defined to lay the foundation for the mathematical model as shown in Fig. 3.37. It comprises half of the product channel and the working channel due to geometrical symmetry.

The wet channel contains wicking material that is evenly saturated with water. As water flow rate by convection is small, water in the wet channel is treated as a thin stagnant film. To simplify the model with reasonable accuracy, the heat and mass transfer by convection mode is considered not significant in comparison to conduction mode. Therefore, the convection mode of heat and mass transfer between water film and plate is considered negligible in the analytical model.

As the hydrodynamic entry length and the thermal entry length are small in comparison to the length of the wet channel, the effect of developing zone is neglected in the proposed model. The airflow in the IEHX is assumed to be fully developed and

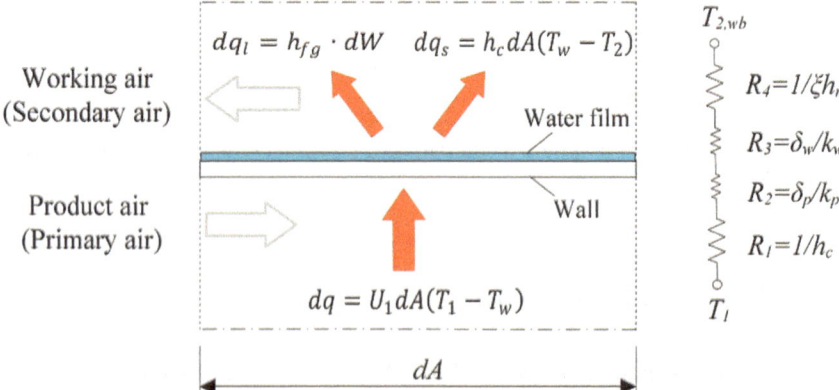

Fig. 3.37 Computational element for mathematical modeling of a counter-flow IEHX

laminar. Nusselt number for appropriate flow geometry, obtained from literature, is conveniently used in the analytical model.

On the dry side, only sensible heat flow occurs between the product air and the water film:

$$dq = U_1 dA(T_1 - T_w) \tag{3.60}$$

where U_1 is the overall heat transfer coefficient that links the product air in the dry channel and the water film in the wet channel, and can be obtained from:

$$U_1 = \frac{1}{\frac{1}{h_c} + \frac{\delta_p}{k_p} + \frac{\delta_w}{k_w}} \tag{3.61}$$

Applying the conservation principle of energy for the product air in the dry channel, we obtain:

$$dq = -m_1 c_{pa} dT_1 \tag{3.62}$$

In the wet channel, the total rate of heat transfer between the water film and the working air is the sum of the sensible heat flow (dq_s) and latent heat flow (dq_l):

$$dq = dq_s + dq_l \tag{3.63}$$

Sensible heat flow is expressed as:

$$dq_s = h_c dA(T_w - T_2) \tag{3.64}$$

where h_c is the convection heat transfer coefficient between the water film and the ambient working air at T_2.

The mass transfer rate between the water film and the moist air is given by

$$dW = h_m dA(w_w - w_2) \tag{3.65}$$

Therefore, the latent heat flow due to water evaporation can be evaluated by

$$dq_l = h_{fg} \cdot dW = h_{fg} h_m dA(w_w - w_2) \tag{3.66}$$

Substituting Eqs. (3.64) and (3.66) into Eq. (3.63), the total heat transfer is written as

$$dq = h_c dA(T_w - T_2) + h_{fg} h_m dA(w_w - w_2) \tag{3.67}$$

The heat transfer coefficient, h_c, is calculated by using the Nusselt number. The Nusselt number for fully developed laminar flow is constant under a specific cross section of the channel [66].

Making the assumption that the Lewis factor for air–water mixture is unity [67], the following expression is written

$$\text{Le} = \frac{h_c}{h_m \cdot c_{pa}} = 1 \tag{3.68}$$

where $c_{pa} = 1.006 + 1.86w$ represents the specific heat of moist air. The mass transfer coefficient, h_m, and the convective heat transfer coefficient, h_c, are correlated by using Lewis factor. As a result, the total heat transfer in wet channel can be expressed as follows:

$$
\begin{aligned}
dq &= h_m \cdot c_{pa} dA (T_w - T_2) + h_{fg} h_m dA (w_w - w_2) \\
&= h_m dA \big[(c_{pa} T_w + h_{fg} w_w) - (c_{pa} T_2 + h_{fg} w_2) \big] \\
&= h_m dA (h_w - h_2)
\end{aligned}
\tag{3.69}
$$

As described in Eq. (3.69), the driving force of total heat transfer in wet channels is determined by the enthalpy difference between the saturated air at water surface and the main moist air stream. It is assumed that the thin moist air layer at the water–air interface is saturated at the water film temperature, T_w. Therefore, the enthalpy of the saturated air at water surface is evaluated as:

$$h_w = c_{pa} T_w + h_{fg} w_w (T_w) \tag{3.70}$$

On the psychometric chart, the wet-bulb temperature lines are nearly parallel with the constant enthalpy lines. In Fig. 3.38, for example, line 1 (l_1) and line 2 (l_2) describe the constant enthalpy line. On the other hand, the wet-bulb temperature along these lines could be approximated as constant. As a result, the enthalpy of the moist air (such as point A) can be approximated by the enthalpy of saturated air at its wet-bulb temperature (such as point B) as

$$h_2(T_2) \approx h_2(T_{2,wb}) = c_{pa} T_{2,wb} + h_{fg} w_2(T_{2,wb}) \tag{3.71}$$

In other words, the enthalpy of air is a function of its wet-bulb temperature. It is reasonable to assume a linear function between the enthalpy and the wet-bulb temperature for small operating range of temperature. For example, the enthalpy difference (Δh) between point B and point C is able to be determined by the wet-bulb temperature difference (ΔT_{wb}) between these two points. To simplify the analysis, a new parameter, ξ, is introduced and is defined as the ratio between the change of enthalpy and the change of wet-bulb temperature and is written as

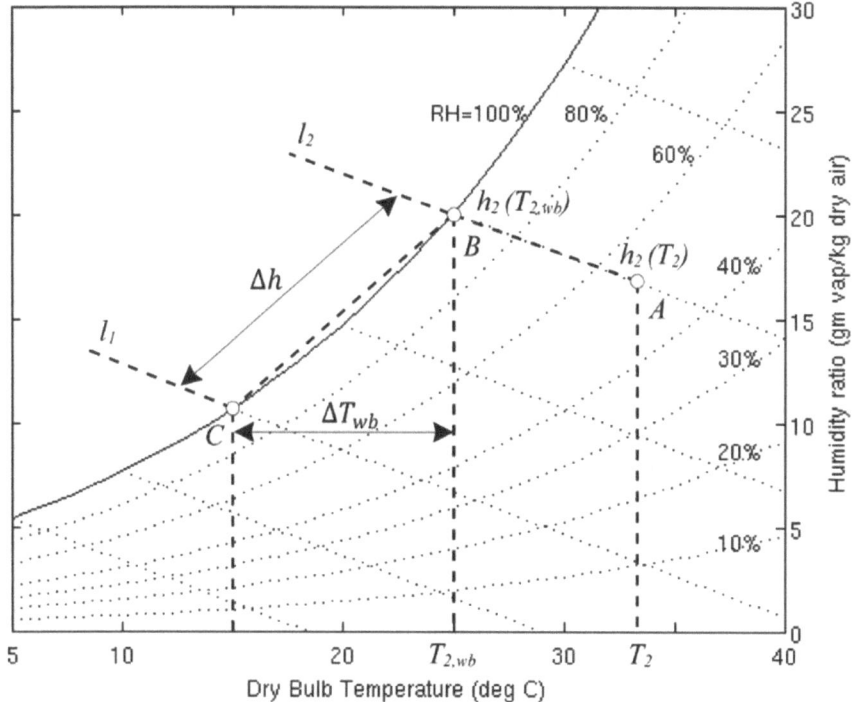

Fig. 3.38 Psychrometric illustration of the correlation between enthalpy and wet-bulb temperature

$$\xi = \frac{\Delta h}{\Delta T_{wb}} \tag{3.72}$$

The value of ξ can be evaluated by the inlet and outlet condition in wet channel. In Eq. (3.69), the heat transfer in wet channel is evaluated by the enthalpy difference $(h_w - h_2)$. Considering Eqs. (3.70), (3.71), and Eq. (3.72), the enthalpy difference, $(h_w - h_2)$, is able to be approximated by the temperature difference between the temperature of the water film and the wet-bulb temperature of the working air as $\xi(T_w - T_{2,wb})$. Therefore, Eq. (3.69) can be rewritten as:

$$dq = \xi h_m dA(T_w - T_{2,wb}) = U_2 dA(T_w - T_{2,wb}) \tag{3.73}$$

where U_2 is the modified overall heat transfer coefficient in wet channels. Thus, the modified thermal resistance in the wet channel is $R_4 = 1/\xi h_m$.

Conservation of energy for the working air within a selected computational element in the wet channel can be evaluated by the change of the wet-bulb temperature as well:

$$dq = -m_2 \cdot dh \approx -m_2 \xi dT_{2,wb} \tag{3.74}$$

Eliminating T_w from Eqs. (3.60) and (3.73):

$$dq = \frac{U_1 U_2}{U_1 + U_2} dA(T_1 - T_{2,\text{wb}}) = U dA(T_1 - T_{2,\text{wb}}) \tag{3.75}$$

where U is the modified overall heat transfer coefficient for combined dry and wet channels:

$$U = \frac{1}{\frac{1}{h_c} + \frac{\delta_p}{k_p} + \frac{\delta_w}{k_w} + \frac{1}{\xi h_m}} \tag{3.76}$$

Equation (3.75) is a key equation since it connects the dry-bulb temperature in the dry channel and the wet-bulb temperature in the wet channel.

The total rate of heat flow over the entire length is the integration of Eq. (3.75) from $A = 0$ to $A = A_{\text{total}}$:

$$Q = \int_0^A U(T_1 - T_{2,\text{wb}}) dA \tag{3.77}$$

From Eqs. (3.62) and (3.74), we obtain:

$$dT_1 = -\frac{dq}{m_1 c_{\text{pa}}} \tag{3.78}$$

$$dT_{2,\text{wb}} = -\frac{dq}{m_2 \xi} \tag{3.79}$$

Subtracting Eq. (3.79) from Eq. (3.78) leaves:

$$d(T_1 - T_{2,\text{wb}}) = -\left(\frac{1}{m_1 c_{\text{pa}}} - \frac{1}{m_2 \xi}\right) dq \tag{3.80}$$

Substituting Eq. (3.75) for dq in Eq. (3.80) yields,

$$\frac{d(T_1 - T_{2,\text{wb}})}{T_1 - T_{2,\text{wb}}} = -U\left(\frac{1}{m_1 c_{\text{pa}}} - \frac{1}{m_2 \xi}\right) dA \tag{3.81}$$

Integrating Eq. (3.81) over the entire surface, we have

$$\int_0^A \frac{d(T_1 - T_{2,\text{wb}})}{T_1 - T_{2,\text{wb}}} = -\int_0^A U\left(\frac{1}{m_1 c_{\text{pa}}} - \frac{1}{m_2 \xi}\right) dA \tag{3.82}$$

$$\ln(T_1 - T_{2,\text{wb}})\Big|_0^A = -U\left(\frac{1}{m_1 c_{\text{pa}}} - \frac{1}{m_2 \xi}\right) A\Big|_0^A \tag{3.83}$$

$$\ln \frac{(T_1 - T_{2,\text{wb}})_A}{(T_1 - T_{2,\text{wb}})_0} = -U\left(\frac{1}{m_1 c_{\text{pa}}} - \frac{1}{m_2 \xi}\right) A \tag{3.84}$$

Take a counter-flow IEHX as an example, Eq. (3.84) can be rewritten as:

$$\ln \frac{T_1'' - T_{2,\text{wb}}'}{T_1' - T_{2,\text{wb}}''} = -U\left(\frac{1}{m_1 c_{\text{pa}}} - \frac{1}{m_2 \xi}\right) A \tag{3.85}$$

Integrating Eqs. (3.62) and (3.74) over the entire length of the channel yields:

$$Q = m_1 c_{\text{pa}}\left(T_1' - T_1''\right) \tag{3.86}$$

$$Q = m_2 \xi \left(T_{2,\text{wb}}'' - T_{2,\text{wb}}'\right) \tag{3.87}$$

Substituting Eqs. (3.86) and (3.87) in Eq. (3.85), and after some rearrangement yields,

$$Q = UA \frac{\left(T_1' - T_{2,\text{wb}}''\right) - \left(T_1'' - T_{2,\text{wb}}'\right)}{\ln \frac{T_1' - T_{2,\text{wb}}''}{T_1'' - T_{2,\text{wb}}'}} \tag{3.88}$$

As a result, we obtain the logarithmic mean temperature difference (*LMTD*) for a counter-flow IEHX:

$$\text{LMTD} = \frac{\left(T_1' - T_{2,\text{wb}}''\right) - \left(T_1'' - T_{2,\text{wb}}'\right)}{\ln \frac{T_1' - T_{2,\text{wb}}''}{T_1'' - T_{2,\text{wb}}'}} \tag{3.89}$$

The LMTD for a parallel-flow IEHX can be derived in the same way as

$$\text{LMTD} = \frac{\left(T_1' - T_{2,\text{wb}}'\right) - \left(T_1'' - T_{2,\text{wb}}''\right)}{\ln \frac{T_1' - T_{2,\text{wb}}'}{T_1'' - T_{2,\text{wb}}''}} \tag{3.90}$$

Figure 3.39 illustrates the changes in temperature in an IEHX that may occur in the air stream for different air flow arrangements. Psychrometric properties of the moist air, such as humidity ratio and wet-bulb temperature, are used in the above equations. It is required to obtain a mathematical expression of these properties for computational analysis.

3.4.3.1 Modelling Process and Flow

The modified LMTD method is able to predict the outlet air temperature under a specific inlet condition. Since the modified overall heat transfer coefficient (U) and

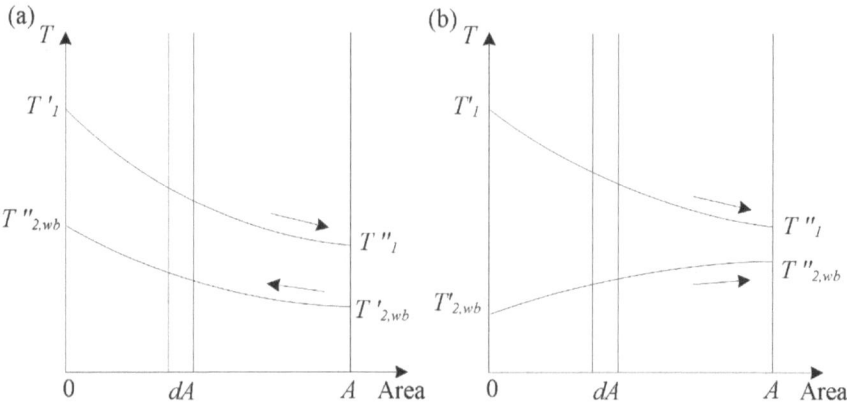

Fig. 3.39 Temperature profiles in indirect evaporative heat exchangers: **a** counter-flow IEHX; and **b** parallel-flow IEHX

the enthalpy to wet-bulb temperature ratio (ξ) are influenced by both inlet and outlet conditions, a judicious given of the initial outlet air temperatures (T_1'', $T_{2,wb}''$) are necessary.

For rating a counter-flow or parallel-sflow IEHX, the calculation procedure can be described as follows. First, the initial values of outlet air temperatures are chosen to determine the enthalpy to wet-bulb temperature ratio (ξ) and the modified overall heat transfer coefficient (U). Second, new values of outlet air temperature (T_1'', $T_{2,wb}''$) are calculated by solving a system of nonlinear equations which include Eqs. (3.85), (3.86), and (3.87). After that, the new value of ξ can be determined based on calculated values of outlet temperature. At last, new values of ξ, T_1'', and $T_{2,wb}''$ are compared with initial values in order to check the convergence. This procedure may be repeated until the desired convergence is achieved.

For rating a cross-flow IEHX, based on the procedure described above, a correction factor (F) is required to modify the LMTD. To obtain the true mean temperature difference for cross-flow arrangement, the LMTD predetermined for counter-flow should be multiplied by the appropriate correction factor so that Eq. (3.88) is modified to

$$Q = \mathrm{UA(LMTD)}(F) \tag{3.91}$$

The correction factor can be determined from the chart produced by Bowman et al. [68]. The dimensionless ratio P and R used in the chart can be calculated as:

$$P = \frac{T_{2,\mathrm{wb}}'' - T_{2,\mathrm{wb}}'}{T_1' - T_{2,\mathrm{wb}}'} \tag{3.92}$$

$$R = \frac{T_1' - T_1''}{T_{2,\mathrm{wb}}'' - T_{2,\mathrm{wb}}'} \tag{3.93}$$

For rating a counter-flow regenerative (M-cycle) IEHX, the determination of $T'_{2,\text{wb}}$ should be considered. In this type of cooler, a part of the product air at its outlet is redirected into the wet channel as working air. Therefore, the inlet wet-bulb temperature of the working air is determined by the outlet temperature of the product air:

$$T'_{2,\text{wb}} = T''_{1,\text{wb}} \tag{3.94}$$

The procedure for rating a counter-flow regenerative IEHX is illustrated in Fig. 3.40.

3.4.3.2   Model Validation

In this part, a validation exercise is first performed on the modified LMTD method. Using a counter-flow regenerative IEHX as an example, we illustrate the application of the model for process design as well as rating the performance of an IEHX. In addition, some suggestions are provided in terms of the thermal resistance analysis.

The outlet air temperature, obtained by the modified LMTD method, is compared with the experimental data as shown in Fig. 3.41. The relative error is calculated as $E = \frac{|T''_{1,\text{m}} - T''_{1,\text{e}}|}{T''_{1,\text{e}}}$. It is evident from the Fig. 3.43 that the modified LMTD method predicts the performance of the IEHX having the largest discrepancy of 5%.

The modified LMTD method has been further validated with experimental data on the cross-flow regenerative IEHX as shown in Fig. 3.42. Comparatively, the simulated results demonstrated good agreement with experimental data to within a maximum discrepancy of 8%.

3.5   Experimental Investigations

The indirect evaporative cooling system is identified as an energy-efficient, sustainable, and environmentally friendly alternative to the conventional vapour compression cooling systems, since no polluting refrigerants and compressors are used. The major limitation of the indirect evaporative cooling systems includes its dependency on the difference between the dry bulb and the wet-bulb temperatures of the ambient air. The cooling of the supply air of indirect evaporative cooling systems is limited to the wet-bulb temperature of the intake air. Carefully designed dew-point evaporative cooling systems are capable of reducing the supply air temperature to below its wet bulb and maintain the comfort space condition in hot climates with high and moderate relative humidity ratios. The performance of the dew-point evaporative cooling systems of counter-flow and cross-flow configurations under different design and operating conditions have been investigated.

Riangvilaikul et al. [13] conducted the parametric study for a counter-flow dew-point evaporative and reported the influences of different operating parameters. The

Fig. 3.40 Flow chart for
rating a counter-flow
regenerative IEHX using the
modified LMTD method

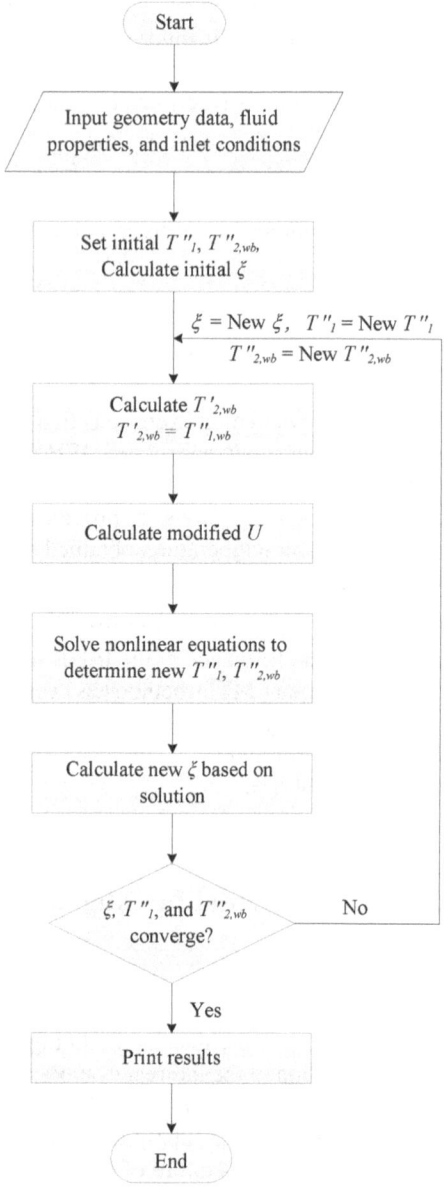

length and width of both dry and wet channels were 1200 mm and 80 mm, respectively
with a 5 mm channel gap and 0.5 mm thickness cotton coated polyurethane sheet as
a wall between two adjacent wet and dry channels. The system consisted of four dry
channels and five wet channels. For the working air to intake air ratio of 0.33 kg/kg
and the water flow rate of 60 g/h, the influences of different operating parameters

Fig. 3.41 Model Validation 1—Comparison between modeled results and experimental data on the counter-flow regenerative IEHX **a** Low flow rate; **b** medium flow rate; and **c** high flow rate

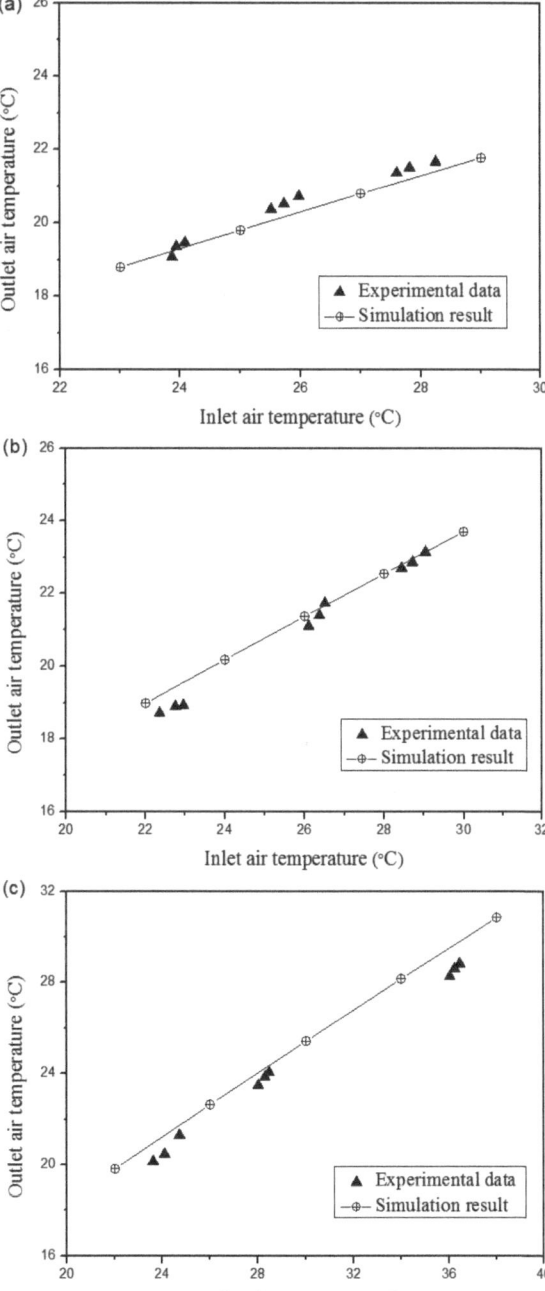

Fig. 3.42 Validation
2—Comparison between
modeled results and
experimental data on the
cross-flow regenerative
IEHX **a** low flow rate;
b medium flow rate; and
c high flow rate

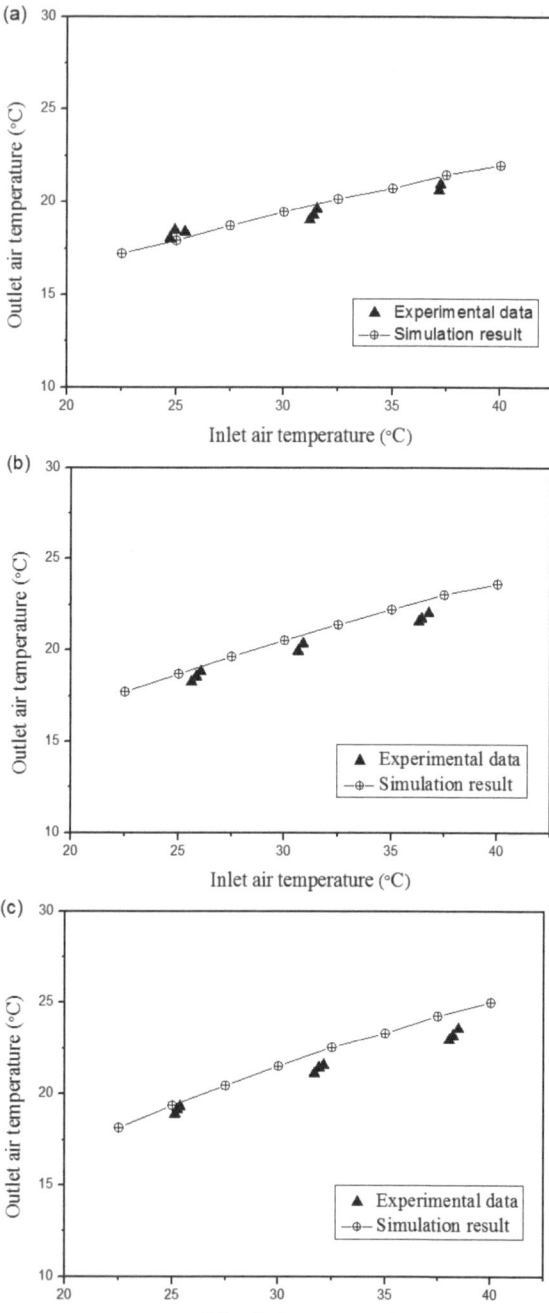

on the cooling effectiveness, such as supply air temperature, wet-bulb effectiveness, and dew-point effectiveness are presented as shown in Fig. 3.43.

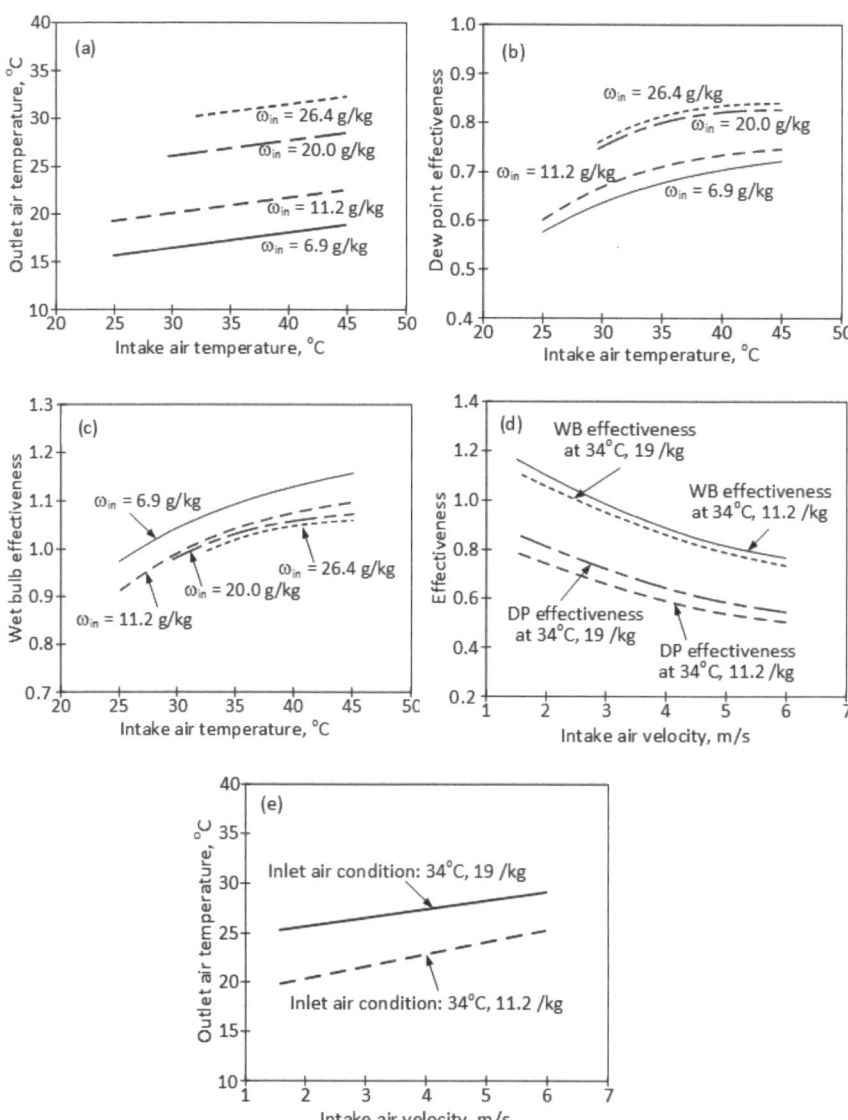

Fig. 3.43 Influences of different operating parameters on the cooling effectiveness **a** outlet air temperature for different inlet air conditions; **b** dew point effectiveness for different inlet air conditions; **c** wet bulb effectiveness for different inlet air conditions; **d** effectiveness for different inlet air velocities; and **e** outlet air temperature for different inlet air velocities

Zhan et al. [41] reported the experimental test results of a commercial M-cycle dew-point cooler with a cross-flow exchanger, which can deliver the dew-point effectiveness of 55–85% and the wet-bulb effectiveness of 110–122% under the operational and ambient test conditions. Jradi et al. [30] tested the cooling performance of a cross-flow dew-point evaporative cooler for the air flow rates of 300, 600, 900, 1200, and 1500 m^3/h. The length of both wet and dry channels was 500 mm with a channel gap of 5 mm. Water was sprayed from the top of the wet channels at about 16 °C. They reported that the cooling system performance characterized by the COP, wet-bulb effectiveness, and dew-point effectiveness were inversely proportional to the flow rates of intake air. Xu et al. [32] designed, constructed and experimentally investigated a novel counter-flow dew-point evaporative cooler with corrugated surfaces and intermittent water supply scheme. Due to the removal of the channel supporting guides and the implementation of the corrugated heat transfer surface, the heat transfer area, heat transfer rate, and the wettability of the wet channel surface were significantly improved. They tested the performance of the cooler under various simulated climatic conditions and achieved a remarkably high COP value of 52.5, wet-bulb effectiveness of 128% and dew-point cooling effectiveness of 95% at the optimal working air ratio of 0.364.

Water-wicking ability, diffusivity, and evaporation ability of wick materials of the wet channels greatly affect the cooling performance of dew-point evaporative coolers. Xu et al. [69] experimentally investigated the performance of seven different fabrics, namely, Coolpass bird eye mesh fabric, Kraft paper, Bamboo charcoal + Coolmax active, Coolpass knitted (double) pique mesh, Topcool spandex single jersey, 320D Supplex + 3 M, and 228 T Supplex + 3 M. A summary of the performance of the tested fabrics is presented in Table 3.6. Coolpass bird eye mesh fabric and Bamboo charcoal + Coolmax active were reported as the most suitable fabrics, which showed increased vertical wicking rates of 182% and 171%, diffusion rates of 14% and 37% and evaporation rates of 20% and 14% respectively, compared to the commonly used Kraft paper. They reported that the cost of the Coolpass bird eye mesh fabric is much lower compared with the Bamboo charcoal + Coolmax active fabric. Therefore, they concluded that Coolpass bird eye mesh fabric would the best choice.

Liu et al. [70] tested the performance of two hybrid cooling systems combining dew-point evaporative coolers with heat pipes for the cooling of a data centre. Corrugated plates were used to fabricate the heat exchanger and the wet surface were covered with a porous fabric to improve the water diffusion. Test results showed that the hybrid systems could achieve the wet bulb effectiveness of 1.25 and the COP up to 34. For the ambient dry bulb temperature of 38.2 °C and the wet-bulb temperature of 22.1 °C, the system could deliver the product air of temperature below 18 °C.

Jia et al. [71] proposed and analysed the performance of two counter-cross-flow dew-point coolers as shown in Fig. 3.44 for residential building air-conditioning applications. One cooler was made of aluminium foil (ALF) and the other one was made of polystyrene board coated with nylon fibres (PS + NL) by electrostatic flocking on the wet channel surface. The PS + NL cooler (core size: 366 × 366 × 250 m^3) delivered the supply air of temperature 0.8–1.2 °C lower than that of the ALF cooler (core size: 866 × 500 × 200 m^3).The test results showed that the dew-point

Table 3.6 Summary of the performance of tested fabrics [69]

Fabric	Performance				Suitability
	Vertical wicking	Diffusion	Evaporation	Mechanical	
1. Coolpass bird eye mesh fabric	Higher	Higher	Higher	Good	Suitable with higher performance
2. Kraft paper	Lowest	Lowest	Lowest	Good	Suitable with lower performance
3. Bamboo charcoal + Coolmax active	Higher	Higher	Higher	Good	Suitable with higher performance
4. Coolpass knitted (double) pique mesh	Higher	Immediate natural diffusion is not good			Not suitable
5. Topcool spandex single jersey	Lower	Immediate natural diffusion is not good			Not suitable
6. 320D Supplex + 3 M	Lower	Highest	Highest	Adhered surface distorted	Suitable with higher performance (not good for adhesive)
7. 228 T Supplex + 3 M	Lower	Highest	Highest	Adhered surface distorted	Suitable with higher performance (not good for adhesive)

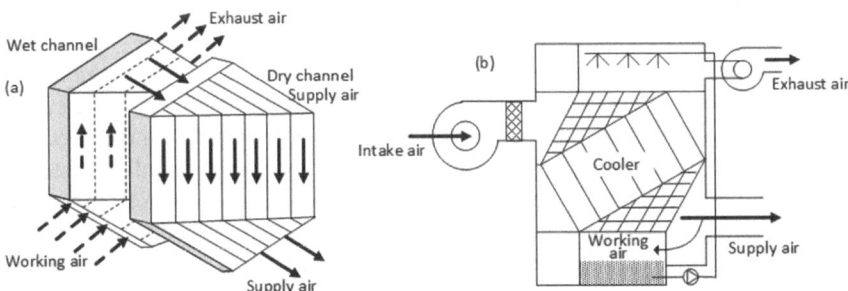

Fig. 3.44 Counter-cross-flow dew point cooler **a** counter-cross-flow pattern; **b** experimental setup

effectiveness varied from 0.467 to 0.786. They reported that the counter-cross-flow configuration of the nylon fibre-coated heat and mass exchanger might reduce the size and weight of the cooler while delivering the same flow rate of air.

The cooling performance of another dew-point evaporative cooler with corrugated plates was experimentally investigated by Liu et al. [72] under the optimized air and water flow arrangements. The test results showed that the cooler could achieve the wet-bulb efficiency of 114%, dew-point efficiency of 68.4%, the cooling capacity of 10,829 W and the maximum COP of 42.8 under the standard operational condition. They reported that the wet-bulb efficiency of the dew-point evaporative cooler was increased by 29.3% and the COP was increased by 34.6% compared with the commercial dew-point cooler Coolerado M30. They claimed that the dew-point evaporative coolers with corrugated plates could achieve the cooling efficiency of more than 10% higher than with the flat plates.

Lin et al. [73] designed, fabricated, and tested the cooling performance of a horizontal counter-flow dew-point evaporative cooler under the transient and steady-state operating conditions. An exponential correlation between the temperature of the supply air and the transient operating time was established. For the intake air temperature of 30–40 °C and the absolute humidity of 10.5–12.0 g/kg, the average achievable product air temperature was 15.9–23.3 °C with a COP spanning 8.6–27.0. The reported pressure drops of the dry and wet channels of length 0.6 m ranged from 16.0 to 29.1 Pa and 19.9 to 52.3 Pa, respectively. They varied the ratio of the working air to the intake air from 0.22 to 0.79 and obtained an optimal cooling capacity at the working air ratio of 0.33.

A summary of the experimental studies conducted by researchers to evaluate the cooling effectiveness of dew-point evaporative coolers of cross-flow and counter-flow configurations under different design and operating parameters are presented in Table 3.7.

3.6 Industrial Status of Dew-Point Evaporative Air Coolers

Dew-point evaporative cooling systems have a distinct advantage over direct and indirect evaporative cooling systems. Hence, the dew-point evaporative cooling system has gained growing attention over the past few years. Compared with the direct and indirect evaporative cooling systems, the dew-point evaporative cooling system is a relatively new technology that is less familiar to cooling industries and manufacturing companies. A small number of evaporative cooling system manufacturing companies are involved in the production of dew-point evaporative coolers of limited capacities and selective configurations. Coolerado Corporation is the largest manufacturer of dew-point evaporative coolers. They have successfully fabricated and installed a number of dew-point evaporative coolers of different designs and cooling capacities across the world for industrial, military, governments, health care, sports and leisure centres, agriculture/plant propagation, food and beverage, retail, schools, residential and data centre use. Coolerado M50 is one of their new energy-efficient

Table 3.7 Summary of flow geometries, operating parameters and cooling effectiveness of dew point evaporative coolers

Study	Flow configuration	Operating parameters	Cooling effectiveness
1. Riangvilaikul et al. [13]	Counter-flow	$T_{i,a} = 25.0–45.0$ °C, $\omega_{i,a} = 7.0–26.0$ g/kg, $v = 2.4$ m/s, $L = 1.0$ m, $H_t = 5$ mm, Working air to intake air ratio = 0.33 kg/kg	$T_{o,a} = 15.6–32.1$ °C, $\varepsilon_{wb} = 0.92–1.14$, $\varepsilon_{dp} = 0.58–0.84$
2. Jradi et al. [30]	Cross-flow	$T_{i,a} = 26.1–41.1$ °C, $\phi_{i,a} = 14.5–38\%$, $\dot{V}_{i,a} = 300–1500$ m³/h, $L = 500$ mm, $H_t = 5$ mm, Working air to intake air ratio = 0.33 kg/kg	$T_{o,a} = 17.3–19.5$ °C, $\varepsilon_{wb} = 0.7–1.169$, $\varepsilon_{dp} = 0.424–0.732$ COP = 5.9–14.2
3. Xu et al. [32]	Counter-flow (corrugated surface)	$T_{i,a} = 26.0–37.8$ °C, $\omega_{i,a} = 19.6–26.2$ g/kg, $\dot{V}_{i,a} = 1350$ m³/h, $L = 1.0$ m, $H_t = 5$ mm, Working air to intake air ratio = 0.44 kg/kg and 0.364 kg/kg (optimal), Height and width of each corrugated wave were 2.8 mm and 11.6 mm	For working air ratio of 0.44: $\Delta T_{o,a} = 4.5–17.1$ °C $\varepsilon_{wb} = 1.0–1.1$, $\varepsilon_{dp} = 0.67–0.76$ COP = 9.7–37.4 For working air ratio of 0.364: $\Delta T_{o,a} = 5.9–19.1$ °C, $\varepsilon_{wb} = 1.04–1.28$, $\varepsilon_{dp} = 0.69–0.95$ COP = 16.5–52.5
4. Liu et al. [70]	Counter-flow (hybrid with heat pipes)	$T_{i,a} = 38.2$ °C, $T_{i,wb} = 22.1$ °C, $\dot{V}_{i,a} = 5400$ m³/h, $L = 1050$ mm, $H_t = 4.3$ mm, Working air to intake air ratio = 0.44 kg/kg	$T_{o,a} = 18$ °C, $\varepsilon_{wb} = 1.25$, COP = 31–34
5. Jia et al. [71]	Counter-cross-flow (ALF and PS + NL)	$T_{i,a} = 29.5–35.5$ °C, $\omega_{i,a} = 13–19$ g/kg $\dot{V}_{i,a} = 35–85$ m³/h, for ALF: $L = 866$ mm and $H_t = 5$ mm, for PS + NL: $L = 366$ mm and $H_t = 3$ mm, Working air to intake air ratio = 0.4 kg/kg	PS + NL: $T_{o,a} = 23.9–26.2$ °C, ALF: $T_{o,a} = 25.9–27.5$ °C PS + NL: $\varepsilon_{dp} = 0.467$ to 0.786 ALF: $\varepsilon_{dp} = 3$ to 7.5% lower than PS + NL PS + NL: COP = 10.1–13.8

(continued)

Table 3.7 (continued)

Study	Flow configuration	Operating parameters	Cooling effectiveness
6. Duan [15]	Counter-flow	$T_{i,a} = 29.6$–35.7 °C, $\omega_{i,a} = 8.7$–9.0 g/kg, $\dot{V}_{i,a} = 0.42$ (m³/h)/ m³ of core volume, $L = 1100$ mm, $H_t = 6$ mm, Working air to intake air ratio = 0.5 kg/kg	$T_{o,a} = 22.4$–24.5 °C, $\varepsilon_{dp} = 0.412$–0.48 COP = 8.3–11.8
7. Liu et al. [72]	Counter-flow (corrugated surface)	$T_{i,a} = 25.4$–38.7 °C, $\phi_{i,a} = 17$–33.9%, $\dot{V}_{i,a} = 3100$–4570 m³/h, $L = 1050$ mm, $H_t = 4.3$ mm, Working air to intake air ratio = 0.360.54 kg/kg	$T_{o,a} = 17.2$–25.4 °C, $\varepsilon_{wb} = 0.749$–1.14, $\varepsilon_{dp} = 0.544$–0.74 COP = 20–42.8
8. Lin et al. [73]	Counter-flow (horizontal)	$T_{i,a} = 30$–40 °C, $\omega_{i,a} = 10.5$–12 g/kg, $v = 1.11$–2 m/s, $L = 600$ mm, $H_t = 3$ mm, Working air to intake air ratio = 0.22–0.79 kg/kg	$T_{o,a} = 15.9$–23.3 °C, $\varepsilon_{wb} = 0.89$–1.44, $\varepsilon_{dp} = 0.6$–0.97, COP = 8.6–27.0
9. Bruno [59]	Counter-flow	$T_{i,a} = 22.5$–40 °C, $\omega_{i,a} = 4.0$–13.1 g/kg, $v = 1.11$–2 m/s, $L = $ mm, $H_t = $ mm, Working air to intake air ratio = kg/kg	$T_{o,a} = 10.2$–21.8 °C, $\varepsilon_{wb} = 0.93$–1.29, $\varepsilon_{dp} = 0.57$–0.83, COP = 4.9–11.8
10. Lee and Lee [24]	Counter-flow regenerative	$T_{i,a} = 27.5$–32.0 °C, $\omega_{i,a} = 9.2$–18.1 g/kg,, $v = 1.0$ m/s, $L = 200$ mm, $H_t = 9.8$ mm, Working air to intake air ratio = 0.3 kg/kg	$T_{o,a} = 16.5$–24.0 °C, $\varepsilon_{wb} = 1.18$–1.22, $\varepsilon_{dp} = 0.75$–0.9,
11. Elberling [27]	Cross-flow	$T_{i,a} = 26.7$–43.8 °C, $T_{i,wb} = 18.1$–23.9 °C, $\dot{V}_{i,a} = 0.53$–1.38 m³/s, Working air to intake air ratio = 0.5 kg/kg	$T_{o,a} = 19.9$–25.6 °C, $\varepsilon_{wb} = 0.81$–0.91 COP = 9.6
12. Zhan et al. [40]	Cross-flow	$T_{i,a} = 25$–40 °C, $T_{i,wb} = 17.9$–30.3 °C, $\dot{V}_{i,a} = 0.036$ m³/s, $L = 0.25$ m, $H_t = 4$ mm, Working air to intake air ratio = 0.5 kg/kg	$T_{o,a} = 11$–30 °C, $\varepsilon_{wb} = 0.5$–0.65

stand-alone products that are stackable side by side to increase the cooling capacity when needed. Their coolers consume as little as 10–20% of the energy consumption of conventional vapour compression air-conditioning systems. Other manufacturing companies such as Aolan (Fujian) Industry Co. Ltd produces dew-point evaporative coolers of limited designs and cooling capacities. The use of the dew-point evaporative cooling systems is still small in terms of its sale volume and the number of building installations.

3.7 Future Research Direction

The dew-point evaporative cooling technology has received considerable attention from researchers and design engineers in recent years due to its potential to reach dew-point temperature of the supply air; resulting in an increase in cooling output, wet-bulb effectiveness, and dew-point effectiveness compared with other conventional indirect evaporative cooling systems. Although the dew-point evaporative cooling systems have achieved a significant increase in the cooling effectiveness, researchers, and design engineers have identified several limitations have motivated them towards conducting future research and development activities. These limitations include (1) flow configuration of wet and dry channels; (2) structure and material of heat and mass exchanger; (3) water distribution and treatment system; (4) hybrid dew-point evaporative cooling systems; and (5) analysis of environmental and economic impacts.

Flow configuration of wet and dry channels: Generally, the heat and mass exchanger of the cross-flow and the counter-flow configurations comprised of pairs of wet and dry channels. As the working air of the counter-flow exchangers travels till the end of the dry channels, the working air is fully cooled before entering the wet channels leading to higher temperature difference and heat transfer between the air streams of dry and wet channels. As a result, the counter-flow exchangers offer around 20% higher cooling capacity, better dew-point and wet-bulb effectiveness compared with cross-flow exchangers under the same operating conditions. However, the pressure drop for the counter-flow exchangers is higher as the volume flow rates of air in the dry and wet channels of the counter-flow exchangers are higher in comparison to the cross-flow exchangers. Cross-flow exchangers have perforated holes widely spreading along the flow paths, which reduce the volume flow rate of air along the length of dry channels and maintain a small volume flow rate in each section of the wet channels. Instead of positioning the holes at the end of the flow channels of the counter-flow exchangers, the perforated holes of variable sizes and numbers can also be aligned along the flow path of the counter-flow exchangers in order to achieve optimum variable flow rates along the length of dry and wet channels. Properly designed guide vanes or spacers can be inserted in both dry and wet channels to promote uniform air flow throughout the exposed surface of the channels. It is worthy to investigate potential flow optimization opportunities of both wet and dry channels.

Structure and material of heat and mass exchanger: The heat and mass exchanger of the dew-point evaporative coolers is usually formed by stacking flat plastic plates, which are easy to fabricate, lightweight, and cost-effective. Supporting guides are inserted between flat plates to maintain the desired gaps for circulating the air in both dry and wet channels. Few researchers [32, 72] have designed based on the corrugated surface exchanger and removed the supporting guides, which are employed to increase the heat transfer area, and thus the cooling effectiveness of the system. The opportunities for developing the modified surfaces to enhance surface area and flow turbulence should be studied with the objective of improving the heat and mass transfer performance of the exchangers. Metals (such as treated copper and aluminium sheets compatible for water-proof coating), fibres and ceramics are the potential material options that have high thermal conductivity and proper shape formation/holding ability. Wick materials such as fibre paper and fabric are generally laminated on the thin water-proof plastic surfaces to manufacture the exchangers. The durability and water wicking ability of these materials were found to be unsatisfactory. Recently, Jia et al. [71] used polystyrene board coated with nylon fibres (PS + NL) by electrostatic flocking and reported the potentials of improving the effectiveness and compactness of the exchangers. Xu et al. [69] experimentally investigated the wicking, diffusion, mechanical and evaporation behaviours of seven different fabrics. The wick made up of copper and aluminium sheets of enhanced surface wettability (hydrophilic performance) could be the most appropriate material/structure for wet channels. Opportunities for developing the new materials with necessary chemical treatments for improved durability and enhanced wicking ability should be further investigated.

Water distribution and treatment system: Distribution of water on the wet surfaces and the treatment of water are another key areas for worthy of investigation. Ahmed et al. [74] experimentally investigated the effect of three water spray modes, namely, external spray, internal spray and mixed internal, and external sprays on the performance of an indirect evaporative cooling system and reported that the mixed-mode performed the best. Presently, water is either sprayed from the top of the vertical wet channels or diffused by capillary action to the horizontal wet channels of the dew-point evaporative cooler. If the flow rate of the spraying water is high, the temperature of the wet surfaces could not drop to the wet-bulb temperature due to the high thermal mass of the flowing water. Instead of spraying water only from the top, the spray nozzles of different flow characteristics can be installed at the top, middle, and bottom planes of larger exchanger blocks. For the horizontal channels, on the other hand, the capillary force of the wick materials is not adequate to distribute the water on the entire surface of wick materials particularly for the larger exchanger blocks. A large horizontal exchanger block can be divided into few small horizontal blocks with water reservoirs installed between two blocks and water can then be fed by capillary action from two sides of the blocks. The water should be carefully treated and filtered to minimize the deposition of the dissolved and suspended solids. The processes of water delivery for different flow configurations of exchangers and the methods of water treatment are not given comprehensive consideration.

Hybrid dew-point evaporative cooling systems: The performance of dew-point evaporative coolers largely depends on the ambient condition. Too humid ambient conditions result in a smaller difference between the dry bulb and dew-point temperature of the air, which hinders the cooling performance of dew-point evaporative coolers and limits their wide applications. To address this problem, researchers and design engineers made some attempts to incorporate the dew-point evaporative coolers with other cooling and dehumidification devices. A few potential hybrid dew-point evaporative cooling systems are shown in Fig. 3.45. A dew-point evaporative cooler can be retrofitted in different arrangements with the conventional cooling coil of central air-conditioning systems or the direct-expansion (DX) air-conditioning units to support the sensible cooling load of buildings in an energy-efficient manner. The approach of incorporating an air dehumidification device to the dew-point evaporative cooler has gained growing attention in recent years from the perspective to enhance the driving potential of water evaporation in wet channels and thereby improve the cooling performance of dew-point evaporative coolers. Commonly used solid desiccants for air dehumidification include silica gel, zeolite, polymer, lithium

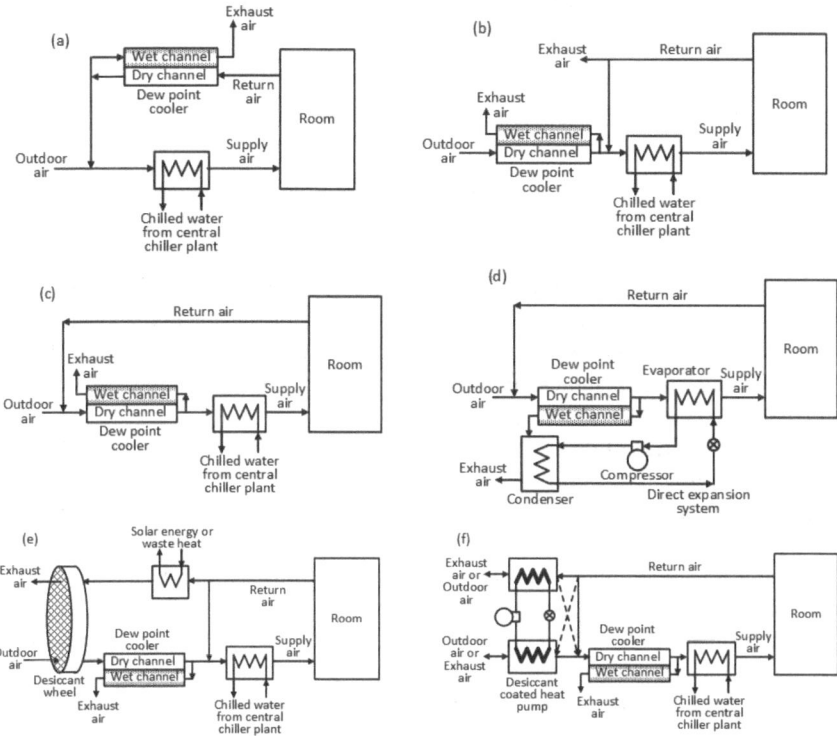

Fig. 3.45 Potential hybrid dew point evaporative cooling systems **a** retrofitted at return duct; **b** retrofitted at fresh air duct; **c** retrofitted at supply duct; **d** integrated with direct expansion unit; **e** desiccant wheel and dew point cooler are retrofitted at fresh air duct; and **f** desiccant coated heat pump and dew point cooler are retrofitted at the fresh air and supply ducts, respectively

chloride salt, titanium silicate etc., while the traditional liquid desiccants are lithium bromide, lithium chloride, and calcium chloride solutions. The use of solar energy or low-grade waste heat to regenerate the desiccants can reduce the demand for fossil fuel. Properly designed hybrid dew-point cooling systems potentially reduce or eliminate the employment of vapour compression air-conditioning systems. The optimization opportunities of the full range of hybrid dew-point evaporative cooling systems are yet to be thoroughly investigated.

Analysis of environmental and economic impacts: Due to the short history of the dew-point evaporative cooling technology, the availability of proper production facilities to manufacture different coolers' sizes and capacities are still relatively immature compared with direct and indirect evaporative cooling systems. The production cost of the dew-point coolers will depreciate considerably once massive production of the coolers get started. The analysis of the potential environmental and economic impacts due to the adoption of the optimized standalone as well as hybrid dew-point cooling systems have received little or no attention by previous researches. Thus far, no clear understanding of the cooler's environmental impact and economic benefits has been established among the users, manufacturers, and professionals. This is another important area that clearly needs investigation.

3.8 Conclusions

An extensive literature review on the potential benefits of the dew-point evaporative cooling system over direct and indirect evaporative cooling systems has been conducted. Key features of the dew-point evaporative cooling system such as the supply air being able to approach to its dew-point temperature without the use of a compressor and any HCFC/CFC refrigerants. This chapter also presents an overview of the dew-point technology in terms of background, working principles, theoretical and experimental studies, and the influences of different design and operating parameters on the cooler's performance.

The development of detailed theoretical models and their numerical and analytical simulations are recognized as an effective tool to understand the coupled heat and mass transfer processes involved and study the influences of different design and operating parameters on the performance of the cooler. Theoretical models of different levels of complexities and their solution techniques are judiciously formulated and presented in this chapter for both cross-flow and counter-flow configurations under the steady-state and transient operating conditions. Different features of the theoretical models are coherently summarized for easy reference.

The construction details and the key findings of a number of experimental prototypes and commercial units are presented. The supply air temperature, dew-point effectiveness, wet-bulb effectiveness, and COP are used as the performance indicators. A summary of the tested design and operating variables for the cross-flow and counter-flow dew-point coolers and their corresponding cooling performances are comprehensibly summarized. Considering the recent development of the technology, potential research directions including the development of new structural materials, wick materials, water distribution systems, and dehumidification opportunities of the circulating air and the optimization of flow configurations of the dry and wet channels are identified and discussed.

References

1. Dudley B (2017) BP statistical review of world energy. London, UK
2. EIA U (2016) Primary energy consumption by source and sector
3. Lior N (2012) Sustainable energy development (May 2011) with some game-changers. Energy 40:3–18
4. Pérez-Lombard L, Ortiz J, Pout C (2008) A review on buildings energy consumption information. Energy Build 40:394–398
5. Yap C, Cai W, Ooi KT, Toh KC, Callavaro G, Pillai EK (2011) Air-con system efficiency primer: a summary. National Climate Change Secretariat and National Research Foundation, Singapore
6. Oh SJ, Ng KC, Thu K, Chun W, Chua KJE (2016) Forecasting long-term electricity demand for cooling of Singapore's buildings incorporating an innovative air-conditioning technology. Energy Build 127:183–193
7. Lin J, Thu K, Bui TD, Wang RZ, Ng KC, Chua KJ (2016) Study on dew point evaporative cooling system with counter-flow configuration. Energy Convers Manage 109:153–165
8. Duan Z, Zhan C, Zhang X, Mustafa M, Zhao X, Alimohammadisagvand B, Hasan A (2012) Indirect evaporative cooling: past, present and future potentials. Renew Sustain Energy Rev 16:6823–6850
9. Glanville P, Kozlov A, Maisotsenko V (2011) Dew point evaporative cooling: technology review and fundamentals. ASHRAE Trans 117
10. Dowdy J, Karabash N (1987) Experimental determination of heat and mass transfer coefficients in rigid impregnated cellulose evaporative media. ASHRAE Trans 93:382–395
11. Mahmood MH, Sultan M, Miyazaki T, Koyama S, Maisotsenko VS (2016) Overview of the Maisotsenko cycle—a way towards dew point evaporative cooling. Renew Sustain Energy Rev 66:537–555
12. Maisotsenko V, Reyzin I (2005) The Maisotsenko cycle for electronics cooling. In: ASME 2005 pacific rim technical conference and exhibition on integration and packaging of MEMS, NEMS, and Electronic Systems collocated with the ASME 2005 heat transfer summer conference. Am Soc Mech Eng, 415–424.
13. Riangvilaikul B, Kumar S (2010a) An experimental study of a novel dew point evaporative cooling system. Energy Build 42:637–644
14. Zhao X, Li JM, Riffat SB (2008) Numerical study of a novel counter-flow heat and mass exchanger for dew point evaporative cooling. Appl Therm Eng 28:1942–1951
15. Z. Duan, Investigation of a novel dew point indirect evaporative air conditioning system for buildings, in, Vol. PhD, University of Nottingham, 2011.
16. Kim MH, Yoon DS, Kim HJ, Jeong JW (2016) Retrofit of a liquid desiccant and evaporative cooling-assisted 100% outdoor air system for enhancing energy saving potential. Appl Therm Eng 96:441–453
17. Kim HJ, Lee SJ, Cho SH, Jeong JW (2016) Energy benefit of a dedicated outdoor air system over a desiccant-enhanced evaporative air conditioner. Appl Therm Eng 108:804–815
18. La D, Dai YJ, Li Y, Tang ZY, Ge TS, Wang RZ (2013) An experimental investigation on the integration of two-stage dehumidification and regenerative evaporative cooling. Appl Energy 102:1218–1228

19. Gao WZ, Cheng YP, Jiang AG, Liu T, Anderson K (2015) Experimental investigation on integrated liquid desiccant—indirect evaporative air cooling system utilizing the Maisotesenko—cycle. Appl Therm Eng 88:288–296
20. Buker MS, Mempouo B, Riffat SB (2015) Experimental investigation of a building integrated photovoltaic/thermal roof collector combined with a liquid desiccant enhanced indirect evaporative cooling system. Energy Convers Manage 101:239–254
21. Oh SJ, Ng KC, Chun W, Chua KJE (2017) Evaluation of a dehumidifier with adsorbent coated heat exchangers for tropical climate operations. Energy
22. Myat A, Thu K, Choon NK (2012) The experimental investigation on the performance of a low temperature waste heat-driven multi-bed desiccant dehumidifier (MBDD) and minimization of entropy generation. Appl Therm Eng 39:70–77
23. Bui TD, Wong Y, Islam MR, Chua KJ (2017) On the theoretical and experimental energy efficiency analyses of a vacuum-based dehumidification membrane. J Membr Sci 539:76–87
24. Lee J, Lee DY (2013) Experimental study of a counter flow regenerative evaporative cooler with finned channels. Int J Heat Mass Transf 65:173–179
25. Dean J, Metzger I (2014) Multistaged indirect evaporative cooler evaluation. National Renewable Energy Laboratory, USA
26. Coolerado C (2014) M30 air conditioner brochure, in. Coolerado Corporation, USA
27. Elberling L (2006) Laboratory evaluation of the Coolerado Cooler™ indirect evaporative cooling unit. PG&E Company, USA
28. Zube D, Gillan L (2011) Evaluating coolerado corportion's heat-mass exchanger performance through experimental analysis. Int J Energy Clean Environ 12
29. Gillan L (2008) Maisotsenko cycle for cooling processes. Int J Clean Environ 9
30. Jradi M, Riffat S (2014) Experimental and numerical investigation of a dew-point cooling system for thermal comfort in buildings. Appl Energy 132:524–535
31. Hsu ST, Lavan Z, Worek WM (1989) Optimization of wet-surface heat exchangers. Energy 14:757–770
32. Xu P, Ma X, Zhao X, Fancey K (2017) Experimental investigation of a super performance dew point air cooler. Appl Energy 203:761–777
33. Riangvilaikul B, Kumar S (2010b) Numerical study of a novel dew point evaporative cooling system. Energy Build 42:2241–2250
34. Kabeel AE, Abdelgaied M (2016) Numerical and experimental investigation of a novel configuration of indirect evaporative cooler with internal baffles. Energy Convers Manage 126:526–536
35. Hasan A (2012) Going below the wet-bulb temperature by indirect evaporative cooling: Analysis using a modified ε-NTU method. Appl Energy 89:237–245
36. Cui X, Chua KJ, Yang WM (2014) Numerical simulation of a novel energy-efficient dew-point evaporative air cooler. Appl Energy 136:979–988
37. Cui X, Chua KJ, Yang WM, Ng KC, Thu K, Nguyen VT (2014) Studying the performance of an improved dew-point evaporative design for cooling application. Appl Therm Eng 63:624–633
38. Cui X, Chua KJ, Islam MR, Ng KC (2015) Performance evaluation of an indirect pre-cooling evaporative heat exchanger operating in hot and humid climate. Energy Convers Manage 102:140–150
39. Cui X, Chua KJ, Islam MR, Yang WM (2014) Fundamental formulation of a modified LMTD method to study indirect evaporative heat exchangers. Energy Convers Manage 88:372–381
40. Zhan C, Duan Z, Zhao X, Smith S, Jin H, Riffat S (2011) Comparative study of the performance of the M-cycle counter-flow and cross-flow heat exchangers for indirect evaporative cooling—paving the path toward sustainable cooling of buildings. Energy 36:6790–6805
41. Zhan C, Zhao X, Smith S, Riffat SB (2011) Numerical study of a M-cycle cross-flow heat exchanger for indirect evaporative cooling. Build Environ 46:657–668
42. Hettiarachchi HDM, Golubovic M, Worek WM (2007) The effect of longitudinal heat conduction in cross flow indirect evaporative air coolers. Appl Therm Eng 27:1841–1848
43. Heidarinejad G, Moshari S (2015) Novel modeling of an indirect evaporative cooling system with cross-flow configuration. Energy Build 92:351–362

44. Anisimov S, Pandelidis D, Danielewicz J (2014) Numerical analysis of selected evaporative exchangers with the Maisotsenko cycle. Energy Convers Manage 88:426–441
45. Anisimov S, Pandelidis D (2015) Theoretical study of the basic cycles for indirect evaporative air cooling. Int J Heat Mass Transf 84:974–989
46. Anisimov S, Pandelidis D, Jedlikowski A, Polushkin V (2014) Performance investigation of a M (Maisotsenko)-cycle cross-flow heat exchanger used for indirect evaporative cooling. Energy 76:593–606
47. Pandelidis D, Anisimov S, Worek WM (2015a) Comparison study of the counter-flow regenerative evaporative heat exchangers with numerical methods. Appl Therm Eng 84:211–224
48. Pandelidis D, Anisimov S, Worek WM (2015b) Performance study of the Maisotsenko Cycle heat exchangers in different air-conditioning applications. Int J Heat Mass Transf 81:207–221
49. Pandelidis D, Anisimov S (2016) Numerical study and optimization of the cross-flow Maisotsenko cycle indirect evaporative air cooler. Int J Heat Mass Transf 103:1029–1041
50. Pandelidis D, Anisimov S (2015a) Numerical analysis of the selected operational and geometrical aspects of the M-cycle heat and mass exchanger. Energy Build 87:413–424
51. Pandelidis D, Anisimov S (2015b) Numerical analysis of the heat and mass transfer processes in selected M-Cycle heat exchangers for the dew point evaporative cooling. Energy Convers Manage 90:62–83
52. Jafarian H, Sayyaadi H, Torabi F (2017) Modeling and optimization of dew-point evaporative coolers based on a developed GMDH-type neural network. Energy Convers Manage 143:49–65
53. Zhu G, Chow TT, Lee CK (2017) Performance analysis of counter-flow regenerative heat and mass exchanger for indirect evaporative cooling based on data-driven model. Energy Build 155:503–512
54. Handbook A (2009) ASHRAE handbook–fundamentals. ASHRAE, Atlanta, GA
55. Handbook A (2013) ASHRAE handbook-Fundamentals. ASHRAE, Atlanta, GA
56. Bergman TL, Incropera FP, Lavine AS (2011) Fundamentals of heat and mass transfer. Wiley
57. Tenne A (2010) Sea water desalination in Israel: planning, coping with difficulties, and economic aspects of long-term risks. State of Israel Desalination Division
58. Burn S, Hoang M, Zarzo D, Olewniak F, Campos E, Bolto B, Barron O (2015) Desalination techniques - A review of the opportunities for desalination in agriculture. Desalination 364:2–16
59. Bruno F (2011) On-site experimental testing of a novel dew point evaporative cooler. Energy Build 43:3475–3483
60. Amer O, Boukhanouf R, Ibrahim HG (2015) A review of evaporative cooling technologies. Int J Environ Sci Dev 6:111–117
61. Lin J, Thu K, Bui TD, Wang RZ, Ng KC, Kumja M, Chua KJ (2016) Unsteady-state analysis of a counter-flow dew point evaporative cooling system. Energy 113:172–185
62. Sohani A, Sayyaadi H, Mohammadhosseini N (2018) Comparative study of the conventional types of heat and mass exchangers to achieve the best design of dew point evaporative coolers at diverse climatic conditions. Energy Convers Manage 158:327–345
63. Baehr HD, Stephan K (2006) Heat and mass transfer. Springer, New York, Berlin
64. Provencher S (1976) A Fourier method for the analysis of exponential decay curves. Biophys J 16:27–41
65. Ogata K (2010) Modern control engineering. Prentice Hall PTR, USA
66. Kays WM, Crawford ME, Weigand B (2005) Convective heat and mass transfer, 4th edn. McGraw-Hill Higher Education, New York
67. Lewis W (1962) The evaporation of a liquid into a gas. Int J Heat Mass Transf 5:109–112
68. Bowman RA, Mueller AC, Nagle WM (1940) Mean temperature difference in design. Trans ASME 62:283–294
69. Xu P, Ma X, Zhao X, Fancey KS (2016) Experimental investigation on performance of fabrics for indirect evaporative cooling applications. Build Environ 110:104–114
70. Liu Y, Yang X, Li J, Zhao X (2018) Energy savings of hybrid dew-point evaporative cooler and micro-channel separated heat pipe cooling systems for computer data centers. Energy 163:629–640

71. Jia L, Liua J, Wangb C, Caoa X, Zhanga Z (2019) Study of the thermal performance of a novel dew point evaporative cooler. Appl Therm Eng 160:114069
72. Liu Y, Akhlaghi YG, Zhao X, Li J (2019) Experimental and numerical investigation of a high-efficiency dew-point evaporative cooler. Energy Build 197:120–130
73. Lin J, Bui DT, Wang R, Chua KJ (2018) The counter-flow dew point evaporative cooler: Analyzing its transient and steady-state behaviour. Appl Therm Eng 143:34–47
74. Ahmed Y, Al-Zubaydi T, Hong G (2019) Experimental study of a novel water-spraying configuration in indirect evaporative cooling. Appl Therm Eng 151:283–293

Chapter 4
Adsorbent-Coated Heat and Mass Exchanger

4.1 Introduction

The cooling load of air-conditioning spaces comprises of sensible cooling load and latent cooling load. Sensible cooling load is generated mainly attributed to the heat gain through the building envelope and the sensible heat rejection by occupants, lamps, computers, printers, photocopy machines, etc. Latent cooling load, on the other hand, is generated from the rejection of water vapour by occupants, the diffusion of moisture through the building envelope and the outdoor fresh air supplied for the ventilation of the air-conditioning spaces. The latent cooling load comprises up to 70% of the total cooling load in tropical climates. The conventional vapour compression chiller systems are commonly employed to support the sensible as well as the latent cooling load. The chiller systems produce chilled water of about 6–7 °C to cool the circulating air below the dew point temperature to condense the moisture of the circulating air and thereby remove the latent cooling load. Due to the generation of low-temperature chilled water, the operation of the conventional vapour compression chiller systems becomes an inefficient energy-intensive process, which consumes significant amount of high-grade electricity as the source of input energy. Several attempts have been made by the researchers to decouple the latent and sensible cooling loads and remove the latent cooling load from the air-conditioning spaces using energy-efficient innovative technologies.

The extensively studied alternative technologies are: (a) electrically activated solid-state systems, (b) electro-mechanical systems, and (c) thermal energy-driven desiccant-assisted heat and mass exchanger systems [1]. Electrically activated solid-state systems and electro-mechanical systems are recognized as the promising energy-efficient alternatives. The magneto caloric effect is an example of the electrically activated solid-state systems. The magneto caloric material is exposed to an externally applied changing magnetic field to repeatedly orient and disorient the magnetic domain. The material absorbs thermal energy to perform the reorientation resulting in a drop in temperature of the surrounding air. Electro-mechanical systems,

© Springer Nature Singapore Pte Ltd. 2021
C. Kian Jon et al., *Advances in Air Conditioning Technologies*, Green Energy and Technology, https://doi.org/10.1007/978-981-15-8477-0_4

such as thermoelectric cooler, use the Peltier effect to create a heat flux at an electrified junction of two different types of conductors. The primary advantages of the electrically activated solid-state systems and electro-mechanical systems compared with the vapour compression type cooling systems are their lack of moving parts, no circulating refrigerant, long life, and small size. However, these technologies are still at the stage of research and development and generally not feasible for practical applications because of their low energy efficiency and higher costs. The latent cooling load can also be reduced by compressing the air to a higher pressure. When the moist air is compressed to a higher pressure, its dew point temperature increases and the water molecules condense by rejecting the latent heat to the surrounding air or cooling water. However, the power consumption of the air compressor is quite high that makes the process infeasible.

The difficulties of decoupling and reducing the latent cooling load are currently addressed by the thermal energy-driven desiccant-assisted dehumidification systems. The desiccants used in the dehumidification systems are broadly divided into the liquid absorbents and solid adsorbents. Despite the extensive use, the main drawback of the liquid absorbent systems is the potential carryover of the absorbent droplets with the circulating air. The liquid absorbents are generally corrosive and may harmful to the occupants who would breathe the air. The solid adsorbents, on the other hand, are environmentally friendly and highly stable. The thermal energy-driven solid adsorbents-assisted dehumidification systems could be designed as fixed-bed, rotary wheel, and adsorbent-coated heat and mass exchangers. As the moist air flows through the exchangers, the dry adsorbents adsorb moisture on its exposed surface, resulting in the decoupling and reduction of latent load in the air. The adsorbed moisture diffuses from the exposed surface to the bulk through the pores of the adsorbents. The moisture adsorption capacity of the adsorbents depends on the thermophysical properties of the adsorbents as well as the moisture content and temperature of the flowing air. The adsorption rate gradually drops as the adsorbent progressively reaches the saturation or equilibrium state. In order to develop a continuously operating system, the adsorbed water molecules are removed from the adsorbents by evaporation using the low-grade waste heat such as solar energy or industrial waste heat. This process of adsorbent regeneration depends largely on the temperature and relative humidity of the circulating air.

The solid adsorbent exchangers are also used extensively in the thermally driven vapour adsorption chiller systems because of the high affinity and capability of adsorbing water vapour, low regeneration temperature, high durability, and minimum maintenance requirements. The major drawbacks of the vapour adsorption chiller systems are the lower coefficient of performance, bulkiness, higher irreversibility and extended payback period [2]. Newly developed high-performance adsorbents can potentially address the inherent challenges of the adsorption chiller systems.

The thermophysical properties of high-performance solid adsorbents, development of different heat and mass exchangers, impacts of the design and operating variables of exchangers on the moisture adsorption and desorption behaviours as well as the opportunities of decoupling and removing the latent cooling load in an energy-efficient manner are presented in this chapter.

4.2 Adsorbents

Adsorbents are generally chemically inert, nontoxic, polar and dimensionally stable solid materials that have the affinity of adsorbing water vapour or other gaseous substances without covalent bonding. The sorption capacity, kinetics and regeneration temperature of adsorbents are carefully analysed to select the adsorbents to achieve the heat and mass transfer requirements of specific applications. In general, the sorption capacity represents the amount of adsorbate (such as water vapour) adsorbed by the adsorbent to reach the equilibrium state corresponding to the partial pressure of vapour or the concentration of liquid at a constant temperature. The quantity of the adsorbate adsorbed is normalized by the mass of the adsorbent to compare the sorption capacity of different adsorbent materials. In the absorption process, the molecules of the adsorbate enter the bulk of the liquid absorbents. However, the adsorption is the adhesion of the molecules of adsorbates on the exposed surface of the adsorbents. The adsorption process is a surface phenomenon that occurs at the interface of two phases, such as solid–gas interfaces, while the absorption involves the whole volume of the absorbent materials. The adsorption and absorption mechanisms of water vapour from the moist air to the solid adsorbents and the liquid absorbents are shown in Fig. 4.1. Within the bulk of an adsorbent, the atoms of adsorbent fill all the bonding requirements for the constituent of the adsorbent material. However, the atoms of the adsorbent are not able to surround the entire exposed surface of the adsorbent. Therefore, the adsorbent materials can attract adsorbate molecules at the exposed surface to reach the equilibrium state. The exposed surface area, surface energy, and available free bonds at the exposed surfaces of the adsorbent control the sorption capacity of the adsorbents.

The exposed surface area and pore structures characterize the porosity of adsorbents. The ratio of the volume of pores and voids to the volume occupied by the solid adsorbent is defined as porosity. The adsorbents with higher porosity are preferable for achieving a higher adsorption capacity. The closed, blind, through and interconnected pores generally form the pore structures by as shown in Fig. 4.2 [3]. The pores are classified according to the internal width as (a) micropore (width less than

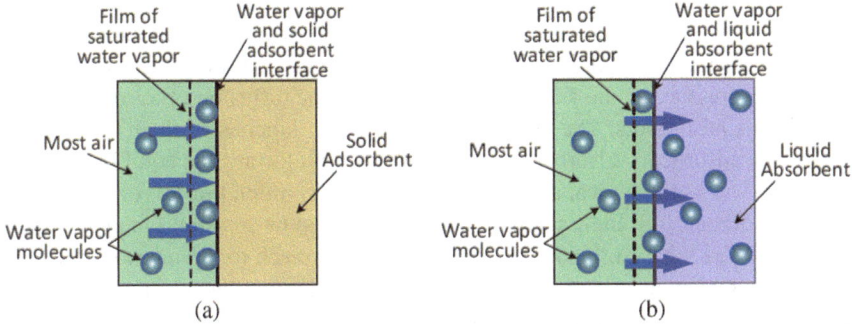

Fig. 4.1 **a** Adsorption process and **b** absorption process of water vapour from moist air

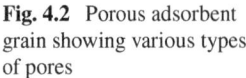

Fig. 4.2 Porous adsorbent
grain showing various types
of pores

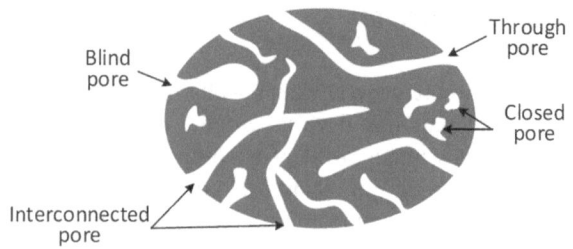

2 nm), (b) mesopore (width between 2 and 50 nm), and (c) macropore (width greater than 50 nm). Micropores and mesopores are particularly important for the adsorption processes. For the pore sizes comparable to the size of the adsorbate molecule, the walls surrounding the pores attract the molecules of adsorbate and fill volumetrically with the adsorbate. Macropores function as the diffusion paths for the adsorbate molecules from the surrounding of adsorbents to the micropores. Very small-sized pores hinder the accessibility of adsorbate molecules to the adsorption surfaces. Therefore, pore size distribution is another important property for characterizing the absorptivity of adsorbents. Polar adsorbents have an affinity to the polar adsorbates such as the molecule of water vapour. Silica gel and zeolite are examples of the polar adsorbents, which are typically hydrophilic. On the other hand, non-polar adsorbents such as polymer adsorbents and silicates are generally hydrophobic.

The adsorption performance of an adsorbent is evaluated using the adsorption isotherm, adsorption isostere, and sorption kinetics. The relationship between the maximum amount of adsorbate adsorbed by an adsorbent of unit mass to reach the equilibrium state corresponding to different concentrations or partial pressures of the adsorbate in the gaseous phase at a constant temperature is called the adsorption isotherm at that temperature. The adsorption isotherms for a wide variety of gas–solid systems have been measured and the majority of the isotherms are classified into six groups [3] as shown in Fig. 4.3. The isotherm profiles should be analysed carefully to evaluate the specific adsorption capacity of the adsorbents at the operating range of concentration or partial pressure of the adsorbates. The performance of an adsorbent is also characterized using the adsorption isostere, which represents the change of the adsorption rate with the variation of operating temperatures and concentrations or partial pressures of the adsorbates. The sorption kinetics, on the other hand, is a measure of the adsorption update with respect to time at a constant temperature and partial pressure of the adsorbates. It measures the diffusion of adsorbate in the pores of the adsorbents. The gradient of the sorption kinetics gradually drops with time as the adsorbents adsorb adsorbates and progress towards an equilibrium state. Figure 4.4 shows the typical adsorption isostere and sorption kinetics of adsorbents.

An adsorbent with high adsorption capacity but slow sorption kinetics is not a good choice as the adsorbate takes too long time to reach to the interior pores. On the other hand, an adsorbent with first sorption kinetics but low adsorption capacity is neither a good choice as a large mass of the adsorbent is required for the required uptakes. In practical operations, the maximum equilibrium adsorption capacity of the

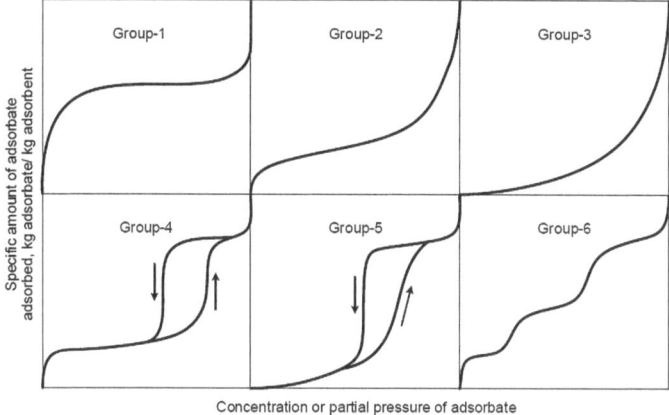

Fig. 4.3 Six groups of adsorption isotherms

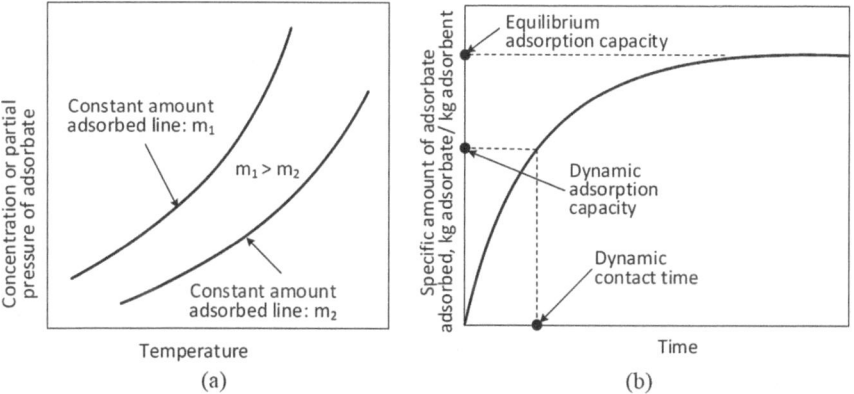

Fig. 4.4 a Adsorption isostere and **b** Sorption kinetics

adsorbents cannot be utilized because of the gradual falling rate of adsorption rate with the uptake of the adsorbates. Therefore, the kinetics and the equilibrium adsorption capacity are judiciously analysed to estimate the practical or dynamic adsorption capacity (shown in Fig. 4.4b) of adsorbents for different contacting processes, contact time ±96, the concentration of adsorbates, and operating temperatures [4]. To regenerate the adsorbent and develop a continuously operating system, the potentials of evaporating the adsorbed adsorbate (water molecules) from the adsorbents at low regeneration temperature using the low-grade waste heat is another important factor for the development of a feasible alternative. Different adsorbents that can potentially be used for the dehumidification of the circulating air of air-conditioning systems and the adsorption of water vapour for the development of vapour adsorption cooling systems are presented in the following section.

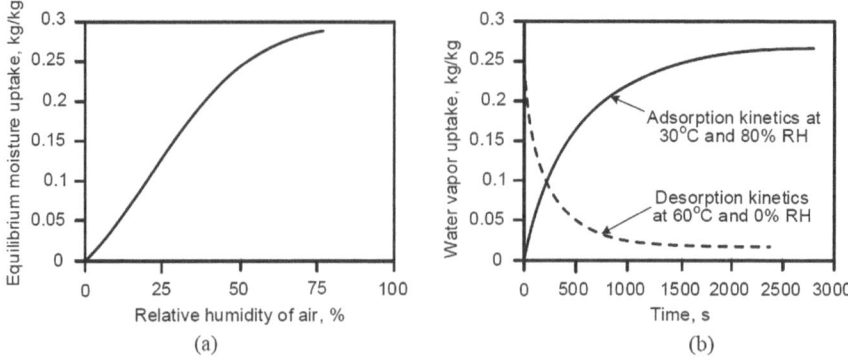

Fig. 4.5 Moisture adsorption performance of silica gel: **a** Moisture adsorption isotherm at 30 °C and **b** moisture adsorption and desorption kinetics

4.2.1   Silica Gel

Silica gel is widely used as an adsorbent in the heat and mass exchangers due to its low cost, high porosity, dimensionally stable, non-toxic, and non-corrosive nature. It is prepared by the reaction between acetic acid and sodium silicate followed by a number of processes such as ageing, pickling, etc. The processes create a microporous structure with a distribution of interconnected pore opening sizes from 0.3 to 6 nm. The interconnected pores form a large exposed surface area that attracts and holds the molecules of water vapour up to 30% of its weight. Silica gel is particularly efficient at temperatures below 30 °C. However, the adsorption capacity drops as the temperature begins to rise. Moisture adsorption performance for silica gel is shown in Fig. 4.5. A 25–45% increase in moisture adsorption capacity can be achieved for lithium chloride (LiCl) and calcium chloride (CaCl2) impregnated composite silica gel. Ge et al. [5] infused 75 w% potassium formate with silica gel and reported two to three times improvement in the moisture adsorption capacity. Moreover, the thermal conductivity of the composite silica gel remarkably increases with the increase of LiCl contents. Zheng et al. [6] reported an increase in thermal conductivity of silica gel of about 90% due to the impregnation of 40 w% LiCl.

4.2.2   Zeolites

Zeolites are natural polar crystalline aluminosilicates. They are also manufactured by hydrothermal synthesis and reforming of different silicate sources and ion exchange processes. The pore diameter of zeolites usually ranges from 0.2 to 0.9 nm. The crystals of zeolite can be pelletized using suitable binders to form mesoporous and macroporous pellets. Aluminium-containing zeolites are dealuminated or aluminium-free silica sources are synthesized to develop the non-polar siliceous zeolites. A new class of high-performance AQSOA zeolites adsorbent materials such as AQSOA-Z01, AQSOA-Z02, and AQSOA-Z05 are developed for dehumidification and adsorption cooling applications. AQSOA zeolites have the distinguished abilities of adsorption and regeneration of moisture at low-temperature ranges compared with silica gels and conventional zeolites. The water vapour adsorption isotherm of AQSOA zeolites is characterized by S-shape with hydrophobic behaviour at low vapour pressure ratios. The adsorption process consists of a monolayer–multilayer adsorption stage at low partial vapour pressure followed by the micropore filling at higher partial vapour pressure. Kayal et al. [7] and Teo et al. [8] measured the water vapour uptakes of AQSOA-Z01, AQSOA-Z02 and AQSOA-Z05 under static and dynamic conditions for the temperatures ranging from 25 to 65 °C and the pressures up to the saturated conditions. They found an extremely sharp rise in water vapour uptakes over a narrow range of vapour pressure ratio as well as a higher influence of temperature on the isotherm shape as shown in Fig. 4.6. Properties and water vapour updates under saturation conditions of AQSOA-Z01, AQSOA-Z02, and AQSOA-Z05 are summarized in Table 4.1.

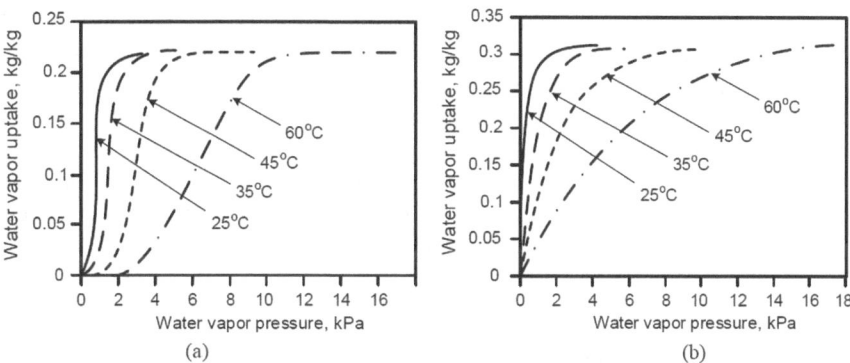

Fig. 4.6 Water vapour uptakes of (**a**) AQSOA-Z01 and (**b**) AQSOA-Z02

Table 4.1 Properties and water vapour updates of AQSOA type Z01, Z02, and Z05 zeolites

Type	Composition	Pore width, nm	Pore volume, cm³/g	Water vapour uptake, kg/kg	Heat of adsorption, kJ/kg
AQSOA-Z01	Iron aluminophosphate	0.74	0.08	0.21	3030
AQSOA-Z02	Silico aluminophosphate	0.37	0.27	0.29	4560
AQSOA-Z05	Aluminophosphate	0.74	0.07	0.22	2800

4.2.3 Polyvinyl Alcohol

Polyvinyl Alcohol (PVA) is considered as another potential adsorbent due to its high
moisture adsorption ability. Anhydrous lithium chloride LiCl is added in different
percentages with the PVA to enhance its dehumidification performance. Zhang et al.
[9] developed a moisture-permeable composite membrane using an active separating
layer of PVA. They added LiCl as an additive to the PVA solution to facilitate the
moisture permeation. The composite membranes exhibit about 70% improvement
of the moisture permeation due to the substantial change of hydrophilicity and crys-
tallinity of PVA. Bui et al. [10] varied the content of LiCl from 0 to 50 w% to develop
hydrophilic composite PVA/LiCl membranes and tested the dehumidification performance. They observed that the addition of LiCl improves both moisture sorption and
diffusion rates, resulting in high moisture permeation. Recently, Vivekh et al. [11]
conducted a comprehensive study on the dehumidification performance of composite
PVA/LiCl polymer desiccants. Silica gel (RD 780), PVA, LiCl and hydroxyethyl
cellulose (HEC) were used to prepare the composite desiccant solutions. A 5 w%
silica gel and 3.3 w% HEC were added to water under continuous stirring at room
temperature to prepare a suitable binding solution of silica gel. They mixed PVA
and appropriate amounts of LiCl with the binding solution to prepare the PVA/LiCl
solutions with 0, 16.7, 33.3, and 50 w% concentrations of LiCl. Experimental results
showed that the composite polymer desiccants with a greater concentration of LiCl
displayed the superior sorption capacity. They reported that the desiccant enabled
high moisture removal rate even at lower regeneration temperatures. The adsorption
and desorption performances of the PVA/LiCl composite polymer desiccants under
different operating conditions are shown in Fig. 4.7.

4.2.4 Superabsorbent Polymer

Superabsorbent polymer (SAP) can absorb a large amount of water vapour relative
to its mass through the hydrogen bonding with the water molecules. Low-density
cross-linked SAPs generally have a higher water absorption capacity. These types of
SAPs swell to a larger degree and become soft when saturated with absorbed water.
Generally, the moisture diffusivity of SAPs is higher than silica gel by 1–2 orders of
magnitude. White et al. [12] tested the dehumidification performance of desiccant
wheel mass exchangers coated with (a) silica gel, (b) ferroaluminophosphate (FAM-
Z01) zeolite with pore size 0.73 nm, and (c) SAP obtained by ion modification from
the polyacrylic acid. They found that the SAP-coated mass exchanger showed the
maximum dehumidification performance at a high relative humidity (>60%) and low
regeneration temperature (50 °C). The temperature of the dehumidified air exiting the
SAP-coated mass exchanger was also lower. Composite SAPs can be developed by
impregnating the hygroscopic salts. The moisture sorption capacity of the composite
SAPs is superior in comparison to that of the pure SAPs and silica gel [13–15].

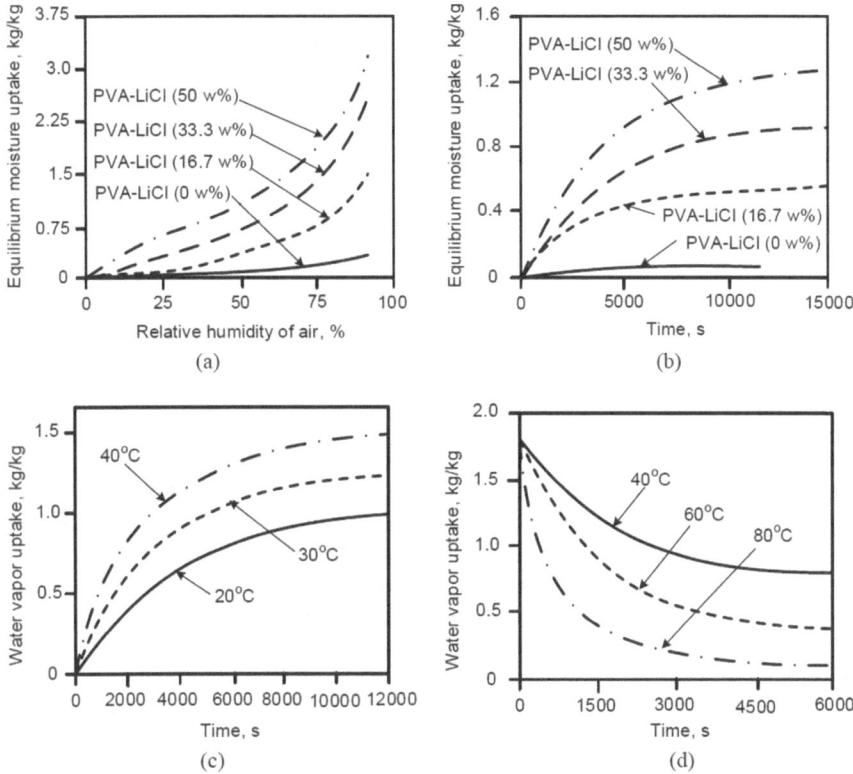

Fig. 4.7 Moisture adsorption performance of PVA with LiCl: **a** moisture sorption isotherm at 30 °C for PVA with 0–50 w% LiCl, **b** sorption kinetics at 30 °C and 80% RH for PVA with 0–50 w% LiCl, **c** sorption kinetics with temperature ranging from 20 to 40 °C at 80% RH for PVA with 50 w% LiCl, and **d** desorption kinetics with temperature ranging from 40 to 80 °C at 0% RH for PVA with 50 w% LiCl

A comparison of the moisture sorption performance of silica gel, pure SAP, and composite SAP–LiCl (30 w%) is shown in Fig. 4.8.

4.2.5 Metal–Organic Framework (MOF)

Metal–organic frameworks (MOFs) are organic–inorganic hybrid materials, which are developed using organic ligands linking metal ions. The composition and structure of MOFs can be changed conveniently to increase their porosity, specific surface area, and adsorption capacity. MOFs have pore volume, specific surface area, and adsorption capacity of about $2 \, cm^3/g$, $6000 \, m^2/g$ and $1.5 \, g/g$, respectively [16, 17]. Kim et al. [18] highlighted the potential of using MOF to extract water from the atmosphere of low humidity levels. They synthesized MOF-801 $\{Zr_6O_4(OH)_4(fumarate)_6\}$ and

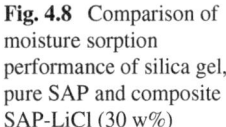

Fig. 4.8 Comparison of moisture sorption performance of silica gel, pure SAP and composite SAP-LiCl (30 w%)

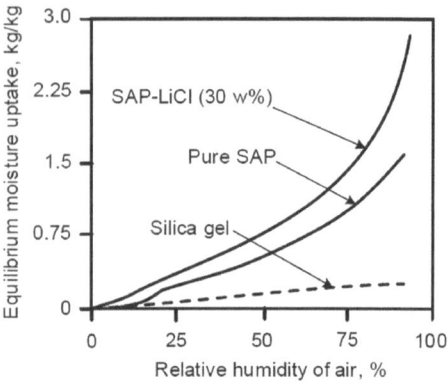

infiltrated into a copper foam to develop a device that can harvest 2.8 L of water per kg of MOF per day. MOFs have a huge potential of using in dehumidification applications. However, limited information on the cyclic performance of MOFs is available in the literature. Generally, MOFs have low hydrothermal stability due to their tendency to decompose in a moist environment. Moreover, limited MOFs are commercially available as the production cost of MOFs is relatively high.

In recent years, silica gel, activated carbon, Y-type zeolite, AQSOA-Z01, AQSOA-Z02, and AQSOA-Z03 are commonly used in industrial dehumidifiers. A comparison of the moisture adsorption performance of these adsorbents is shown in Fig. 4.9.

Adsorbents are usually chosen based on their adsorption capacity, regeneration temperature, adsorption kinetics, corrosion, and durability. Greater adsorption capacity helps to increase the cycle time and reduce the mass of required adsorbent. Lower regeneration temperature allows effective utilization of low-grade heat sources and increases the efficiency of moisture adsorption processes.

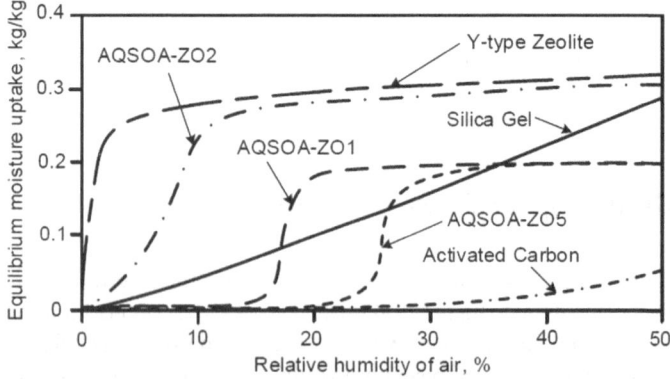

Fig. 4.9 Moisture adsorption performance of adsorbents commonly used in industrial dehumidifiers

4.3 Solid Adsorbent Dehumidification Systems

Solid adsorbent dehumidification systems are broadly classified as (a) fixed-bed, (b) rotary wheel, and (c) adsorbent coated heat and mass exchangers.

4.3.1 Fixed-Bed Dehumidifier

In fixed-bed dehumidifiers, the adsorbent materials are packed in stationary cylinders. The moist air and the hot regeneration air flow alternatively through the stationary bed of adsorbent to achieve the dehumidified process air and the regeneration of the adsorbent. Since the dehumidification process is not continuous, many beds of adsorbent are installed to achieve the continuous dehumidified process air. While the beds of adsorbent adsorb moisture from the stream of moist air and produce the dehumidified air, the stream of hot airflows through the other beds to regenerate them for the next cycle of dehumidification. Although the fixed-bed dehumidifiers are easy to fabricate, the overall heat and mass transfer coefficients are low due to the poor contact between the adsorbent matrix and the stream of air. Consequently, a large volume of adsorbent materials is required to generate the necessary exposed surface area of the adsorbent matrix for the adsorption of the unit mass of moisture.

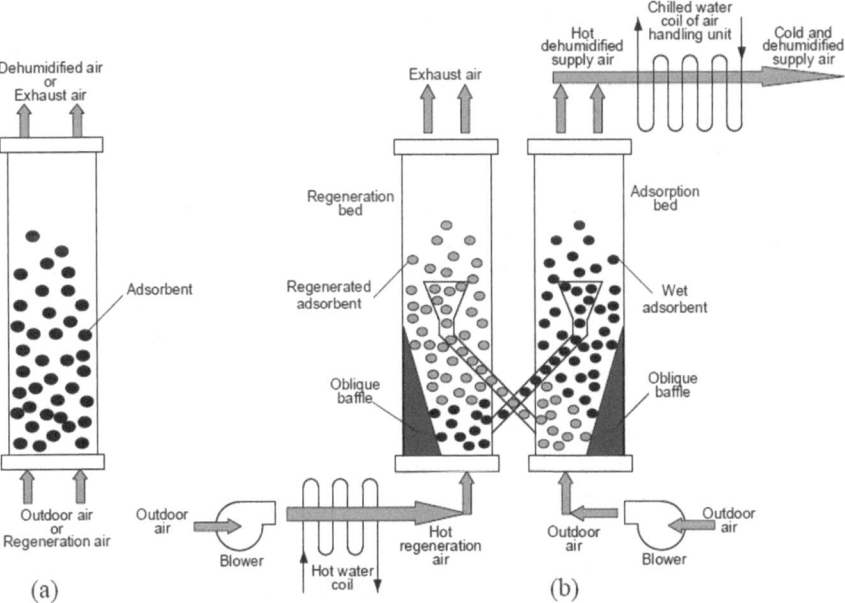

Fig. 4.10 Schematics of (**a**) conventional fluidized bed and (**b**) circulating fluidized bed system

The conventional fluidized bed (Fig. 4.10a) and the circulating fluidized bed system (Fig. 4.10b) are developed to overcome the limitations of fixed-bed dehumidifiers [19]. Two blowers are used to form two fluidized beds of the circulating fluidized bed system. Two funnels and piping systems are designed to collect the wet and regenerated falling adsorbent particles from the two fluidized beds and continuously transfer them to the other fluidized bed. An oblique baffle can be added at the bottom of the fluidized beds to eliminate the formation of the dead space in the fluidized bed and increase the circulation effect. The circulating fluidized bed system exhibits good heat and mass transfer performance and low pressure drop. Experimental results show that the dehumidification and regeneration capacities of a conventional single-cylinder fluidized bed (Fig. 4.10a) are increased by about 22.7% and 19.6%, respectively, compared with a fixed-bed system. The circulating fluidized bed with the oblique baffle system (Fig. 4.10b) further enhances the dehumidification performance by about 14%.

4.3.2 Rotary Wheel Dehumidifiers

Rotary wheel dehumidifiers are generally fabricated by rolling up the corrugated porous ceramic fibre sheet and glass fibre sheet coated with adsorbents. One piece of long fibre sheet coated with adsorbent is formed to the corrugated shape using the corrugating machine. The corrugated sheet is then bonded to another adsorbent coated flat sheet to maintain its corrugated shape. Finally, the sheets are rolled to form the cylindrical wheel that contains a large number of parallel air channels. The wheel is rotated slowly by a small motor. The process air and the regeneration hot air flow through the channels of the desiccant wheel as shown in Fig. 4.11 for continuous dehumidification of the process air and the regeneration of the wheel. The pressure drop and dehumidification performance of various shapes of air channels such as triangular, square, hexagonal, sinusoidal and circular (Fig. 4.11) are studied [20]. Study results show that the triangular channel adsorbent wheel adsorbs about 11 and 42% more moisture than the circular and square channel adsorbent wheels. However, the mean pressure drop in the triangular air channel is 73.5 and 69% higher than that in the circular and square air channels. The ratio of the required blower work to the dehumidification work, known as the figure of merit, is used to evaluate the overall energy efficiency of air channel with different shapes as presented in Table 4.2. The circular air channel shows the best energy performance. The small values of the figure of merit indicate that the required blower work to overcome the channel pressure drop is very small. Hence, the performance of heat and mass transfer is the key performance indicator that largely depends on the thermophysical properties of adsorbent and substrate materials. The effective utilization rate of the adsorbent walls depends on the rotational speed of the wheel. If the thickness of adsorbent walls exceeds the critical value under the optimum rotational speed of the wheel, only a thin layer of the adsorbent wall takes part in the moisture adsorption process.

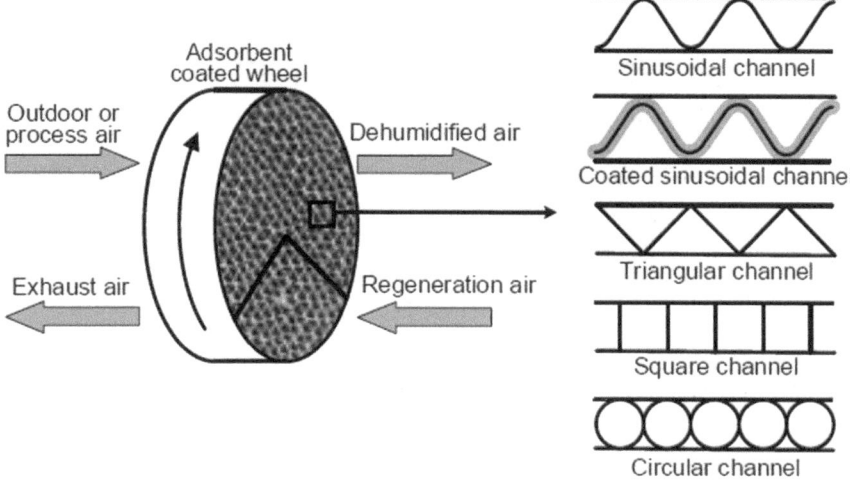

Fig. 4.11 Rotary wheel dehumidifiers with different shapes of air channels

Table 4.2 The values of work ratio for air channel of different shapes [20]	Variable	Triangular channel	Square channel	Circular channel
	Figure of merit	8.27×10^{-8}	6.29×10^{-8}	4.07×10^{-8}

The adsorption and the regeneration processes of the rotary adsorbent wheel are adiabatic. Because of the low thermal conductivity and specific heat capacity of the substrate materials, the heat of adsorption generated during the adsorption process increases the temperature of the adsorbent materials. As a result, the potential of dehumidification drops and the temperature of the dehumidified air increases. The specific enthalpy of the process air remains almost constant at the inlet and outlet of the adsorbent wheel, which represents the conversion of latent heat of the absorbed moisture to the sensible heat. If the adsorbent wheel system is retrofitted with the air handling units of central air-conditioning systems as shown in Fig. 4.12, the cooling load for the central air-conditioning systems remains almost constant. However, the central air-conditioning systems can be operated at a higher chilled water supply temperature of about 15 °C to support the sensible cooling load of the circulating air resulting in an improvement of the energy efficiency and the reduction of electrical power consumption of about 25% of the central air-conditioning systems. Flat plate solar thermal collectors can be used to generate the hot water and supply to the heating coil to regenerate the adsorbent.

The overall dehumidification and cooling performance of the hybrid adsorbent coated wheel and conventional air handling unit can be enhanced by installing a heat recovery thermal wheel in a series configuration between the adsorbent-coated wheel and the cooling coils of the conventional air-handling unit as shown in Fig. 4.13. The

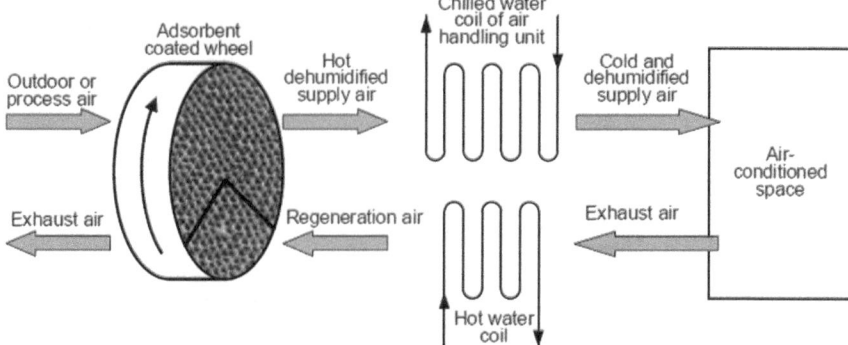

Fig. 4.12 Adsorbent-coated wheel is retrofitted with the air handling unit of the air-conditioning system

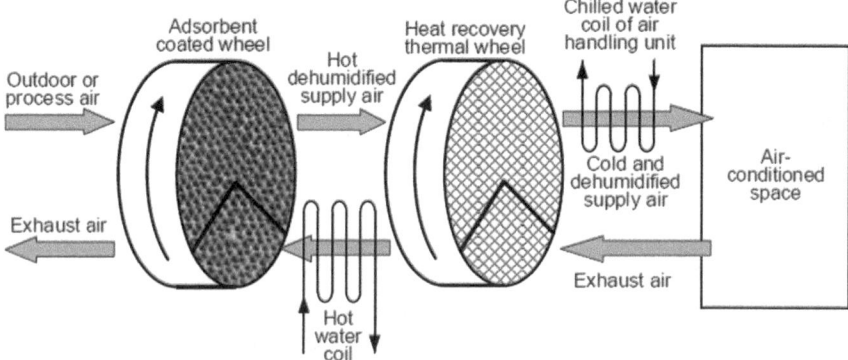

Fig. 4.13 Adsorbent-coated wheel and thermal wheel are retrofitted with the air handling unit of the air-conditioning system

circular, square, triangular and honeycomb matrix of uncoated metal of relatively high specific heat capacity and thermal conductivity are commonly used to fabricate the heat recovery thermal wheels. The thermal wheel absorbs heat from the hot dehumidified supply air and then transfer to the stream of exhaust air for the regeneration of the adsorbent-coated wheel. As a result, the cooling load of the chilled water coil of air handling units as well as the heating load of the hot water coils drop.

4.3.3 Adsorbent-Coated Heat and Mass Exchanger

During the adsorption process, the kinetic energy of the adsorbed gaseous molecules of moisture is released into the adsorbents in the form of heat of adsorption. Hence, the adsorption process becomes exothermic. The adsorbent first adsorbs the heat of adsorption and finally transfers the heat to the flowing air. As a result, the temperature of the adsorbent and the flowing air increases during the dehumidification processes of the fixed-bed and the rotary wheel dehumidifiers. The adoption capacity of the adsorbent decreases with the increase of temperature [21, 22]. The heat of adsorption equivalent to the latent heat of the adsorbed moisture is turned into the sensible heat of the dehumidified air. Therefore, the dehumidified hot air is required to flow over the subsequent cooling coils to release the generated sensible heat and reduce the temperature of the dehumidified air. The total cooling load of the subsequent cooling coils remains almost unchanged.

Adsorbent-coated heat and mass exchangers (ACHMEs) are developed to address the limitations of the fixed-bed and the rotary wheel dehumidifiers. In ACHMEs, appropriately selected adsorbents are coated on the fins or extended surfaces of the exchangers and the moist process air flows over the coated extended surfaces. A cooling fluid is passed through the channels or tubes during the dehumidification process to effectively remove the heat of adsorption. As the thermal conductivity of the metallic fins or extended surfaces and the channels or tubes is quite high, the cooling fluid effectively removes a significant portion of the heat of adsorption and the total cooling load of the subsequent cooling coils is reduced. Due to the effective heat transfer to the cooling fluid, the temperature of the adsorbent remains low during the moisture adsorption process, which enhances the moisture adsorption capacity of the adsorbents [23].

There are two major thermal resistances for transferring the regeneration heat to the adsorbent of the fixed-bed and the rotary wheel dehumidifiers. The first thermal resistance is from the heat source to the bulk of the regeneration air and the second one is from the bulk of the regeneration air to the adsorbent. The thermal resistances are high due to the low convective heat transfer coefficient of the regeneration air. Consequently, a high-temperature heat source is required to overcome the thermal resistances and transfer the required regeneration heat to the adsorbent. The regeneration performance of the ACHMEs improves as the heating fluid of the high convective heat transfer coefficient internally heats the adsorbent. The metallic fins or channels of high thermal conductivity reduce the thermal resistances and enhances regeneration efficiency. Finned tube heat exchangers are predominantly used in ACHMEs. Figure 4.14a shows an uncoated heat exchanger and Fig. 4.14b shows a silica gel-coated heat and mass exchanger [24]. Fins are made of thin aluminium (Al) and tubes are made of copper (Cu). Jeong et al. [25] investigated the dehumidification performance of an internally cooled fixed-bed dehumidification system as shown in Fig. 4.15. Silica gel was coated on the internal aluminium fins to develop the heat and mass exchanger. The return air is dehumidified as well as cooled and the regeneration air in the regeneration process is heated up directly by the internal heat exchanger as shown in the Psychrometric chart (Fig. 4.15b). Cooling and heating water temperatures of 16.1 and 34.7 °C were used. The outlet absolute humidity and temperature of the process air of 7.9 g/kg and 18 °C were achieved. Key results show that the system can be operated using renewable and low-grade waste energy sources such as solar thermal and industrial waste heat. To study the effect of internal cooling, Ge et al. [26] tested the dehumidification performance of a silica gel coated heat

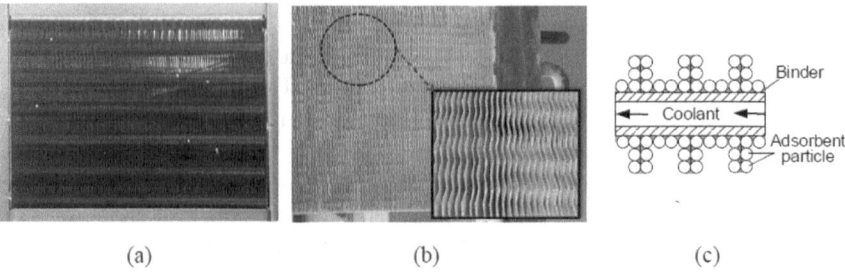

(a)   (b)   (c)

Fig. 4.14 a Uncoated finned tube heat exchanger, **b** silica gel-coated heat and mass exchanger, and **c** schematic of adsorbent coating on the fin surface [24]

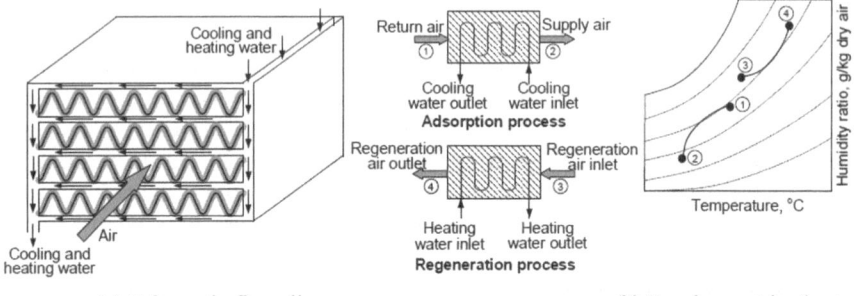

(a) Schematic flow diagram   (b) Psychrometric chart

Fig. 4.15 Schematic flow diagram and Psychrometric chart of fixed-bed dehumidification system with internal heat exchanger

and mass exchanger with and without cooling water supply. The inlet air temperature and humidity ratio of 30 °C and 14.3 g/kg, cooling water inlet temperature of 25 °C, regeneration water temperature of 60 °C, and cycle time of 600 s were maintained in the experimental study. They reported an improvement of the outlet air humidity ratio by about 2.5 g/kg when the cooling water was supplied through the heat exchanger. Saeed et al. [27] conducted a detailed review and highlighted the superiority of the ACHMEs to handle sensible and latent loads simultaneously. The advantages of the ACHMEs over the fixed-bed and the rotary wheel dehumidifiers are summarized in Table 4.3 [28]. The higher moisture adsorption capacity, the efficient removal of the heat of adsorption, and the potentials of using the low-grade waste heat for the regeneration have created the opportunities of integrating the ACHMEs for the performance enhancement of several thermally driven technologies such as adsorption chillers, heat pumps, and atmospheric water harvesters.

Table 4.3 Advantages of adsorbent coated heat and mass exchangers over the fixed-bed and the rotary wheel dehumidifiers [28]

Properties	Fixed-bed dehumidifiers	Rotary wheel dehumidifiers	Adsorbent-coated heat and mass exchangers
Heat transfer efficiency	Low, because of higher multiple thermal resistances	Low, because of higher multiple thermal resistances	High, because of high thermal conductivity of metallic fins and internal heating
Pressure drop	High	Low	Low
Adsorption capacity	Low	Low	High because of efficient cooling
Total cooling load	Total cooling load remains the same. Latent heat is converted to sensible heat	Total cooling load remains the same. Latent heat is converted to sensible heat	Total cooling load is reduced as the cooling fluid takes away a portion of the latent heat
Continuous dehumidification	Possible with two beds	Possible using a single rotary wheel	Possible with two coated exchangers
Adsorbent material utilization	Low	High	High

4.4 Binders

The selection of binder and coating method determines the overall heat and mass transfer effectiveness of ACHMEs. Binders must have high thermal conductivity as well as good adhesive strength to glue the adsorbent to the heat exchanger fin within the operating temperature range. However, binders may fill the available pore spaces of the adsorbents and impede the mass adsorption process. Therefore, binders should be selected carefully to achieve optimal heat transfer and mass adsorption rates. Polyvinyl alcohol (PVA), aluminium hydroxides, polyaniline, polytetrafluoroethylene (PTFE), graphite powders, and metallic foams are commonly used for binding zeolites and silica gel on metallic fin and tube surfaces. Although these binders of relatively high thermal conductivity reduce the thermal resistance and improve the heat transfer rates, they may slightly reduce the mass adsorption performance. Chang et al. [29] have confirmed that PVA enhances heat transfer performance of silica gel and the percentage of silica gel not involved in adsorption due to PVA binder is insignificant. Li et al. [30] reported that hydroxyl cellulose (3.3 w%) showed the superior adhesive strength to bind silica gel onto aluminium fins as well as higher Brunauer–Emmett–Teller (BET) surface area available for adsorption. Moreover, silica gel with hydroxyl cellulose binder coating achieved the maximum heat transfer performance compared with the granular packed bed. The critical properties that evaluate the feasibility and the suitability of binders are illustrated below.

4.4.1   Adhesive Ability

Adhesive ability represents the strength of the binder to glue the adsorbents on heat exchanger fins within the range of operating temperatures. Li et al. [30] presented a detailed list of binders. Organic binders such as hydroxy cellulose, polyvinyl alcohol (PVA), and gelatin exhibited better binding strength on the fin surface while inorganic binders such as sepiolite and bentonite displayed poor adhesive ability. Nitrogen adsorption technique is generally used to measure the BET surface area of adsorbents. Studies have shown that the BET surface area of the organic binders is higher than that of the inorganic epoxy binders.

4.4.2   Influence on the Adsorption Uptake

Binders have the affinity to fill the pore spaces of adsorbents and hinder the mass adsorption process. The optimum weight ratio of a binder that has a minimum effect on the adsorption performance of an adsorbent should be determined. Li et al. [30] conducted a detailed study with 1.67, 3.3, 6.67 and 10% weight ratio of hydroxy cellulose binder and measured the effect of binder contents on the BET surface and pore volumes of 3A and RD types of silica gel. The weight ratio of 3.3% was reported as the optimum that provided the maximum value of the BET surface area and pore volume. Similar tests are necessary for selecting appropriate binders for other adsorbent materials.

4.4.3   Heat Transfer Performance

In conjunction with the mass adsorption performance of an adsorbent, the thermal resistance and cycle time play a key role on the overall performance of an adsorption system. Transient heat transfer performance of silica gel with hydroxy cellulose binder is three to four times higher when compared with the granular silica gel packing fixed-bed system [30]. Moreover, the time required to attain the peak heat transfer rate of the fixed-bed granular silica gel packing is reduced from 70 s to about 10 s for the hydroxy cellulose binder silica gel-coated adsorption system resulting in higher heat and mass transfer performances.

4.5   Adsorbent Coating Techniques

An appropriate coating technique is essential to achieve a durable and uniform coating of adsorbents on the metal surface of ACHMEs. Studies have shown that researchers have adopted three main coating techniques, namely, dip-coating, electrostatic spray coating, and direct synthesis. The advantages and limitations of these techniques are discussed in the following sections.

4.5.1 Dip Coating

Dip coating is the simplest and least expensive technique. Oh et al. [24] illustrated the steps of dip coating. The finned tube heat exchanger is carefully cleaned by immersing it in an alkaline solution followed by washing using clean water and drying. An optimum concentration and weight ratio of the solution of adsorbent and binder is then prepared and the mixture is magnetically stirred to ensure a homogeneous mixing without any lump formation. The cleaned and dried finned tube heat exchanger is then dipped in the solution several times. After dipping, the heat exchanger is heated in an oven at about 120 °C for 1–2 h depending on the thickness of the coating. Usually, the dipping and drying processes are repeated for five to six times until the desired coating thickness is formed. The optimum weight ratio of the binder and adsorbent solution should be used to minimize the blocking of the pores of adsorbent that negatively affect the adsorption capacity [31]. Test results showed that the mass diffusivity of dip-coated silica aero gel is similar to that of solid silica aerogel [32]. Aimjaijit et al. [33] presented the silica gel synthesizing technique to enhance the adhesive properties of the silica gel. Although the dip-coating technique is widely used, repeated dipping may lead to an uneven distribution of the adsorbent resulting in the blockage of the fin gaps.

4.5.2 Electrostatic Spray Coating

Electrostatic spray coating is a specialized process. It involves the preparation of the mixture of binder and adsorbent and the electrostatic spraying process. The optimum ratio of the binder and adsorbent is mixed thoroughly and the resulting powder is loaded into the charging barrel of the electrostatic spray coating device. The power is sprayed under the influence of a powerful electrostatic field through the gun of the coating device on the exposed surfaces of the finned tube heat exchanger. The spray coating together with the impregnation produces a good uniform coating thickness of about 120 μm without any cracks, which enhances the adsorption performance [34]. The initial investment for the electrostatic spray coating process is high compared with the dip-coating technique. Moreover, the spray coating device has a few sensitive components that need regular maintenance.

4.5.3 Direct Synthesis

The direct synthesis of zeolites on the finned tube heat exchangers and various metal substrates such as aluminium and copper foams have been extensively investigated. The synthesis process involved hydrothermal zeolitization in a closed system with varying temperature, pressure, and time depending on the types of zeolites. The required synthesis time could be a few hours to several days. The hydrothermal experiments established that the zeolite phase laumontite is formed in the temperature interval between 30 and 450 °C at 1 kbar H_2O pressure and an experiment duration of 6 weeks. Bonaccorsi et al. [35] reported that direct synthesis coating of zeolite 4A and zeolite Y on aluminium form good adhesive strength, coating thickness, and other properties. The industrial application of direct synthesis coating is limited as the synthesis yields a coating thickness of about 10–50 µm. Among all types of coating processes, dip-coating and electrostatic spray coating requires impeller operating steps and can be easily implemented in industries for mass production compared with the direct synthesis method. Dip coating has an advantage over the electrostatic spraying coating as the use of sensitive devise and gun for spraying raises the investment, operating and maintenance costs. Moreover, heat exchangers of complex geometries can be coated uniformly using the dip coating. The adhesive strength of dip coating can be enhanced by employing appropriate synthesis methods. A comparison of the three types of coating processes is summarized in Table 4.4.

4.6 Dynamic Performance of Adsorbent Coated Heat and Mass Exchangers

The dynamic performance of Adsorbent-Coated Heat and Mass Exchangers (ACHMEs) depends on a number of parameters such as dry bulb temperature ($T_{bd,in}$), humidity ratio ($\omega_{a,in}$) and flow rate (m_a) of inlet air, cooling water temperature (T_{cool}), regeneration or hot water temperature (T_{reg}), face velocity of the exchanger (V_a), cycle time (t_{cycle}) and adsorbent used. The humidity ratio and temperature of the

Table 4.4 Comparison of dip, electrostatic spray, and direct synthesis coating processes

Parameters	Dip coating	Electrostatic spray coating	Direct synthesis coating
Achievable coating thickness, µm	120	120	10–50
Intensities of reactions	Mild	Mild	Severe
Industrial mass production	Easy	Easy	Difficult
Capital and maintenance costs	Cheap	Slightly expensive	Expensive

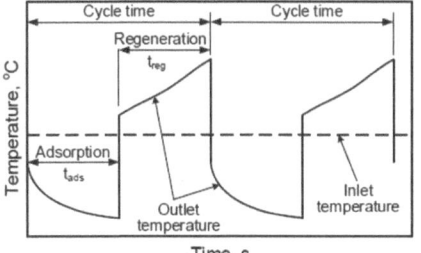 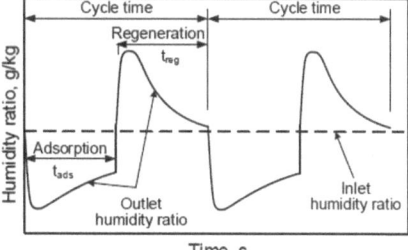

Fig. 4.16 Typical variations of the temperature and humidity ratio of the circulating air during the adsorption and regeneration processes

circulating air are continuously measured at the inlet and outlet of the exchanger to evaluate the dynamic adsorption and regeneration performance. The typical variations of the humidity ratio and temperature of the circulating air during the adsorption and regeneration processes are schematically presented in Fig. 4.16.

During the dehumidification process, the humidity ratio of the outlet air first decreases sharply and then gradually increases as the adsorbent approaches its saturation. The regeneration process is opposite to the adsorption process. The humidity ratio of the outlet air increases sharply at the initial stage of the regeneration process and then steadily decreases until it is on par with the inlet humidity ratio. Dynamic performances of ACHMEs are calculated using the following parameters.

4.6.1 Average Moisture Removal Capacity

Transient moisture removal (D_{tran}) performance of ACHMEs is determined by measuring the humidity ratio of the circulating air at the inlet and outlet of the exchanger as:

$$D_{tran} = \omega_{a,in} - \omega_{a,out} \tag{4.1}$$

where D_{tran} (g/kg) represents the transient moisture removal, $\omega_{a,in}$ (g/kg) is the humidity ratio of the circulating air at the inlet, and $\omega_{a,out}$ (g/kg) is the humidity ratio of the circulating air at the outlet of the exchanger. As the adsorbent approaches its saturation during the adsorption process, the transient value of $\omega_{a,out}$ increases resulting in a steady drop of D_{tran} with time. Hence, an effective adsorption time (t_{ads}) should be chosen such that the transient moisture removal remains significant. The average moisture removal capacity (MRC) of ACHMEs during the effective adsorption time (t_{ads}) is calculated as:

$$\text{MRC} = \frac{1}{t_{\text{ads}}} \int_0^{t_{\text{ads}}} D_{\text{tran}} \mathrm{d}t \qquad (4.2)$$

where MRC (g/kg) represents the average moisture removal capacity and t_{ads} (s) is the effective adsorption time. Some researchers use dehumidification capacity (W_{d}) to compare the dehumidification performance of ACHMEs. The dehumidification capacity is defined as the total amount of moisture adsorbed from the circulating air by the ACHMEs in its cycle time as:

$$W_{\text{d}} = (\text{MRC}) \times m_{\text{a}} \qquad (4.3)$$

where W_{d} (g/s) represents the dehumidification capacity and m_{a} (kg/s) is the mass flow rate of the circulating air.

4.6.2 Thermal Coefficient of Performance

Thermal coefficient of performance (COPth) of ACHMEs is defined as the ratio of the average cooling capacity (Q_{cooling}) of dehumidified air produced during the adsorption process to the average heat exchange rate (Q_{heating}) of hot water during the regeneration process of the adsorbent and is expressed as:

$$\text{COP}_{\text{th}} = \frac{Q_{\text{cooling}}}{Q_{\text{heating}}} \qquad (4.4)$$

Theoretically, the thermal energy required for regenerating the adsorbent should translate into an equivalent cooling capacity of the produced dehumidified air. However, due to the thermal mass of ACHMEs and the losses involved in the system, the cooling capacity of the dehumidified air generated by the ACHME system is lower than its regeneration energy requirement. The COP_{th} largely depends on the temperature and flow rate of the cooling water. If the temperature of the cooling water is lower than the dew-point temperature of the inlet air, the value of the COP_{th} could become greater than one. In such a situation, the cooling capacity of the generated dehumidified air is achieved due to the combination of the moisture condensation and the moisture adsorption by the desiccant.

The average cooling capacity (Q_{cooling}) of dehumidified air generated during the adsorption process is calculated as:

$$Q_{\text{cooling}} = \frac{m_{\text{a}}}{t_{\text{ads}}} \int_0^{t_{\text{ads}}} (h_{\text{a,in}} - h_{\text{a,out}}) \mathrm{d}t \qquad (4.5)$$

where $Q_{cooling}$ (kW) represents the average cooling capacity of dehumidified air, m_a (kg/s) is the mass flow rate of the circulating air, t_{ads} (s) is the effective adsorption time, $h_{a,in}$ (kJ/kg) is the enthalpy of the circulating air at the inlet, and $h_{a,out}$ (kJ/kg) is the enthalpy of the circulating air at the outlet of the exchanger.

Similarly, the average heat exchange rate ($Q_{heating}$) of the hot water during the regeneration process of the adsorbent is calculated as:

$$Q_{heating} = \frac{m_w c_{p,w}}{t_{reg}} \int_0^{t_{reg}} \left(T_{w,in} - T_{w,out}\right) dt \tag{4.6}$$

where $Q_{heating}$ (kW) represents the average heat exchange rate of the hot water, m_w (kg/s) is the mass flow rate of the hot water, t_{reg} (s) is the effective regeneration time, $c_{p,w}$ (kJ/kg K) is the specific heat capacity of hot water, $T_{w,in}$ (°C) is the temperature of hot water at the inlet, and $T_{w,out}$ (°C) is the temperature of hot water at the outlet of the exchanger.

4.7 Parametric Study of ACHMEs

The effects of different operating parameters on the dehumidification and thermal performance of ACHMEs are discussed in the ensuing sections.

4.7.1 Effect of Moisture Content of Air

The change in the inlet air humidity ratio or the inlet air relative humidity at constant temperature represents the change of moisture content in the air. The higher moisture content of air results in the greater partial pressure of moisture in the air, which improves the driving potential for moisture transfer from the stream of air to the exposed surface of adsorbents. As a result, the dehumidification performance of ACHMEs increases with the increase of moisture content in the air. The COP_{th} of ACHMEs also improves with the higher moisture content of air because of the increase in cooling capacity. Table 4.5 shows the effect of the inlet air moisture content on the dehumidification and thermal performance of ACHMEs.

4.7.2 Effect of Inlet Air Dry Bulb Temperature

The effect of the variation of inlet air dry bulb temperature on the dehumidification and thermal performance of ACHMEs is studied for two different scenarios

Table 4.5 Effect of inlet air moisture content on the dehumidification and thermal performance of ACHMEs.

Adsorbent material	Moisture content in air	Constant operating parameters	Dehumidification performance	Thermal performance
Pure silica gel [26]	$\omega_{a,in}$ increased from 10 to 17.5 g/kg	$T_{bd,in} = 30\,°C$ $T_{cool} = 25\,°C$ $T_{reg} = 60\,°C$ $t_{cycle} = 10$ min $V_a = 1.0$ m/s	MRC increases from 1.0 to 4.1 g/kg	COP$_{th}$ increases from 0.28 to 0.52
Polymer [26]			MRC increases from 0.5 to 2.0 g/kg	COP$_{th}$ increases from 0.15 to 0.3
Pure silica gel [36]	RH$_{in}$ increased from 50% to 70%	$T_{bd,in} = 30\,°C$ $T_{cool} = 15\,°C$ $T_{reg} = 50\,°C$ $t_{cycle} = 10$ min $V_a = 1.54$ m/s	MRC increases from 2.1 to 3.2 g/kg	COP$_{th}$ increases from 0.6 to 0.9
Composite silica gel with 38 w% LiCl [36]			MRC increases from 2.3 to 3.9 g/kg	COP$_{th}$ increases from 0.94 to 1.2
Pure silica gel [37]	RHin increased from 40 to 70%	Tbd,in $= 30\,°C$ Tcool $= 20\,°C$ Treg $= 50\,°C$ Va $= 1.7$ m/s	Wd increases from 150 g/h to 900 g/h	Qcooling increases from 780 to 930 W
Composite silica gel with 30 w% LiCl [37]			Wd increases from 250 to 1100 g/h	Qcooling increases from 820 to 1000 W

depending on whether the humidity ratio or the relative humidity of the inlet air is maintained constant. The humidity ratio represents the actual moisture content of air, which does not change with the change of dry bulb temperature. At constant humidity ratio of air, when the dry bulb temperature of the inlet air is increased, the relative humidity of the inlet air decreases. On the other hand, when the dry bulb temperature of the inlet air is increased keeping the relative humidity constant, the humidity ratio of the inlet air increases. The increase of the humidity ratio increases the moisture content and the partial vapour pressure of air, which acts as a driving potential for moisture transfer from the bulk air to the exposed surface of the adsorbent. As a result, the dehumidification performance of ACHMEs improves. At higher inlet air dry bulb temperature, more thermal energy is transferred from the bulk air to the coolant, thereby increases the cooling capacity and COPth of the ACHMEs.

At constant humidity ratio of the inlet air, higher dry bulb temperature of air increases the temperature of adsorbent results in the decrease of its adsorption capacity because of the reduction in the relative humidity at the exposed surface of the adsorbent. As the adsorption capacity of the adsorbent is a function of relative humidity, the *MRC* of the ACHMEs decreases. Table 4.6 represents the effect of inlet air dry bulb temperature on the dehumidification and thermal performance of ACHMEs.

Table 4.6 Effect of inlet air dry bulb temperature on dehumidification and thermal performance of ACHMEs

Adsorbent material	Inlet air dry bulb temperature, °C	Constant operating parameters	Dehumidification performance	Thermal performance
Pure silica gel [36]	Increased from 25 to 35	$RH_{in} = 60\%$ $T_{cool} = 15\,°C$ $T_{reg} = 50\,°C$ $t_{cycle} = 10$ min $V_a = 1.7$ m/s	MRC increases from 1.6 to 3.0 g/kg	COP_{th} increases from 0.4 to 1.2
Composite silica gel with 38 w% LiCl [36]			MRC increases from 2.1 to 3.8 g/kg	COP_{th} increases from 0.56 to 1.6
Pure silica gel [26]	Increased from 25 to 35	$\omega_{a,in} = 14.3$ g/kg $T_{cool} = 25\,°C$ $T_{reg} = 60\,°C$ $t_{cycle} = 10$ min $V_a = 1.0$ m/s	MRC decreases from 3.1 to 2.8 g/kg	COP_{th} increases from 0.4 to 0.45
Polymer [26]			MRC decreases from 1.7 to 1.55 g/kg	COP_{th} increases from 0.31 to 0.35

4.7.3 Effect of Face Velocity of Exchanger

The face velocity of an exchanger represents the velocity of the air passing through the heat and mass transfer surfaces of the exchanger. The interaction time for the moisture transfer process between the circulating air and the exposed surface of ACHMEs decreases with the increase in the face velocity. Consequently, both MRC and COP_{th} of the ACHMEs decrease with higher face velocity. However, the dehumidification capacity (W_d) of the ACHMEs increases due to the higher mass flow rate of the inlet air. The effect of face velocity on the dehumidification and thermal performance of ACHMEs is presented in Table 4.7. The MRC and the COP_{th} of the composite silica gel with 38w% LiCl-coated exchanger improved between 18 and 40% and 20 and 25%, respectively, compared with the pure silica gel coated exchanger for the operating range of face velocities.

Table 4.7 Effect of face velocity of inlet air on dehumidification and thermal performance of ACHMEs [36]

Adsorbent material	Face velocity, m/s	Constant operating parameters	Dehumidification performance	Thermal performance
Pure silica gel	Increased from 0.86 to 1.54	$T_{bd,in} = 30\,°C$ $RH_{in} = 60\%$ $T_{cool} = 15\,°C$ $T_{reg} = 50\,°C$ $t_{cycle} = 10$ min	MRC decreases from 3.9 to 2.3 g/kg	COP_{th} decreases from 1.0 to 0.8
Composite silica gel with 38 w% LiCl			MRC decreases from 4.6 to 3.2 g/kg	COP_{th} decreases from 1.2 to 1.0

4.7.4 Effect of Cooling Water Temperature

The adsorption of moisture in an exothermic process. The moisture adsorption capacity of adsorbents decreases with the increase of temperature. The dehumidification performance of ACHMEs improves when the heat of adsorption is effectively removed from the adsorbent using the cooling water of low temperature. The COPth of ACHMEs also improves with lower cooling water temperature because of the increase in cooling capacity. Table 4.8 represents the effect of the cooling water temperature on the dehumidification and thermal performance of ACHMEs. The results indicate that the pure silica gel-coated exchangers need much lower cooling water temperature than the composite desiccant to achieve the same dehumidification effect.

4.7.5 Effect of Regeneration Temperature

Regeneration temperature has a significant impact on the dehumidification performance as well as the cooling capacity ($Q_{cooling}$) of ACHMEs. Higher regeneration temperature facilitates the deep drying of adsorbents. As a result, the capacity of adsorbing moisture during the dehumidification process improves remarkably. However, the energy efficiency of the regeneration process decreases with increasing regeneration temperature due to the higher heat loss when the regeneration process is switched to the dehumidification process. Moreover, some portion of the cooling effect is offset by the thermal energy stored in the adsorbents during the regeneration process. The decrease in cooling capacity and the drop in energy efficiency of the regeneration process are attributed to the thermal mass of the ACHME and the heat loss to the surroundings. With the increase in regeneration temperature, the reduction in the cooling capacity of the ACHME becomes more pronounced. Zheng et al. [38] tested the dehumidification performance of silica gel, SAPO-34 and FAPO-34 by increasing the regeneration temperature from 50 to 80 °C. They maintained inlet air dry bulb temperature at 30 °C, inlet air humidity ratio at 16 g/kg and cooling water temperature at 20 °C. Due to the increase in regeneration temperature, the dehumidification capacities of silica gel, SAPO-34, and FAPO-34 were enhanced by 28%, 7%, and 69%, respectively. Hu etal. [37] experimentally investigated the change of cooling capacity of the silica gel coated and the composite silica gel with 30 w% LiCl-coated ACHMEs by regulating the regeneration temperature from 30 to 70 °C while maintaining the inlet air dry bulb temperature at 30 °C, inlet air relative humidity at 60% and cooling water temperature at 20 °C. They reported that the cooling capacities of the silica gel coated and the composite silica gel-coated ACHMEs were dropped from 625 to 425 W and 675 W to 525 W, respectively. Table 4.9 represents the effect of changing the regeneration temperature on the MRC and the COP_{th} of ACHMEs.

Table 4.8 Effect of cooling water temperature on the dehumidification and thermal performance of ACHMEs

Adsorbent material	Cooling water temperature, °C	Constant operating parameters	Dehumidification performance	Thermal performance
Pure silica gel [37]	T_{cool} decreased from 25 to 10 °C	$T_{bd,in} = 30$ °C $T_{reg} = 50$ °C $V_a = 1.7$ m/s	For $RH_{in} = 55\%$: W_d increases from 100 to 700 g/h; For $RH_{in} = 65\%$: W_d iincreases from 200 to 1100 g/h	For $RH_{in} = 55\%$: $Q_{cooling}$ increases from 200 to 1200 W; For $RH_{in} = 65\%$: $Q_{cooling}$ increases from 250 to 1350 W
Composite silica gel with 30 w% LiCl [37]	T_{cool} decreased from 25 to 10 °C	$T_{bd,in} = 30$ °C $T_{reg} = 50$ °C $V_a = 1.7$ m/s	For $RH_{in} = 55\%$: W_d increases from 300 to 980 g/h; For $RH_{in} = 65\%$: W_d increases from 425 to 1300 g/h	For $RH_{in} = 65\%$: $Q_{cooling}$ increases from 250 to 1350 W For $RH_{in} = 65\%$: $Q_{cooling}$ increases from 270 to 1400 W

Table 4.9 Effect of changing the regeneration temperature on the dehumidification and thermal performance of ACHMEs

Adsorbent material	Regeneration temperature, °C	Constant operating parameters	Dehumidification performance	Thermal performance
Pure silica gel [36]	T_{reg} increased from 40 to 60 °C	$T_{bd,in} =$ 30 °C $RH_{in} = 60\%$	MRC increases from 2.2 to 2.5 g/kg	COP_{th} decreases from 1.7 to 0.7
Composite silica gel with 38 w% LiCl [36]		$T_{cool} =$ 15 °C $t_{cycle} = 10$ min $V_a = 1.54$ m/s	MRC increases from 2.6 to 3.2 g/kg	COP_{th} decreases from 1.6 to 0.5
Pure silica gel [26]	T_{reg} increased from 60 to 80 °C	$T_{bd,in} = 30$ °C $\omega_{a,in} = 14.3$ g/kg $T_{cool} = 25$ °C	MRC increases from 3.1 to 5.0 g/kg	For 60–70 °C: COP_{th} increases from 0.43 to 0.48 For 70–80 °C: COP_{th} decreases from 0.48 to 0.38
Polymer [26]		$t_{cycle} = 10$ min $V_a = 1.0$ m/s	MRC increases from 1.4 to 2.9 g/kg	COP_{th} decreases from 0.33 to 0.22

4.7.6 Effect of Cycle Time

A complete cycle comprises of adsorption and regeneration process as illustrated in Fig. 4.16. At the initial period of the effective adsorption time (t_{ads}), the humidity ratio of the supply air decreases sharply as the unsaturated walls surrounding the pores of adsorbent rapidly attract the molecules of moisture. The walls are started to fill volumetrically with the molecules of moisture. The adsorbed molecules of moisture then diffuse into the pores of the adsorbent and the moisture adsorption rate is controlled by the diffusion coefficient of the pores. The opposite phenomenon occurs during the effective regeneration time (t_{reg}). The humidity ratio of the exhaust air increases sharply at the initial period of the effective regeneration time (t_{reg}) as the saturated walls surrounding the pores of adsorbent rapidly release the molecules of moisture. As a result, both MRC and COP_{th} of ACHMEs increase at the initial period of switching until a maximum value is reached and then starts dropping. Zhao et al. [21] studied the effect of increasing the cycle time on silica gel coated heat exchangers for the operating parameters summarized in Table 4.10. They reported that the optimum MRC and COP_{th} performances were achieved for the cycle time of 600 s. The adsorption and regeneration profiles shown in Fig. 4.16 exhibit that the cycle time should be shortened to reduce the humidity ratio of the supply air.

The effects of different operating parameters on the dehumidification and thermal performance of ACHMEs are documented in Table 4.11.

Table 4.10 Operating parameters to study the effect of cycle time [21]

Cycle time, s	Dry bulb temperature, °C	Humidity ratio, g/kg	Cooling water temperature, °C	Regeneration temperature, °C
300–1200	29 and 31	15–16	24 and 28	80 and 85

Table 4.11 Summary of the effects of different operating parameters on the dehumidification and thermal performance of ACHMEs

Operating parameters	Moisture removal capacity (MRC)	Thermal coefficient of performance (COP$_{th}$)
Increase in moisture content of air	MRC increases as the driving potential is enhanced	COP$_{th}$ improves as the cooling capacity increases
Increase in dry bulb temperature of air	For constant RH$_{in}$: MRC increases as the partial vapour pressure of air increases	COP$_{th}$ improves as more thermal energy transfers to coolant
	For constant $\omega_{a,in}$: MRC decreases as the temperature of adsorbent increases	
Higher face velocity of exchanger	MRC reduces as the contact time between the adsorbent and air decreases	COPth drops as the contact time between the adsorbent and air decreases
Decrease in Cooling water temperature	MRC increases as the heat of adsorption is effectively removed	COPth improves as more thermal energy transfers to coolant
Increase in regeneration temperature	MRC increases as adsorbent experiences a deep drying	COP$_{th}$ depends on regeneration temperature and thermal mass of exchanger
Cycle time	MRC increases at the initial period of switching until a maximum value is reached and then starts dropping	COP$_{th}$ increase at the initial period of switching until a maximum value is reached and then starts dropping

4.8 Hybrid Applications of Heat and Mass Exchangers

Adsorbent-coated heat and mass exchanger (ACHME) systems are identified as an energy-efficient alternative of the conventional vapour compression central air-conditioning chiller systems. The internal cooling and heating features of the ACHMEs remarkably enhance the moisture adsorption and regeneration performance. Due to the efficient heat transfer characteristics of the ACHMEs, the low-grade industrial waste heat and solar thermal energy are used to regenerate the adsorbents during the regeneration process. Similarly, the heat of adsorption generated during the dehumidification process is conveniently transferred to the cooling tower. The ACHMEs are retrofitted in many innovative ways to improve the overall energy performance of the dehumidification and cooling processes.

4.8.1 ACHMEs Coupled with Solar Thermal System

The schematic diagram of the ACHMEs coupled with the solar thermal collector and cooling tower systems are shown in Fig. 4.17. Based on the types of adsorbent used in the ACHMEs, the flat plate or the evacuated tube solar thermal collectors are used to generate hot water for the regeneration of adsorbent. Valve 1, 2, 3, and 4 are opened and valve 5, 6, 7, and 8 are closed to operate ACHME-1 and ACHME-2 in the adsorption and regeneration modes, respectively. Motorized dampers D-1 and D-2 are opened, and D-3 and D-4 are closed to circulate the dehumidified cold air in the air-conditioned spaces. A chilled water coil could be installed in the supply air duct to maintain the required temperature of the supply air. After the pre-set cycle time, the modes of operation of ACHME-1 and ACHME-2 are changed by opening the valve-5, 6, 7, and 8, and closing the valve-1, 2, 3, and 4. At the same time, the motorized dampers D-1 and D-2 are closed, and D-3 and D-4 are opened to maintain a constant supply of the conditioned air to the spaces. A major portion of the heat of adsorption generated in the adsorption process of moisture is transferred to the cooling tower. As a result, the cooling load of the chiller plant drops significantly. The chilled water coils need to support only the sensible cooling load of the supply air. As the moisture of the supply air is adsorbed by the adsorbent of ACHMEs, it is not necessary to supply the chilled water to the chilled water coil below the dew-point temperature of

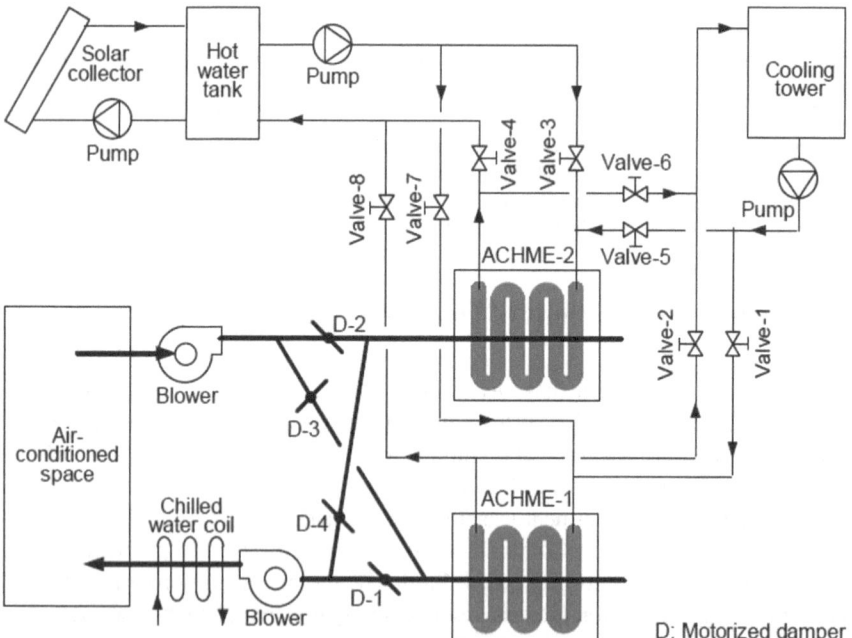

Fig. 4.17 Schematic diagram of the ACHMEs coupled with the solar thermal collector and cooling tower systems

the supply air. The chiller plant can be operated to generate the chilled water supply temperature of 12–15 °C, which further improves the energy efficiency of the chiller plants. Zhao et al. [39] installed a heat exchanger to recover the heat from the exhaust air of regenerating ACHME. They reported that the dehumidification performance of the ACHME system improved over 2% and the requirement of the regeneration energy dropped from 4485 to 2016 kJ due to the installation of the heat recovery system. Consequently, the *COPth* of the ACHME with the heat recovery system was almost doubled.

4.8.2 Multi-bed ACHME System

The airflow resistance of ACHMEs depends on the face velocity. Oh et al. [24] highlighted that the airflow in a perpendicular direction across the ACHME increased the face velocity and resulted in higher energy consumption of the blower of overcome the flow resistance of the circulating air. Multiple beds of ACHME can be installed in a V-shaped configuration to reduce the face velocity and the flow resistance. The photograph and schematic diagrams of a two-bed ACHME system are shown in Fig. 4.18. The ACHMEs are installed at an angle of 30°. This inclined configuration of ACHMEs increases the frontal interaction area between the adsorbent and the moist air, which eventually reduces both face velocity and flow resistance, and increases the contact time between the desiccant and the moist air. The notable advantage of the ducting and damper system design is that the dehumidification and regeneration processes are continuous. Motorized damper D-1 and D-2 are opened and D-3 and D-4 are closed as shown in Fig. 4.18b to operate the beds of ACHME-1 and ACHME-2 in the dehumidification and regeneration modes, respectively. After the pre-set cycle time, the modes of operation of the beds of ACHME-1 and ACHME-2 are changed by opening the damper D-3 and D-4 and closing the damper D-1 and D-2 as shown in Fig. 4.18c. The hot water required for the regeneration of the adsorbent can be produced by extracting the low-grade waste heat of industrial processes. Depending on the required temperature of the supply air, the cold water from the cooling tower or the chiller system can be supplied to the ACHME involved in the dehumidification and cooling processes. A number of the solenoid valves are installed in the hot and cold water pipes to regulate the flow of the hot and cold water in the beds of ACHME-1 and ACHME-2 depending on their modes of operation.

4.8.3 Adsorbent-Coated Direct Expansion System

The evaporator and condenser coils of the direct expansion air-conditioning units can be replaced with ACHMEs to utilize the cooling effect of the evaporator coil and the heating effect of the condenser coil and enhance the overall energy performance of the system. The evaporator coil of the conventional direct expansion air-conditioning

(a) Photograph of a two-bed ACHME system

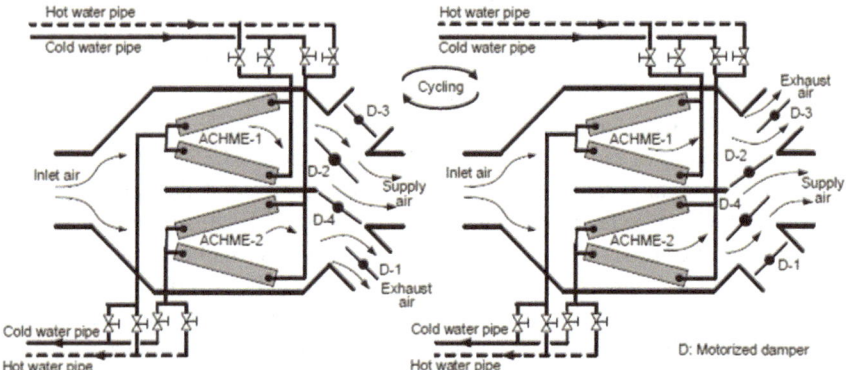

(b) ACHME-1 is in dehumidification   (c) ACHME-2 is in dehumidification
and ACHME-2 is in regeneration mode   and ACHME-1 is in regeneration mode

Fig. 4.18 Photograph and schematic diagrams of a two-bed ACHME system

units is cooled below the dew-point temperature of circulating air to handle both
sensible and latent cooling loads. As a result, the compressor is operated at a higher
pressure difference of refrigerant across the evaporator and condenser coils resulting
in higher power consumption of the compressor. However, the excessive cooling of air
for the condensation of the moisture can be avoided by installing the ACHMEs in the
direct expansion air-conditioning units as shown in Fig. 4.19 and operating the system
at a higher evaporator temperature. The ducting layout with the arrangement of the
motorized dampers enables the system to deliver dehumidified cold air continuously
to the air-conditioned space. For the mode displayed in Fig. 4.19, the dehumidification
and cooling of the supply air are carried out in the evaporator coil of ACHME-1 while
the heat of the condenser coil ACHME-2 is utilized to regenerate the adsorbent. A
four-way valve is installed to reverse the flow direction of the refrigerant and change
the modes of operation of ACHME-1 and ACHME-2. After the pre-set cycle time,
the positions of the motorized dampers together with the four-way valve are altered
to reverse the mode of operation of ACHMEs and the flow configurations of the

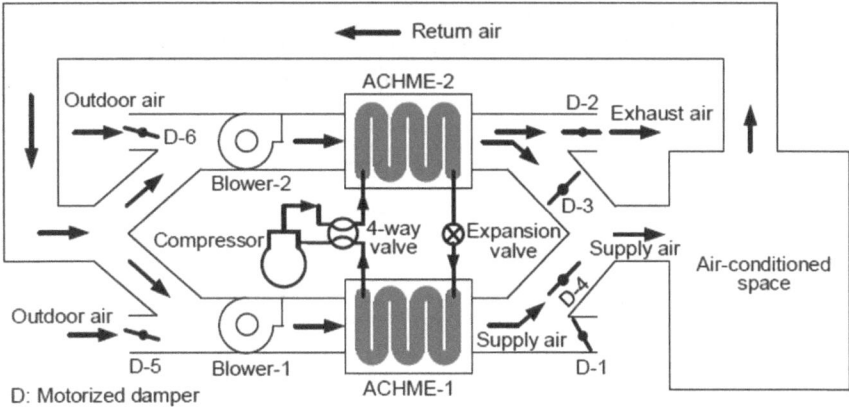

Fig. 4.19 Schematic diagram of an adsorbent coated direct expansion system

supply and exhaust air. The opening of the motorized damper D-5 and D-6 can be modulated to regulate the outdoor airflow rate through the system. Tu et al. [40] tested the performance of an adsorbent coated direct expansion system for the standard room air-conditioning conditions and reported that the average supply air temperature of 20 °C and the humidity ratio of 8.5 g/kg were achieved with a system COP of about 6.2, which is nearly double compared to conventional room air conditioners.

4.8.4 ACHMEs Coupled with Adsorption Chiller

The cooling capacity and the energy efficiency of vapour adsorption chillers largely depend on the water vapour adsorption and regeneration performance of the adsorption and desorption beds. Packed beds of adsorbent are commonly used in the vapour adsorption chillers. Properly selected pure or composite adsorbent coated heat and mass exchangers can be used to enhance the cooling capacity as well as the energy performance of the vapour adsorption chiller systems. Good isothermal adsorption capacity and high durability are two key properties of adsorbents. The working principle and the cycling process of the vapour adsorption chiller systems are illustrated in Fig. 4.20. The water vapour condensed in the condenser coils can be extracted and used as potable water. Otherwise, the condensed water can flow by gravity to the evaporator to continue the process. Kayal et al. [7] studied the adsorption characteristics of Z01 and Z02 AQSOA zeolites and reported that Z01 is more suitable for adsorption chillers as the pore volume of Z01 is high, and a heat source of about 65 °C can be used for the regeneration of Z01. The novel MOF desiccants are also identified in the literature as the potential adsorbents for the vapour adsorption chiller systems due to their faster adsorption kinetics and excellent long-term stability.

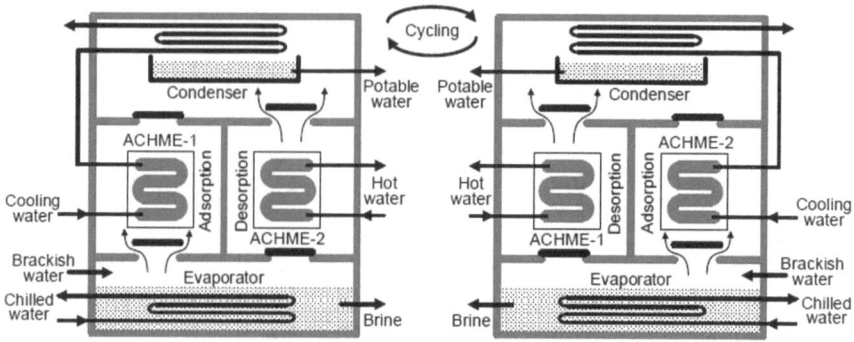

Fig. 4.20 Working principle and cycling process of the vapour adsorption chiller systems

4.9 Conclusions

Conventional vapour compression air-conditioning systems are energy intensive that uses high-grade electrical energy. Adsorbent-coated heat and mass exchangers have been identified as a potential energy-efficient alternative for generating the dehumidified cold air using low-grade waste heat and supporting the cooling load of air-conditioning spaces. This chapter presented the adsorption and desorption characteristics of different pure and composite adsorbents. To enhance the heat and mass transfer performance, different techniques for coating the adsorbents on the heat exchanger surfaces have been discussed. Thereafter, the theory for evaluating the performance of exchanger systems and the influence of different design and operating parameters on the dehumidification and cooling performance are presented. Finally, various hybrid applications of the adsorbent coated exchangers as well as the concept design of the systems are illustrated.

References

1. Goetzler W, Zogg R, Young J, Johnson C (2014) Energy savings potential and RD&D opportunities for non-vapour-compression HVAC technologies. Energy Effi Renew Energy
2. Demir H, Mobedi M, Ülkü S (2008) A review on adsorption heat pump: problems and solutions. Renew Sustain Energy Rev 12:2381–2403
3. Rouquerol F, Rouquerol J, Sing K (1999) Adsorption by powders and porous solids. In: Principles, methodology and applications. Academic Press, London NW1 7DX, UK, pp 6–25
4. Suzuki M (1990) Adsorption engineering. Elsevier, Amsterdam, The Netherlands, pp 2–93
5. Ge TS, Zhang JY, Dai YJ, Wang RZ (2017) Experimental study on performance of silica gel and potassium formate composite desiccant coated heat exchanger. Energy 141:149–158
6. Zheng X, Ge TS, Jiang Y, Wang RZ (2015) Experimental study on silica gel-LiCl composite desiccants for desiccant coated heat exchanger. Int J Refrig 51:24–32
7. Kayal S, Baichuan A, Saha BB (2016) Adsorption characteristics of AQSOA zeolites and water for adsorption chillers. Int J Heat Mass Transfer 92, 1120–1127

8. Teo HWB, Chakraborty A, Fan W (2017) Improved adsorption characteristics data for AQSOA types zeolites and water systems under static and dynamic conditions. Microporous Mesoporous Mater 242:109–117

9. Zhang LZ, Wang YY, Wang CL, Xiang H (2008) Synthesis and characterization of a PVA/LiCl blend membrane for air dehumidification. J Membr Sci 308:198–206

10. Bui DT, Nida A, Ng KC, Chua KJ (2016) Water vapour permeation and dehumidification performance of poly (vinyl alcohol)/lithium chloride composite membranes. J Membr Sci 498:254–262

11. Vivekh P, Bui DT, Wong Y, Kumja M, Chua KJ (2019) Performance evaluation of PVA- LiCl coated heat exchangers for next generation of energy-efficient dehumidification. Appl Energy 237:733–750

12. S.D. White, M. Goldsworthy, R. Reece, T. Spillmann, Gorur A, D.Y. Lee, Characterization of desiccant wheels with alternative materials at low regeneration temperatures. Int J Refrig 34 (2011) 1786–91

13. Yang Y, Rana D, Lan CQ (2015) Development of solid super desiccants based on a polymeric superabsorbent hydrogel composite. RSC Adv 5:59583–59590

14. Lee J, Lee DY (2012) Sorptioncharacteristicsofanovelpolymericdesiccant. Int J Refrig 35:1940–1949

15. Chen CH, Hsu CY, Chen CC, Chen SL (2015) Silica gel polymer composite desiccants for air conditioning systems. Energy Build 101:122–132

16. Furukawa H, Ko N, Go YB, Aratani N, Choi SB, Choi E, Yazaydin AO, Snurr RQ, O'Keeffe M, Kim J, Yaghi OM (2010) Ultrahigh porosity in metal-organic frameworks. Science 329(5990):424–428

17. Farha OK, Eryazici I, Jeong NC, Hauser BG, Wilmer CE, Sarjeant AA, Snurr RQ, Nguyen ST, Yazaydın AO, Hupp JT (2012) Metal−organic framework materials with ultrahigh surface areas: Is the sky the limit? J Am Chem Soc 134:15016–15021

18. Kim H, Yang S, Rao SR, Narayanan S, Kapustin EA, Furukawa H, Umans AR, Yaghi OM, Wang EN (2017) Water harvesting from air with metal-organic frameworks powered by natural sunlight. Science 356(6336):430–434

19. Chen CH, Ma SS, Wu PH, Chiang YC, Chen SL (2015) Adsorption and desorption of silica gel circulating fluidized beds for air conditioning systems. Appl Energy 155:708–718

20. Al-Sharqawi HS, Lior N (2008) Effect of flow-duct geometry on solid desiccant dehumidification. Ind Eng Chem Res 47(5):569–585

21. Zhao Y, Ge TS, Dai YJ, Wang RZ (2014) Experimental investigation on a desiccant dehumidification unit using fin-tube heat exchanger with silica gel coating. Appl Therm Eng 63:52–58

22. Ismail AB, Li A, Thu K, Ng KC, Chun W (2013) On the thermodynamics of refrigerant + heterogeneous solid surfaces adsorption. Langmuir 29, 14494–14502

23. Pesaran AA (1993) A review of desiccant dehumidification technology. In: EPRI's electric dehumidification: energy efficient humidity control for commercial and institutional buildings conference, New Orleans, Louisiana

24. Oh SJ, Ng KC, Thu K, Ja MK, Islam MR, Chun W, Chua KC (2017) Studying the performanceof a dehumidifier with adsorbent coated heat exchangers for tropical climate operations. Science and Technology for the Built Environment 23(1):127–135

25. Jeong J, Yamaguchi S, Saito K, Kawai S (2011) Performance analysis of desiccant dehumidification systems driven by low-grade heat source. Int J Refrig 34:928–945

26. Ge TS, Dai YJ, Wang RZ, Peng ZZ (2010) Experimentalcomparisonandanalysisonsilica gel and polymer coated fin-tube heat exchangers. Energy 35:2893–2900

27. Saeed A, Al-Alili A (2017) A review on desiccant coated heat exchangers. SciTechnol Built Environ 23(1):136–150

28. Vivekh P, Kumja M, Bui DT, Chua KJ (2018) Recent developments in solid desiccant coated heat exchangers–a review. Appl Energy 229:778–803

29. Chang KS, Chen MT, Chung TW (2005) Effects of the thickness and particle size of silica gel on the heat and mass transfer performance of a silica gel-coated bed for air-conditioning adsorption systems. Appl Therm Eng 25(14):2330–2340

30. Li A, Thu K, Ismail AB, Shahzad MW, Ng KC (2016) Performance of adsorbent-embedded heat exchangers using binder-coating method. Int J Heat Mass Transf 92:149–157
31. Freni A, Bonaccorsi L, Calabrese L, Caprì A, Frazzica A, Sapienza A (2015) SAPO-34 coated adsorbent heat exchanger for adsorption chillers. Appl Thermal Eng 82:1–7
32. Nawaz K, Schmidt SJ, Jacobi AM (2014) Mass diffusion coefficient of desiccants for dehumidification applications: silica aerogels and silica aerogel coatings on metal foams. In: 15th international refrigeration and air conditioning conference, vol 2432, pp 1–10
33. Aimjaijit PCW, Chuangchote S (2016) Synthesis of silica gel and development of coating method for applications in a regenerative air dehumidifier. Appl Mech Mater 839:70–74
34. Fang Y, Zuo S, Liang X, Cao Y, Gao X, Zhang Z (2016) Preparation and performance of desiccant coating with modified ion exchange resin on finned tube heat exchanger. Appl Therm Eng 93:36–42
35. Bonaccorsi L, Proverbio E, Freni A, Restuccia G (2007) In situ growth of zeolites on metal foamed supports for adsorption heat pumps. J Chem Eng Japan 40(13):1307–1312
36. Jiang Y, Ge TS, Wang RZ, Hu LM (2015) Experimental investigation and analysis of composite silica-gel coated fin-tube heat exchangers. Int J Refrig 51:169–179
37. Hu LM, Ge TS, Jiang Y, Wang RZ (2015) Performance study on composite desiccant material coated fin-tube heat exchanger. Int J Heat Mass Transfer 90:109–120
38. Zheng X, Wang RZ, Ge TS, Hu LM (2015) Performance study of SAPO-34 andFAPO-34 desiccants for desiccant coated heat exchanger systems. Energy 93:88–94
39. Zhao Y, Dai YJ, Ge TS, Wang HH, Wang RZ (2016) A high performance desiccant dehumidification unit using solid desiccant coated heat exchanger with heat recovery. Energy Build 116:583–592
40. Tu YD, Wang RZ, Ge TS, Zheng X (2017) Comfortable, high-efficiency heat pump with desiccant-coated, water-sorbing heat exchangers. Sci Rep 7(40437):1–10

Chapter 5
Liquid Desiccant Air-Conditioning Systems

List of Symbols

A	Heat transfer area, m^2
c_p	Specific heat capacity, J/(kg·K)
d_h	Hydraulic diameter, m
D_s	Diffusivity of water in desiccant solution, m^2/s
D_a	Diffusivity of vapour in air, m^2/s
g	Acceleration due to gravity, m/s^2
h_a	Convective heat transfer coefficient of air, W/(m^2·K)
h_i	Convective heat transfer coefficient between solution bulk and wall, W/(m^2·K)
h_o	Convective heat transfer coefficient between interface and solution bulk, W/(m^2·K)
h_w	Convective heat transfer coefficient of coolant, W/(m^2·K)
i	Specific enthalpy, J/kg
i_{st}	Specific enthalpy of moisture at the exposed surface of desiccant film, J/kg
$i_{par,w}$	Partial enthalpy of absorbed moisture at the exposed surface of desiccant film, J/kg
Ja_s	Jacobs Number of liquid desiccant, dimensionless
$k_{m,a}$	Mass transfer coefficient of air, kg/(m^2 s)
$k_{m,s}$	Mass transfer coefficient of solution, m/s
k	Thermal conductivity, W/(m·K)
Ka_s	Kapitza Number of solution, dimensionless
LHR	Latent heat ratio, dimensionless
$m_{v,a}$	Mass fraction of vapour at a-surface, dimensionless
$m_{v,s}$	Mass fraction of vapour at s-surface, dimensionless
M	Mass flow rate, kg/s
$\overline{M}$	Molecular weight, kg/kmol
MMR	Moisture removal rate, kg/(m^2·h)
Nu	Nusselt Number, dimensionless

© Springer Nature Singapore Pte Ltd. 2021
C. Kian Jon et al., *Advances in Air Conditioning Technologies*, Green Energy and Technology, https://doi.org/10.1007/978-981-15-8477-0_5

P	Pressure, bar
P_o	Reference pressure, 1 bar
Pr	Prandtl Number, dimensionless
Q_a	Rate of heat transfer from interface to air, W
Q_i	Rate of heat transfer from solution bulk to water, W
Q_{lat}	Latent energy transfer rate, W
Q_o	Rate of heat transfer from interface to solution bulk, W
Q_{sen}	Sensible energy transfer rate, W
Re	Reynolds Number, dimensionless
Sc	Schmidt Number, dimensionless
Sh	Sherwood Number, dimensionless
T	Temperature, °C
U_{b-w}	Overall energy transfer coefficient from desiccant solution to water, W/(m²·K)
X	Length segment along the direction of the channel, m
W	Width of plate, m

Greek Symbols

δ_{wall}	Thickness of wall, m
μ	Dynamic viscosity, Pa·s
ρ	Density, kg/m³
ϕ	Relative humidity of air, percentage
ω_a	Absolute humidity of air, kg/kg
ω_s	Concentration of desiccant in the solution, kg/kg

Subscripts

a	Air, a-surface
b	Bulk
if	Solution interface
in	Inlet condition
l	Desiccant
out	Outlet conditions
s	Desiccant aqueous solution, s-surface
v	Vapour, moisture
w	Water
wall	Wall

5.1 Introduction

Air-conditioning systems of buildings consume a major fraction of the world's total energy consumption. The projected worldwide electric energy consumption for the air-conditioning system will reach to about 4,000 TWh in the year 2050 and will further increase up to 10,000 TWh in the year 2100 [1]. To achieve the targeted worldwide energy efficiency goals, the reduction of the electric energy consumption in air-conditioning systems is, therefore, a major concern. The conventional vapour compression air-conditioning systems are energy-intensive process, where the circulating air is cooled below the dew point temperature to dehumidify the air and maintain the comfort condition of air-conditioning spaces. To meet the low relative humidity requirements of various industrial production areas, the supply temperature of the chilled water is further reduced to cool and dehumidify the supply air to the required low moisture levels and then reheat the air using duct heaters before supplying to the production spaces. The compression lift and the resulting energy consumption of the vapour compression systems increase with the decrease of chilled water supply temperature. Moreover, the duct heaters consume electric energy, which is finally added as an incremental heat load for the air-conditioning systems.

Researchers have made significant attempts to develop energy-efficient alternative air-conditioning systems, which include desiccant assisted air-conditioning systems, latent and sensible cooling load decoupling systems, indirect evaporative cooling systems, direct evaporative cooling systems, etc. In hot and humid climates, the desiccant assisted air-conditioning systems are recognized as energy-efficient alternatives to the conventional vapour compression systems due to their advantage in removing the latent load of spaces using the low-grade energy. Extensive research works have been conducted on the dehumidification of air using liquid desiccants, solid desiccants, and membrane-based liquid desiccant systems. Each of these dehumidification systems has its own set of challenges. Attempts have been made by the researchers and design engineers to understand the moisture transfer processes and develop energy-efficient design alternatives. The liquid desiccant systems possess several advantages [2], namely, (1) excellent dehumidification performance, (2) low-grade energy sources can be used for the regeneration of desiccant solutions, (3) low-pressure drop of the process air stream, (4) multiple absorbers and regenerators of liquid desiccant systems can conveniently be connected by pumping, and (5) preventing the growth of bacteria, moulds, viruses, and microbes.

One of the key advantages of the liquid desiccant dehumidification systems is the opportunity to install the absorbers and regenerators at strategic locations. For instance, the absorbers of the liquid desiccant dehumidification systems can be retrofitted with the conventional chilled water coil of air handling units to dehumidify the circulating air, and the regenerators can be installed to the rooftop, where solar energy or other low-grade energy is available. The desiccant solution can conveniently be circulated through the beds of absorber and regenerator. Moreover, the switching of air, like the solid desiccant dehumidification systems, is not required, which enables the liquid dehumidification system to supply dehumidified air continuously with negligible fluctuation of its moisture contents. However, the liquid desiccant dehumidification systems have few inherent challenges that need to handle carefully to eliminate the possibility of contaminating the indoor air and enhance the energy performance and capacity of the systems. Major challenges include carryover of the liquid desiccant with supply air, optimization of local heat and mass transfer performance of the beds of absorbers and regenerators, enhance the exposed surface area of liquid desiccant by improving the wettability of bed packing materials, and matching the dehumidification and cooling capacity of the hybrid liquid desiccant air-conditioning systems. All these challenges will be systematically discussed in the ensuing sections.

5.2 Principles and Features of the Liquid Desiccant Air-Conditioning Systems

The control systems divide the cooling load of buildings into the sensible cooling load and the latent cooling load to maintain pre-set indoor air temperature and relative humidity of air-conditioned spaces. The conventional vapour compression air-conditioning chiller plants support both latent and sensible cooling load of spaces by producing chilled water below dew point temperature to condense the moisture and cool the circulating air. The vapour compression chiller systems can support the sensible cooling load efficiently but the systems become less efficient to handle the latent cooling load due to the generation of low-temperature chilled water. The conventional systems can preciously control the sensible cooling load but become inefficient as well as challenging while controlling the latent cooling load and relative humidity of spaces. Moreover, moulds and bacteria grow on the wet surfaces of the chilled water coils of the air handling units during the condensation of moisture, which can affect the indoor air quality and causes health issues. The liquid desiccant air-conditioning system is a feasible and cost-effective alternative to overcome the above-mentioned issues of the conventional vapour compression air-conditioning systems. The main components of the liquid desiccant air-conditioning systems are absorber, regenerator, liquid desiccant storage tank, and sensible cooling units. As shown in Fig. 5.1, the liquid desiccant dehumidifier and the sensible cooling unit separately handle the latent cooling load and the sensible cooling load, respectively [3]. The latent cooling load and the relative humidity can be controlled preciously by the absorption of moisture from the humid air to the liquid desiccant surface.

The absorbed moisture is evaporated by heating the dilute solution of liquid desiccant in the regenerator (not shown in Fig. 5.1) using low-grade heat. The indirect evaporative cooling systems and the auxiliary coolers are installed to support the sensible cooling load. The liquid desiccant air-conditioning systems are more energy efficient and provide effective control of the humidity and temperature of air-conditioned spaces. The individual control of humidity by the liquid desiccant

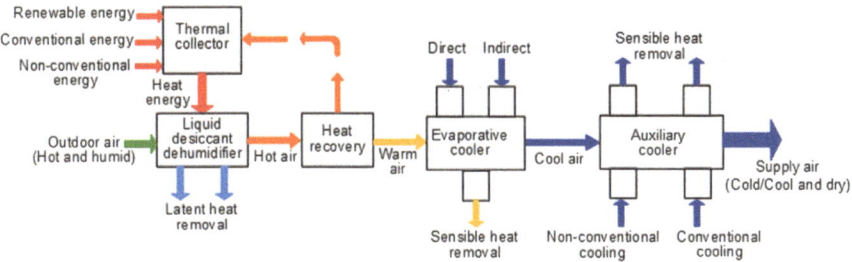

Fig. 5.1 Working principle of liquid desiccant air-conditioning systems

dehumidifier and regulation of the temperature by the cooler increase the overall performance of the system. The liquid desiccant air-conditioning systems eliminate the moisture condensation issue of the cooling coils, which causes the growth of different fungi viruses, and provide a better indoor air quality due to the sanitizing effects of desiccant materials. Jiang et al. [4] compared the relative advantages and limitations of the conventional vapour compression air-conditioning systems and the liquid desiccant air-conditioning systems, which are presented in Table 5.1.

As shown in Fig. 5.2, the absorber, regenerator, and liquid desiccant storage tank are the main components of a liquid desiccant dehumidifier. The beds of the absorber and generator are generally filled with packing materials. The precooled concentrated solution of liquid desiccant is sprayed from the top of the packed bed of absorber. The solution flows downwards due to the gravity over the surface of the packing materials as a thin falling film. The circulating air with high humidity flows over the exposed surfaces of the falling film of liquid desiccant in parallel, counter or cross-flow configuration. Due to the difference of the partial vapour pressures at the exposed surface of the film of liquid desiccant and the bulk of air stream, the moisture of the flowing air migrates to the exposed surface of the film of liquid desiccant. Finally, the absorption of moisture takes place at the film–air interface. Consequently, the concentration of the liquid desiccant solution drops. The dilute solution eventually accumulates in the basin located at the bottom of the absorber.

Table 5.1 Relative advantages and limitations of the conventional vapour compression air-conditioning systems and the liquid desiccant air-conditioning systems [4]

Parameter	Vapour compression air-conditioning system	Liquid desiccant air-conditioning system
Cost of operation	High	Save about 40%
Driving source of energy	Electricity, natural gas, vapour	Low grade energy, e.g. solar energy, waste heat etc.
Humidity control	Average	Accurate
Quality of indoor air	Average	Good
System instalment	Average	Slightly complicate
Capacity for storage of energy	Average	Good

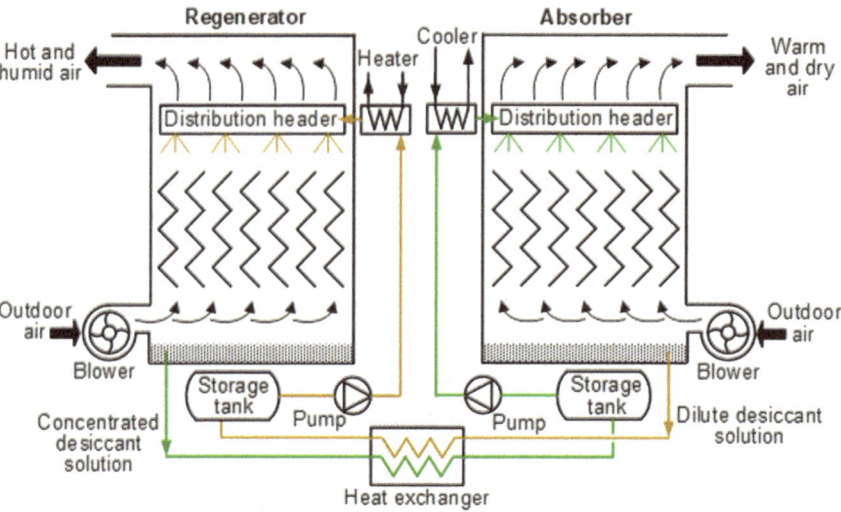

Fig. 5.2 The schematic of an adiabatic liquid desiccant dehumidifier

A filter or drift eliminator is installed at the air outlet of the absorber to eliminate the liquid desiccant carryover with the flowing air. The concentrated solution of the liquid desiccant is precooled before spraying at the top of the absorber to reduce the partial vapour pressure of the desiccant film compared with that of the moisture of the flowing air.

The structure of the regenerator is similar to the absorber; however, the moisture migration direction in the regenerator occurs in the opposite. The dilute desiccant solution of the basin of the absorber is heated to about 65–80 °C before spraying at the top of the regenerator to increase the partial vapour pressure at the exposed surface of the desiccant film compared to that of the moisture of the circulating air. As a result, water evaporates from the film of liquid desiccant in the regenerator and migrates to the flowing air. Finally, the concentrate solution of liquid desiccant accumulates in the basin of the regenerator. Low-grade heat is used to preheat the dilute solution. The concentrated solution is again precooled and sprayed at the top of the absorber to continue the process. A cooling tower is generally employed to precool the concentrated solution. Two desiccant storage tanks are installed for storing the concentrated and the dilute solutions of liquid desiccant to support the variable latent cooling load and continue the dehumidification process for the period when the sources of heat energy are not available.

The processes in liquid desiccant dehumidifiers are commonly classified as adiabatic and internally cooled [5]. Figure 5.2 shows the schematic of an adiabatic liquid desiccant dehumidifier. The absorption of moisture in the liquid desiccant is an exothermic process. The liquid desiccant absorbs the heat generated during the moisture absorption process. Hence, the temperature and partial vapour pressure of the

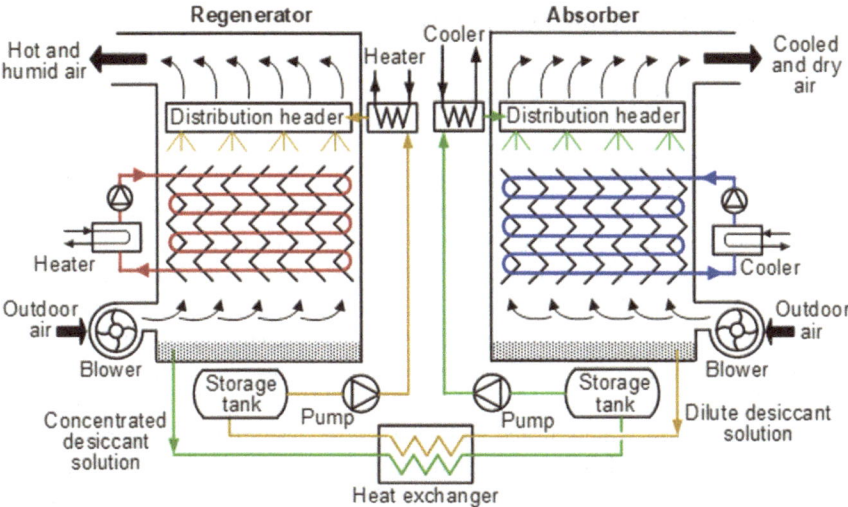

Fig. 5.3 The schematic of an internally cooled/heated liquid desiccant dehumidifier

desiccant film increase, which eventually decreases its moisture absorption performance and increases the temperature of flowing air. The adiabatic dehumidifiers are still widely used for the dehumidification applications of the residential as well as commercial buildings because of its simple configuration and the large contact area between the desiccant film and the flowing air. The flow rate of desiccant solution is increased in the adiabatic dehumidifier to minimize the rise of temperature of the desiccant solution and flowing air. The high flow rate solution also helps to wet the packing materials completely.

To overcome the limitations of the adiabatic dehumidifier and to remove the heat generated in the desiccant film, the internally cooled dehumidifier, as shown in Fig. 5.3, has been a key focus. The desiccant films are cooled in the absorber and heated in the regenerator while the films of liquid desiccant flow over the pickling materials to enhance the rates of moisture absorption and evaporation, respectively. The condition of complete wetting of the packing materials can be achieved at low flow rates of desiccant by adding the surfactants to the liquid desiccant or using the wick attained packing materials of improved surface wettability. The low flow rate of desiccant reduces the film thickness, heat transfer resistance, and the desiccant droplet carryover problems.

5.3   Liquid Desiccants Solutions

5.3.1   Single Desiccant Solutions

The overall dehumidification performance of the liquid desiccant dehumidification systems largely depends on the characteristics of the desiccant solutions. The key considerations for the selection of desiccant solutions include high moisture absorption capacity, low regeneration temperature, availability, and cost. The preferred properties of a good desiccant solution include low surface vapour pressure, high thermal conductivity and moisture diffusivity, low viscosity, non-corrosive, non-volatile, non-toxic, non-flammable, odourless, and chemically stable. Lithium chloride (LiCl), lithium bromide (LiBr), calcium chloride ($CaCl_2$), and triethylene glycol are among the most widely tested and used single desiccants. The partial vapour pressure at the exposed surface of the desiccant solutions at low temperature and high concentration becomes lower than that of the circulating humid air. The desiccant solutions absorb moisture at the exposed surface. The absorbed moisture slowly diffuses from the exposed surface to the bulk of the desiccant solution, which depends on the diffusivity of the desiccant solution. The moisture absorption is an exothermic process resulting in the increase of temperatures of the desiccant solution as well as the surrounding air. The temperature rise of a desiccant solution depends on its thermal conductivity and specific heat capacity.

The partial vapour pressure of calcium chloride at a given temperature and concentration is relatively high. Moreover, it is unstable at specific ranges of concentrations and inlet air conditions, which limit its widespread application. Calcium chloride is likely to crystallize during application in internally cooled dehumidifiers. However, calcium chloride is the cheapest and most readily available desiccant. Experiments and investigations found that lithium chloride is the most stable desiccant with remarkably low partial vapour pressure. The cost of lithium chloride is slightly higher compared with other desiccants. Attempts have been made by researchers to mix the lithium chloride and calcium chloride in different weight percentages to get cost-effective mixtures for specific dehumidification applications. The partial vapour pressure and cost of lithium bromide are intermediate. Both lithium chloride and lithium bromide corrode the metal whether used alone or in combination. Triethylene glycol was mainly used in the early stages of the liquid desiccant dehumidification systems. The surface vapour pressure triethylene glycol (TEG) is very low which may cause the triethylene glycol to evaporate and flow with the circulating air to the air-conditioned spaces. Potassium formate is identified as a good alternative desiccant for many applications. It is a relatively weaker desiccant as compared to the lithium bromide or lithium chloride but it has the ability to dry the circulating air and maintain the comfort condition of spaces if the operating parameters are properly controlled. A comparative assessment of LiCl, LiBr, $CaCl_2$, and TEG solutions based on their relevant thermodynamic properties is presented in Table 5.2 [6].

The partial vapour pressure at the exposed surface of the liquid desiccants often varies due to the difference in their purities. Conde [7] reviewed the sources of

Table 5.2 Comparative assessment of conventional desiccant solutions [6]

Property at 25 °C/75 °C	Unit	LiCl 40%	LiBr 60%	CaCl$_2$ 45%	TEG 95%
Vapour pressure	Pa	593/9379	265/4779	952/14,707	0.528/9.355
Density	Kg/m^3	1251/1224	1716/1684	1445/1413	1118/1079
Dynamic viscosity	10^{-3}pa s	8.33/2.90	7.4/2.8	14.0/4.81	54.0/6.5
Isobaric specific heat capacity	KJ/(kg k)	2.692/2.874	1.876/1.946	2.269/2.503	2.270/2.549
Surface tension	10^{-3} N/m	95.8/90.0	94.5/89.4	96.5/91.3	45.55/43.70
Thermal conductivity	W/(mK)	0.53/0.59	0.41/0.45	0.54/0.60	0.25/0.22
Cost	Euro/kg	21.91	27.57	1.88	2.93

measured data from the year 1850 onwards and proposed calculation model for the vapour pressure, differential enthalpy of dilution, solubility boundary, density, surface tension, dynamic viscosity, thermal conductivity, and specific heat capacity of aqueous solutions of the chlorides of lithium and calcium, particularly suited for use as a desiccant in sorption based air conditioning equipment. Chaudhari and Patil [8] measured the vapour pressure of an aqueous solution of lithium chloride of + 99% purity in the concentration range of 12.9–44.2% weight and in the temperature range of 30–100 °C. Experimental data show that the vapour pressure drops from 163.03 mm Hg to 4.28 mm Hg when the solution of 44.186% weight of lithium chloride is cooled from 100 to 30 °C. Test results highlighted the suitability of lithium chloride solution for the dehumidification of air and the opportunity to regenerate the solution using low-grade waste heat. They fitted the experimental data to the Antoine type equation. The average deviation is less than 1.0%. They employed the Haltenberger method to calculate the enthalpies of the solution via experimental vapour pressure and heat capacity data. The enthalpy data were fitted to a polynomial equation by the least-squares method within the range of temperature 0–120 °C. They plotted the enthalpy–concentration isotherms and showed the limitations imposed by the crystallization points of the solutions. By using the isotherms, one can evaluate the performance of a dehumidifier and the absorption heat pumps using the lithium chloride—water solution. The relationship between the percentage concentration by weight of salt and the density was also expressed using a polynomial equation incurring an average error of 0.04%.

McNelly [9] presented the thermodynamic properties of the aqueous solution of lithium bromide. Kaita [10] developed the equations for calculating the vapour pressure and the enthalpy of lithium bromide solution of concentrations 40–65 wt% in the temperature range of 40–210 °C. To study the effects of nanoparticles, solution concentration and temperature on the saturated vapour pressure and the water vapour absorption rate, Wang et al. [11] measured the saturated vapour pressure and the mass transfer rate of lithium bromide aqueous solution in the presence of 40 nm CuO nanoparticles with the concentrations of 0.0, 0.01, 0.05, 0.1, and 0.12 wt%. They conducted the tests under the temperatures between 20 and 60 °C and concentrations between 56 and 60%. They reported that the saturated vapour pressure of the solution decreased first and then increased with the higher nanoparticle mass fraction. The saturated vapour pressure reached the minimum value at 0.1 wt% of the CuO nanoparticles. They observed that the effective absorption ratio had the maximum values of 2.20 and 2.85 in 58% LiBr solution with nanoparticle mass fractions of 0.05% and 0.1%, respectively.

5.3.2 Solution of Mixed Desiccants

To evaluate the opportunities of developing cost-effective desiccant solutions, Ertas et al. [12] conducted experimental works on the thermal properties of different mixtures comprising lithium chloride and calcium chloride solutions. They measured the vapour pressures of five mixtures of lithium chloride and calcium chloride solutions with ratios of 1.0:0, 0.7:0.3, 0.5:0.5, 0.3:0.7, and 0:1.0 under the total electrolyte mass concentration of 20%. They reported that the 100% lithium chloride solution showed the lowest vapour pressure under different temperatures. They concluded that the lithium chloride and calcium chloride of ratio 1:1 should be a cost-effective choice.

To improve the solubility and, thereby, overcome the crystallization issue of lithium bromide water solution, Krolikowska et al. [13] added a small amount of ethylene glycol, diethylene glycol, triethylene glycol, and glycerol as additives to the lithium bromide water solution (additive to LiBr mass fraction of 0.3) and measured the phase equilibria, liquid density, and dynamic viscosity over a wide range of temperatures and compositions. Experimental results showed that adding a small amount of glycol to the binary solution of lithium bromide and water enabled a significant increase in the solubility of lithium bromide in water, which could potentially reduce the crystallization issues of lithium bromide water solution. Due to the lowest values of the viscosity and the highest values of the water activity coefficient for the lithium bromide, glycerol and water solution, the glycerol could be considered as the best additive.

Tsai et al. [14] added triethyleneglycol and propylene glycol to a solution mix of lithium bromide and lithium chloride to obtain a composite liquid desiccant solution and measured the thermal properties of the mixed solution. Chen et al. [15] and Chen et al. [16] also combined diethylene, triethylene, and tetraethylene glycol with magnesium chloride solution. They measured the vapour pressure and reported the enhancement of mass transfer performance of the mixed solutions.

Wen et al. [17] prepared a newly mixed liquid desiccant solution of 25% lithium chloride, 39% hydroxyethyl urea, and 36% water to study the potential of reducing its causticity in a metal-based regenerator. They measured its basic thermal properties including density, viscosity, and conductivity and compared it with a 35% lithium chloride solution. The regeneration performances of the solutions were experimentally studied. An average increase of the regeneration performances of 14.1% was reported because of the larger wetting ratio and greater fluctuation of falling film.

The wetting ratio increased from 81.5 to 87.8%, and the standard deviation of the film thickness increased from 25.441 to 31.672 mm with more rigorous fluctuations. The new solution significantly reduced causticity due to the addition of hydroxyethyl urea and a reduction in lithium chloride concentration.

Donate et al. [18] prepared the solutions of lithium bromide, lithium bromide + sodium formate, lithium bromide + potassium formate, lithium bromide + potassium acetate, and lithium bromide + sodium lactate, and measured their vapour pressures and water vapour absorption performances in a vacuum chamber. They also measured density, viscosity, enthalpies of dilution and solubility of the mixtures. Their results suggested that the formula of $LiBr + CHO_2Na + water$ ($LiBr/CHO_2Na = 2$ by mass) could be a promising candidate for the LiBr solution. The four organic salts in combination with lithium bromide exhibited the requirement of low regeneration temperature compared with the pure lithium bromide. The lithium bromide and organic salt mixtures showed the optimum properties such as low vapour pressures, less crystallization temperature, and latent heat of absorption.

5.4 Packing Materials

The moisture absorption performance of the liquid desiccant dehumidifiers also depends on the flow configurations and the exposed surface generation characteristics of the packing materials. Packing materials can be broadly classified as random packing materials and structured packing materials. Rosette ring, pall ring, and ladder ring of irregular geometry are widely used as random packing materials, which are placed randomly in the bed of the dehumidifiers. The potential of uneven distribution of the liquid desiccant particularly at the higher flow rate is a practical issue for the random packing materials. Furthermore, wall flow and channel flow with poor wetting of the random packing materials may occur when the flow rate of liquid desiccant is low. Random packing materials are generally used in adiabatic bed dehumidifiers. The structured packing materials, on the other hand, have fixed geometric forms, which are installed sequentially to form the designed geometric patterns and the flow direction of the desiccant solution. Grid packing materials, wavy plate packing materials, and silk net packing materials are structured packing materials, which enhance the desiccant-air exposed surface area to some extent and lower the desiccant-air flow resistance. Structured packing materials are used for the fabrication of the internally cooled dehumidification beds. The key features of the structured packing materials include low-pressure drop, high exposed surface area of desiccant film, high moisture absorption and evaporation rate, and flexible manipulation. Generally, the designed profiles of the structured packing materials are formed by pressing the thin metal sheets, which are vulnerable to corrosion and deformation. For the internally cooled dehumidifiers, the liquid desiccant may crystallize and clog the narrow channels. The design of the wavy packing material is a better alternative that reduces the potential of clogging due to the generation of flow turbulence of the liquid desiccant.

The desiccant-air contacting area per unit volume of the packing materials (known as volumetric area), the void volume per unit volume of the packing materials (void ratio), and the spacing intervals between packing material layers are the major factors to be considered during the design and selection of packing materials. The pressure drop of the stream of circulating air is another major criterion for the design or selection of packing materials since the pressure drop of the air stream is directly related to the energy consumption of the blower. The airflow resistance decreases with the increase in the void ratio. Pressure drops in both random and structured packing materials have been reported in the literature. Gandhidasan et al. [19] developed a rigorous model to predict the pressure drops of different packing materials and summarized them with some parameters of the packing materials as shown in Table 5.3. The superficial desiccant velocity of 0.001 m/s and the superficial air velocity of 2.5 m/s were used in the study. The sheet-type structured packing (Mellapak 250Y) has the lowest while the small size of random packing shows the highest value dry pressure drop. An optimum spacing interval between the layers imposes less resistance on the circulating air.

Many researchers have conducted comparative experiments on the performance of random and structured packing materials and identified that structured packing

Table 5.3 The physical characteristics and pressure drops of selected random and structured packing materials [19]

Packing	Type/Size (mm)	Surface area per unit volume(m^2/m^3)	Void fraction	Dry pressure drop, (Pa/m)	Wet pressure drop, (Pa/m)	Flooding condition pressure drop, (Pa/m)
Random ceramic packings						
Raschig rings	15	264	0.698	2926	–	–
	30	137	0.775	915	1556	3168
	35	126	0.773	904	1513	3223
Berl saddles	15	300	0.561	–	–	–
	35	133	0.750	625	977	3161
Intalox saddles	20	300	0.672	–	–	–
	25	183	0.732	–	–	–
	35	135	0.760	603	938	3162
Pall rings	25	219	0.740			
	35	165	0.760	682	1108	3020
Structured packings						
Sulzer	BX-gauze type	450	0.860	429	686	2546
Sulzer	Mellapack 250 Y	250	0.850	177	244	3104
Gempack	2A	394	0.920	265	392	2747

materials had the characteristics of higher heat and mass transfer capacity, higher efficiency, and lower pressure drop compared with the random packing materials [20, 21]. The moisture absorption performance of packing materials drops if the spacing between the layers is unsuitably small or big. The spacing interval of 6–8 mm is widely used by researchers and design engineers. The moisture absorption and regeneration effectiveness, and the specific moisture absorption and regeneration effectiveness of few packing materials are summarized in Tables 5.4 and 5.5, respectively [22]. The wetting ratio is also an essential factor that affects the mass and heat

Table 5.4 Moisture absorption and regeneration effectiveness of structured and random packing materials [23]

Structured materials	Effectiveness, %	Random materials	Effectiveness, %
Moisture transfer effectiveness during the absorption process			
Cross-corrugated cellulose	63–68	Tripack No. 1/2 PE spheres	10–42
Cross-corrugated PVC	65–72	Tri Packs No. 1/2 PP	35–83
Z-type gauze packing	25–42	1.6 cm PP Flexi rings	50–70
Arrays of aluminium plate stack	20–48	Random packing	90–95
Celdek packing	40–75	2.54 cm PP RauschertHiflow rings	72–95
PC panels (10 mm, 30 deg)	40–58	1 inch PP RauschertHiflow rings	55–78
Wire mesh packing (75 layers)	25–65	25 mm plastic pall Rings (LiBr)	35–94
Gauze packing	35–88	25 mm plastic pall Rings (LiCl)	35–95
Cross corrugated ceramic packing	25–58	25 mm plastic pall Rings (KCOOH)	35–94
Moisture transfer effectiveness during the regeneration process			
Structured materials	Effectiveness, %	Random materials	Effectiveness, %
Arrays of wood plate stack	58–100	1 inch ceramic Rasching ring	50–68
Cellulose rigid media pads	25–38	2.54 cm PP RauschertHiflow rings	60–98
Cross corrugated ceramic packing	12–28	1 inch PP RauschertHiflow rings	78–98
Cross corrugated plate	34–75	25 mm plastic pall Rings (LiCl)	20–80
Z-type gauze packing (2)	5–15	25 mm plastic pall Rings (LiBr)	22–80
Cellulse corrugated packing	20–40	25 mm plastic pall Rings (KCOOH)	22–80

Table 5.5 Specific moisture absorption and regeneration effectiveness of structured and random packing materials [23]

Structured materials	Specific effectiveness, g/m³ s	Random materials	Specific effectiveness, g/m³ s
Specific moisture transfer effectiveness during the absorption process			
Cross-corrugated cellulose	12–28	Tripack No. 1/2 PE spheres	6–38
Cross-corrugated PVC	15–30	Tri Packs No. 1/2 PP	22–56
Celdek 7090	15–30	1.6 cm PP Flexi rings	10–32
Cross corrugated ceramic packing	10–28	2.54 cm PP RauschertHiflow rings	8–26
Z-type gauze packing (2)	18–36	1 inch PP RauschertHiflow rings	8–20
Plant fibre packing	25–48	25 mm plastic pall Rings (LiCl)	1–10
Cellulose corrugated packing	26–55	25 mm plastic pall Rings (LiBr)	2–12
Wire mesh packing (75 layers)	10–12	25 mm plastic pall Rings (KCOOH)	2–10
Specific moisture transfer effectiveness during the regeneration process			
Arrays of wood plate stack	18–54	1 inch ceramic Rasching ring	36–52
Cellulose rigid media pads	32–81	Tripack No. 1/2 PE spheres (tray)	2–14
Cross corrugated ceramic packing	6–22	Tripack No. 1/2 PE spheres (nozzle)	10–22
Celdek packing	5–8	Polypropylene Tripack	18–16
Wire mesh packing (6 layers)	8–9	2.54 cm PP RauschertHiflow rings	15–42
Cross corrugated plate	10–35	1 inch PP RauschertHiflow rings	50–92
Z-type gauze packing (2)	5–20	Z Shaped plastic packing	5–18
Plant fibre packing	10–36	25 mm plastic pall Rings (LiCl)	2–8
Cellulose corrugated packing	25–48	25 mm plastic pall Rings (KCOOH)	2–10

transfer performance of packing materials. Because of the flow channels and surface characteristics, the wetting performance of different packing materials differs from each other. Theoretical models have been developed to calculate the effective interfacial area and wetting ratio of packing materials [23]. The quality of the packing materials should also be carefully considered. The packing materials should not be corroded or distorted due to the soaking of liquid desiccant. The material should be able to withstand the pressure of the circulating air.

5.5 Theoretical Study on Liquid Desiccant Dehumidification Systems

Several theoretical models of different levels of complexities have been developed by researchers to study the heat and mass transfer behaviours of adiabatic and internally cooled liquid desiccant dehumidifiers of the cross, counter and parallel flow configurations. In the most detailed numerical models, the conservation equations of mass, momentum, and energy are solved simultaneously. They include the least number of assumptions and often do not require a great amount of knowledge on the heat and mass transfer coefficients. However, some assumptions need to be made about the nature of the liquid film and the flow regime. Several models that use formulations similar to that of traditional heat exchanger analysis have also been developed. These models require the variation of the heat and mass transfer coefficient for the falling film, cooling fluid, and circulating air as inputs.

Ren et al. [24] assumed the equilibrium humidity ratio of the desiccant solution as a linear function of its temperature and concentration and formulated a set of one-dimensional differential equations to represent the heat and mass transfer process of liquid desiccant dehumidifiers. The governing equations of momentum, energy and mass diffusion for the air and desiccant solution were numerically solved by Ali et al. [25] and Dai and Zhang [26] to investigate the heat and mass transfer between the air and the falling solution film of cross-flow configuration. Emhofer et al. [27] neglected the coupling between air and desiccant solution and assumed a constant film thickness for developing the mathematical model for the cross-flow configuration. Nada [28] assumed a constant tube surface temperature and solved the governing equations for the parallel, counter and cross-flow using finite-difference technique. Simulation results demonstrated that the parallel flow provided more cooling and dehumidification due to the smaller relative velocity between air and desiccant film.

Tu et al. [29] used the finite-difference model to study the adiabatic dehumidification process, while Luo et al. [30] established the Computational Fluid Dynamics (CFD) technology for the adiabatic and the internally cooled dehumidifiers and demonstrated the importance of considering variable local properties of the desiccant solution. Huang et al. [31] used the finite volume approach to solve the governing equations of a water-cooled membrane-based liquid desiccant dehumidifier and reported that the fully developed Nusselt and Sherwood numbers for the

Table 5.6 Flow considerations for the film of liquid desiccants and adopted solution techniques

Desiccant solution flow consideration	Solution techniques	Iteration	Accuracy	Application
Without consideration of film thickness	ε-NTU [33]; Simplified [34]	Simple/no iteration	Relatively less accurate	Predict performance
Uniform film thickness	Finite difference [29]; Finite volume [35]; ε-NTU [36]	Effective iterative process	Good accuracy	Design and optimization
Variable film thickness	Finite difference [25]; Finite volume [31]; CFD technology [30]	Complicated iterative process	Best accuracy	Design and optimization
Variable mass flow rate resulting in variable film thickness	Stable fourth order Runge–Kutta scheme. Heat and mass transfer coefficients were extracted [37] from the detailed model with variable film thickness [38]	Simple iteration	Very good accuracy	Design and optimization

membrane-based system were 2–3% smaller than those for an adiabatic one. The review of various mathematical models reveals that finite difference model, effectiveness Number of Transfer Unit (NTU) model, and simplified models are mainly employed to study the adiabatic dehumidifier [32]. As the moisture absorption rate for the adiabatic dehumidifiers is less compared with the internally cooled dehumidifiers, the film thickness of the desiccant solution is generally considered as constant for the adiabatic dehumidifiers. A summary of different flow considerations for the film of liquid desiccants and adopted solution techniques is presented in Table 5.6.

5.6 Development of the Theoretical Model

As discussed in Sect. 5.2, the liquid desiccant dehumidifiers are commonly designed to operate as an adiabatic dehumidifier or an internally cooled dehumidifier. In conventional liquid desiccant dehumidifier systems, the dehumidifier consists of several vertical plates, inclined plates, horizontal tubes, or vertical tubes. The concentrated solution of the liquid desiccant, also known as strong solution, is sprayed on the surfaces of the plates or tubes at the top as shown in Fig. 5.4. The desiccant solution usually flows under gravity as a thin falling film. The cooling fluid flows

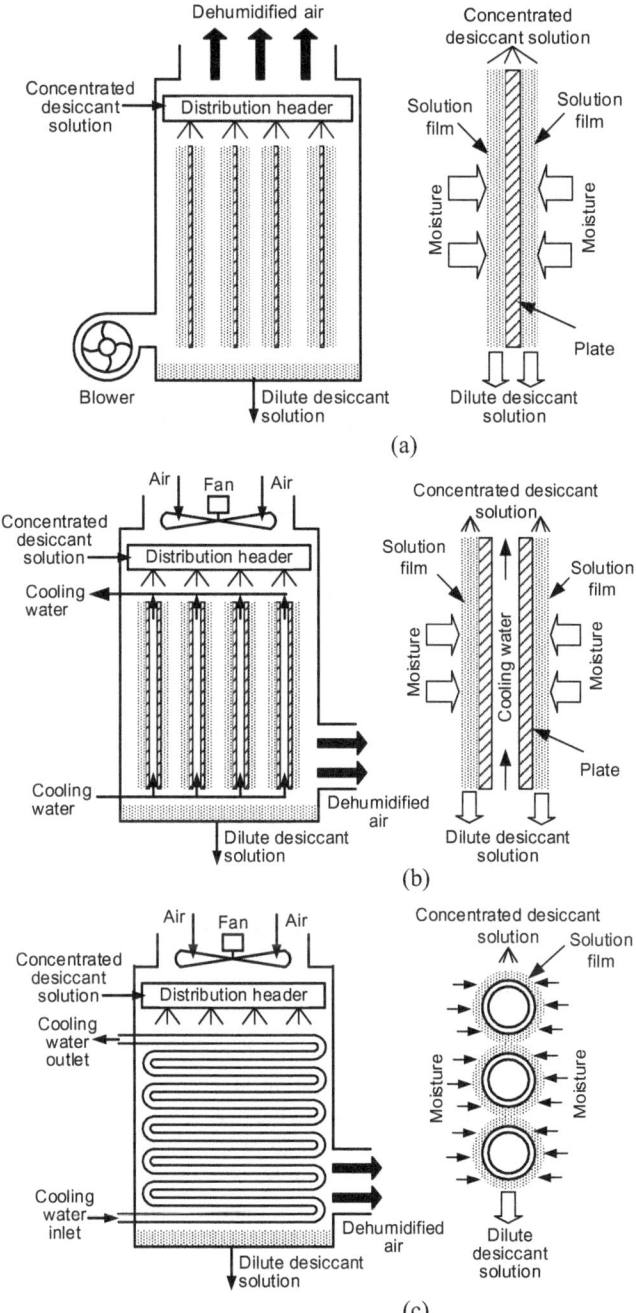

Fig. 5.4 Schematic diagram of flow configuration **a** vertical wall adiabatic dehumidifier, **b** vertical wall internally cooled dehumidifier, and **c** horizontal tube internally cooled dehumidifier

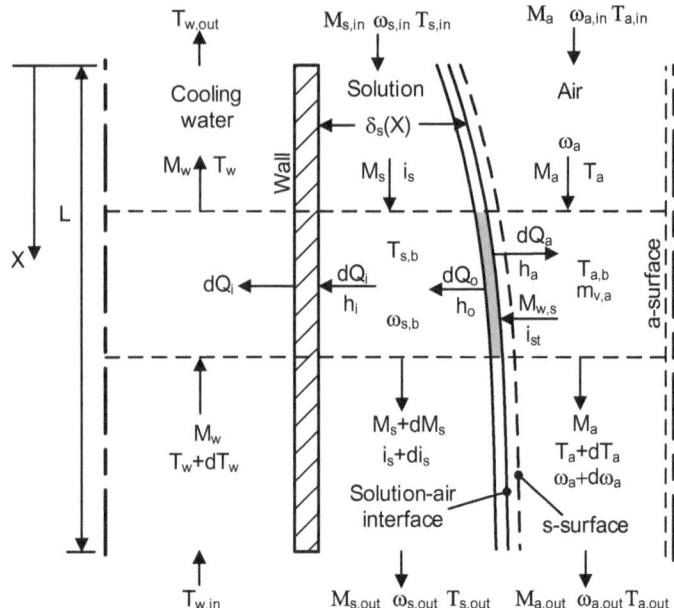

Fig. 5.5 Schematic of the control volume of mathematical model

in a direction that is counter or parallel to the solution flow direction. Air is circulated over the falling film of the liquid desiccant in the counter flow, parallel flow, or cross-flow configurations. If the detailed shape and surface structures are ignored, the dehumidifier can be idealized by a surface with a falling film and air on one side and the coolant on the other side.

In the physical model shown schematically in Fig. 5.5, a thin film of strong desiccant solution flows down over a vertical flat plate. The cooling water flows in the opposite direction to form a counter flow heat and mass exchanger. Air is circulated over the falling film in the parallel flow configuration. The exposed surface of the falling film of desiccant solution is in contact with the flowing air. At the inlet, the strong solution of desiccant is at a uniform temperature of $T_{s,in}$ and concentration $\omega_{s,in}$ (mass of desiccant per unit mass of the solution) corresponding to an equilibrium vapour pressure of $P_{v,s,in}$ that is different from the partial vapour pressure of moisture of the circulating air $P_{v,a}$. Because of this vapour pressure difference, mass transfer of the moisture takes place at the solution–air interface. This absorbed moisture diffuses into the film of liquid desiccant. The heat generated in the absorption process, known as the heat of absorption, flows through the desiccant film as well as transfers to the air stream by convection. For internally cooled dehumidifiers, the major fraction of the heat of absorption flows through the film to the cooling medium. Due to symmetry, only a single falling film of the solution and one-half of the streams of air and cooling water are considered in developing the theoretical model. The variation in liquid desiccant flow rate due to the absorption of moisture is incorporated in the model.

The following assumptions are made in developing the governing equations: (i) The desiccant solution is a Newtonian fluid and its physical properties are function of the temperature and concentration, (ii) The flow of the film of desiccant solution is laminar and non-wavy, (iii) No shear forces are exerted on the solution by the air, (iv) Vapour pressure equilibrium exists between the air and the solution at the interface, (v) Heat transfer by conduction and mass transfer by diffusion in the direction of solution flow is negligible, (vi) The system is in a steady-state condition, and (vii) There are no chemical reactions.

Considering a small control volume as shown in Fig. 5.5, the mass conservation equation for the desiccant can be written as:

$$M_1 = {}_{s,b}M_s = {}_{s,in}M_{s,in} \tag{5.1}$$

Since the flow rate of desiccant salt is constant, differentiation of Eq. (5.1) gives:

$$dM_1 = d_{s,b}M_s + {}_{s,b}dM_s \tag{5.2}$$

From the mass balance of the desiccant solution and the moisture of the air:

$$dM_s = dM_{w,s} \tag{5.3}$$

The energy conservation equation for the control volume at steady state can be expressed as:

$$M_s i_s + i_{st} dM_{ws} = (M_s + dM_s)(i_s + di_s) + dQ_i + dQ_a \tag{5.4}$$

The rate of heat transfer from the interface of the desiccant film to the air can be written as:

$$dQ_a = h_a \big(T_{s,if} - T_{a,b}\big) dA = M_a c_{p,a} dT_{a,b} \tag{5.5}$$

The heat transfer from the bulk of the desiccant film at $T_{s,b}$ to the wall at T_{wall} can be expressed as:

$$dQ_i = h_i \big(T_{s,b} - T_{wall}\big) dA \tag{5.6}$$

The energy equation for the coolant gives:

$$dQ_i = h_w (T_{wall} - T_w) dA = -M_w c_{p,w} dT_w \tag{5.7}$$

The heat transfer from the bulk of the desiccant film at $T_{s,b}$ to the cooling water at T_w can be expressed in term of the overall heat transfer coefficient U_{b-w} as:

$$dQ_i = U_{b-w}(T_{s,b} - T_w)dA = -M_w c_{p,w} dT_w \tag{5.8}$$

The energy balance for a small control element at the interface of the desiccant film gives:

$$dM_{w,s} i_{st} = dM_{w,s} i_{par,w} + dQ_o + dQ_a \tag{5.9}$$

where $h_{par,w}$ is the partial enthalpy of the absorbed moisture, which is a function of the concentration and temperature desiccant of the exposed surface.

The rate of energy transfer from the film–air interface to the bulk of the desiccant film can be expressed as:

$$dQ_o = h_o(T_{s,if} - T_{s,b})dA \tag{5.10}$$

Considering an a-surface in the bulk of air and an imaginary s-surface just outside the exposed surface of the desiccant film (shown in Fig. 5.5), the moisture transfer rate from the a-surface to the s-surface and finally from the s-surface to the bulk of the desiccant film can be written as:

$$dM_{w,s} = k_{m,ss}(s_{,b} - s_{,if})dA = k_{m,a}(m_{v,a} - m_{v,s})dA \tag{5.11}$$

The a-surface represents the local conditions of the bulk air stream. However, the s-surface is very close to the exposed surface of the desiccant film. It is assumed that the temperatures of the s-surface and the exposed surface of the desiccant film are equal. The s-surface is considered to be in equilibrium corresponding to the temperature and the concentration of the solution of exposed surface of the desiccant film.

The mass fraction of moisture at the bulk (i.e. a-surface) of the air stream can be expressed as:

$$m_{v,a} = \frac{v,a}{a} = \frac{P_{v,a}\overline{M}_v}{P_{v,a}\overline{M}_v + (1 - P_{v,a})\overline{M}_a} \tag{5.12}$$

The partial pressure of moisture at the bulk of the air stream depends on the relative humidity and the temperature of the bulk air, which can be calculated as:

$$P_{v,a} = \left\{ \emptyset e^{\left[27.0214 - \frac{6887}{T_{a,b}+273.15} - 5.311\log\frac{T_{a,b}+273.15}{273.15}\right]} \right\} / 100000 \tag{5.13}$$

The mass fraction of moisture at the s-surface close to the exposed surface of the desiccant film can be written as:

$$m_{v,s} = \frac{v,s}{s} = \frac{P_{v,s}\bar{M}_v}{P_{v,s}\bar{M}_v + (1 - P_{v,s})\bar{M}_a} \tag{5.14}$$

The partial pressure of moisture at the exposed surface (i.e. s-surface) of the film depends on the thermophysical properties and temperature of the desiccant solution. LiCl and LiBr are two extensively used desiccant. For the LiCl solution, the partial pressure of moisture can be calculated using the below empirical correlation [39]:

$$\log(100P_{v,s}) = H + \frac{J}{(T_{s,if} + 273.15)} + \frac{K}{(T_{s,if} + 273.15)^2} \tag{5.15}$$

where, $H = 7.323355 - 0.062366\,L + 0.0061613\,L^2 - 0.0001043\,L^3$.

$J = -1718.157 + 8.2255*1 - 2.2131*1^2 + 0.02461^3$

$K = -97{,}575.68 + 3839.979\,L - 421.429\,L^2 + 16.731\,L^3$ and $L = \frac{s,if}{(1-s,if)\bar{M}_{LiCl}}$

Similarly, the partial pressure of moisture at the exposed surface (i.e. s-surface) of the film of LiBr solution is determined using the below empirical correlation [40]:

$$\log(100P_{v,s}) = E + \frac{F}{(T'_{s,if} + 273.15)} + \frac{G}{(T'_{s,if} + 273.15)^2} \tag{5.16}$$

where, $T'_{s,if} = \frac{T_{s,if} - \sum_{n=0}^{3} Y_n(100_{s,if})^n}{\sum_{n=0}^{3} Y_n(100_{s,if})^n}$

$E = 7.05, F = -1596.49, g = -104{,}095.5$

$X_o = -2.00755, X_1 = 0.16976, X_2 = -3.133362 \times 10^{-3}, X_3 = 1.97688 \times 10^{-5}$

$Y_o = 124.937, Y_1 = -7.71649, Y_2 = 0.152286, Y_3 = -7.95090 \times 10^{-4}$

The enthalpies of LiCl and LiBr solutions can be expressed as a function of the solution temperature and concentration as [40, 41]:

$$i_s = 1000\sum_{i-1}^{5}(A_i + T_{s,b}B_i + T_{s,b}^2 C_i)_{s,b}^{i-1} \tag{5.17}$$

where, A_i, B_i and C_i are the coefficients of Eq. (5.17). Values of the coefficients for LiCl and LiBr solutions are presented in Tables 5.7 and 5.8, respectively.

Table 5.7 Values of coefficients used in Eq. (5.17) for LiCl [41]

i	A_i	B_i	C_i
1	−66.2324	4.5751	−8.09689 × 10⁻⁴
2	11.2711	−0.146924	2.18145 × 10⁻⁴
3	−0.79853	6.307226 × 10⁻³	−1.36194 × 10⁻⁵
4	2.1534 × 10⁻²	−1.38054 × 10⁻⁴	3.20998 × 10⁻⁷
5	−1.66352 × 10⁻⁴	1.06690 × 10⁻⁶	−2.64266 × 10⁻⁹

Table 5.8 Values of coefficients used in Eq. (5.17) for LiBr [40]

i	A_i	B_i	C_i
1	-2024.33	18.2829	$-3.7008214 \times 10^{-2}$
2	163.309	-1.1691757	2.8877666×10^{-3}
3	-4.88161	3.248041×10^{-2}	$-8.1313015 \times 10^{-5}$
4	6.302948×10^{-2}	-4.034184×10^{-4}	9.9116628×10^{-7}
5	-2.913705×10^{-4}	1.8520569×10^{-9}	$-4.4441207 \times 10^{-9}$

The reduction of the moisture content of the air stream is equal to the amount of moisture absorbed by the desiccant film:

$$dM_{w,s} = K_{m,s}\rho_s\big(s_{s,b} - s_{s,if}\big)dA = -M_a d_a \tag{5.18}$$

The heat balance for a small element at the interface of the desiccant film gives:

$$K_{m,s}\rho_s\big(i_{st} - i_{par,w}\big)\big(s_{s,b} - s_{s,if}\big) = h_o\big(T_{s,if} - T_{s,b}\big) + h_a\big(T_{s,if} - T_{a,b}\big) \tag{5.19}$$

Equation (5.19) can be simplified to obtain an auxiliary equation from the equilibrium condition at the interface as:

$$s_{,if} = \psi_1 + \psi_2 T_{s,if} \tag{5.20}$$

where, $\psi_1 = \frac{K_{m,s}\rho_s i_{vs,b} + h_o T_{s,b} + h_a T_{a,b}}{K_{m,s}\rho_s i_v}$; $\psi_2 = -\frac{h_o + h_a}{K_{m,s}\rho_s i_v}$; $i_v = i_{st} - i_{par,w}$

Equations (5.1) to (5.11) and (5.18) to (5.20) form the governing equations for the theoretical model.

The following are the boundary conditions for the solution:

(a) For the air stream: At $X = 0$, $\omega_a = \omega_{a,in}$, $T_a = T_{a,in}$ and mass flow rate of dry air $= M_a$
(b) For the desiccant solution: At $X = 0$, $M_s = M_{s,in}$, $T_s = T_{sb,in}$ and $\omega_s = \omega_{s,in}$
(c) At the interface of the falling film: $T_s = T_{s,if}$ and $\omega_s = \omega_{s,if}$
(d) For the cooling water: At $X = 0$, $T_w = T_{w,out;}$; at $X = L$, $T_w = T_{w,in}$

Equation (5.2) can be rearranged as:

$$M_s \frac{d s_{s,b}}{dA} + s_{s,b}\frac{dM_s}{dA} = 0 \tag{5.21}$$

Combining Eq. (5.21) and (5.1) gives:

$$\frac{d s_{s,b}}{dA} = -\frac{M_1}{M_s^2}\frac{dM_s}{dA} \tag{5.22}$$

Differentiating Eq. (5.17) gives:

$$\frac{di_s}{dA} = 1000\sum_{i-1}^{5}\left(B_i + 2T_{s,b}C_i\right)_{s,b}^{i-1}\frac{dT_{s,b}}{dA}$$
$$+ 1000\sum_{i-1}^{5}\left(A_i + T_{s,b}B_i + T_{s,b}^2 C_i\right)(i-1)_{s,b}^{i-2}\frac{d\omega_{s,b}}{dA}$$

$$\frac{dT_{s,b}}{dA} = \frac{\frac{di_s}{dA} - 1000\sum_{i-1}^{5}\left(A_i + T_{s,b}B_i + T_{s,b}^2 C_i\right)(i-1)_{s,b}^{i-2}\frac{d\omega_{s,b}}{dA}}{1000\sum_{i-1}^{5}\left(B_i + 2T_{s,b}C_i\right)_{s,b}^{i-1}} \qquad (5.23)$$

Combining Eqs. (5.1), (5.22) and (5.23) gives:

$$\frac{dT_{s,b}}{dA} = \frac{\frac{di_s}{dA} + 1000\frac{M_1}{M_s^2}\frac{dM_s}{dA}\sum_{i-1}^{5}\left(A_i + T_{s,b}B_i + T_{s,b}^2 C_i\right)(i-1)\left(\frac{M_1}{M_s}\right)^{i-2}}{1000\sum_{i-1}^{5}\left(B_i + 2T_{s,b}C_i\right)\left(\frac{M_1}{M_s}\right)^{i-1}}$$

$$\frac{dT_{s,b}}{dA} = \alpha + \beta \qquad (5.24)$$

where, $\alpha = \dfrac{\frac{di_s}{dA}}{1000\sum_{i-1}^{5}\left(B_i + 2T_{s,b}C_i\right)\left(\frac{M_1}{M_s}\right)^{i-1}}$ and $\beta = \dfrac{1000\frac{M_1}{M_s^2}\frac{dM_s}{dA}\sum_{i-1}^{5}\left(A_i + T_{s,b}B_i + T_{s,b}^2 C_i\right)(i-1)\left(\frac{M_1}{M_s}\right)^{i-2}}{1000\sum_{i-1}^{5}\left(B_i + 2T_{s,b}C_i\right)\left(\frac{M_1}{M_s}\right)^{i-1}}$

Again, combining Eqs. (5.4), (5.5) and (5.8), and neglecting the small term $di_s dM_s$ gives:

$$M_s i_s + i_{st}dM_{ws} = (M_s + dM_s)(i_s + di_s) + U_{b-w}\left(T_{s,b} - T_w\right)dA + h_a\left(T_{s,if} - T_{a,b}\right)dA$$

$$M_s i_s + i_{st}dM_{ws} = M_s i_s + M_s di_s + i_s dM_s + U_{b-w}\left(T_{s,b} - T_w\right)dA + h_a\left(T_{s,if} - T_{a,b}\right)dA \qquad (5.25)$$

Combining Eqs. (5.25) and (5.3) gives:

$$M_s di_s = (i_{st} - i_s)\frac{dM_s}{dA} - U_{b-w}\left(T_{s,b} - T_w\right) - h_a\left(T_{s,if} - T_{a,b}\right)$$

$$\frac{di_s}{dA} = \frac{(i_{st} - i_s)\frac{dM_s}{dA} - U_{b-w}\left(T_{s,b} - T_w\right) - h_a\left(T_{s,if} - T_{a,b}\right)}{M_s} \qquad (5.26)$$

Manipulating and simplifying Eqs. (5.1) to (5.11), (5.24) and (5.26) form the following set of governing equations:

$$\frac{dM_s}{dA} = K_{m,s}\rho_s\left(s_{,b} - s_{,if}\right) \qquad (5.27)$$

$$\frac{d_{s,b}}{dA} = -\frac{M_1}{M_s^2}K_{m,s}\rho_s\left(s_{,b} - s_{,if}\right) \qquad (5.28)$$

$$\frac{dT_{s,b}}{dA} = \alpha + \beta \tag{5.29}$$

$$\frac{di_s}{dA} = \frac{(i_{st} - i_s)\frac{dM_s}{dA} - U_{b-w}(T_{s,b} - T_w) - h_a(T_{s,if} - T_{a,b})}{M_s} \tag{5.30}$$

$$\frac{dT_{a,b}}{dA} = \frac{h_a}{M_a c_{p,a}}(T_{s,if} - T_{a,b}) \tag{5.31}$$

$$\frac{da}{dA} = -\frac{K_{m,s}\rho_s}{M_a}\left(_{s,b} - _{s,if}\right) \tag{5.32}$$

$$\frac{dT_w}{dA} = \frac{U_{b-w}}{M_w c_{p,w}}\left(T_w - T_{s,b}\right) \tag{5.33}$$

The overall heat transfer coefficient between the water and the desiccant solution can be written as:

$$U_{b-w} = \left(\frac{1}{h_w} + \frac{1}{h_i} + \frac{\delta_{wall}}{K_{wall}}\right) \tag{5.34}$$

where, $h_w = \frac{Nu_w K_w}{d_{h,w}}$ and [42], $Nu_w = 0.023 Re_w^{0.8} Pr_w^{0.3}$, for $Re_w \geq 2300$;
$Nu_w = 3.185$, for $Re_w < 2300$;
The convective heat transfer coefficients of the solution film are calculated as [43]:

$$h_i = Nu_i K_s \left(\frac{\mu_s^2}{\rho_s^2 g}\right)^{-1/3} \tag{5.35}$$

$$h_o = Nu_o K_s \left(\frac{\mu_s^2}{\rho_s^2 g}\right)^{-1/3} \tag{5.36}$$

where, $Nu_i = 0.138 Re_s^{-0.132} Pr_s^{0.351}$ and

$$Nu_o = 1.064 \times 10^{-3} Re_s^{-0.093} Pr_s^{0.45} Ja_s^{0.55} Ka_s^{0.6} \left(\frac{P_{v,s}}{P_o}\right)^{1.3}$$

The mass transfer coefficient of the desiccant solution is expressed as:

$$K_{m,s} = Sh_s D_s \left(\frac{\rho_s^2 g}{\mu_s^2}\right)^{1/3} \tag{5.37}$$

where, Sherwood number [45]:

$$Sh_s = 1.064 \times 10^{-3} Re_s^{-0.093} Sc_s^{0.45} Ja_s^{0.55} Ka_s^{0.6} \left(\frac{p_{v,s}}{p_0} \right)^{1.3}$$

The convective heat and mass transfer coefficients of air can be determined as [42]:

$$h_a = \frac{Nu_a K_a}{d_{h,a}} \tag{5.38}$$

$$K_{m,a} = \frac{Sh_a D_a \rho_a}{d_{h,a}} \tag{5.39}$$

where, $Nu_a = 0.023 Re_a^{0.8} Pr_a^{0.3}$, for $Re_a \geq 2300$;
$Nu_a = 3.185$, for $Re_a < 2300$;
$Sh_a = 0.023 Re_a^{0.8} Sc_a^{0.3}$, for $Re_a \geq 2300$;
$Sh_a = 3.185$, for $Re_a < 2300$;

Seven coupled differential Eqs. (5.27) to (5.33) form the governing equations that represent the heat and mass transfer processes of the internally cooled dehumidifier. The governing equations are solved simultaneously with appropriate boundary conditions using a fourth-order Runge–Kutta (R.K.) scheme. A MATLAB computer code was written to solve the governing equations. The model is also solved for the adiabatic wall condition by solving Eqs. (5.27) to (5.32) and setting the overall heat transfer coefficient from the bulk of the film of desiccant solution to the cooling water as zero. The flow diagram, shown in Fig. 5.6, describes the sequence of the various steps of calculation for the internally cooled and adiabatic dehumidifiers. Initially, the necessary parameters and fluid properties are introduced. The cooling water inlet temperature is taken as a known input. The calculation starts with a guessed value of the cooling water outlet temperature. Therefore, the properties of the desiccant solution, air and cooling medium at the first control element that is located at the entry of the dehumidifier are known. The set of governing equations is then solved using the fourth-order R.K. scheme for the first control element to determine the parameters of the desiccant solution, air and coolant at the second control element. The calculation process is continued until the parameters in the last control element are determined. If the calculated temperature of the coolant at the last control element is not within ±0.01 °C of the actual inlet temperature of the coolant, the entire calculation is repeated with a new guessed value of the outlet temperature of the coolant. The by-section method is used to guess the outlet temperature of the coolant. The process is continued until the convergence criteria are satisfied.

Simulation results from the theoretical model are validated with the dehumidification and cooling performance data of the parallel flow water-cooled dehumidifier that are available in the published literature. Ren et al. [24] developed a comprehensive model for similar flow configuration and presented the variation of moisture content of the stream of air and the temperatures of the desiccant solution and the stream of air with the dimensionless air-side Number of Transfer Unit (*NTU*) in

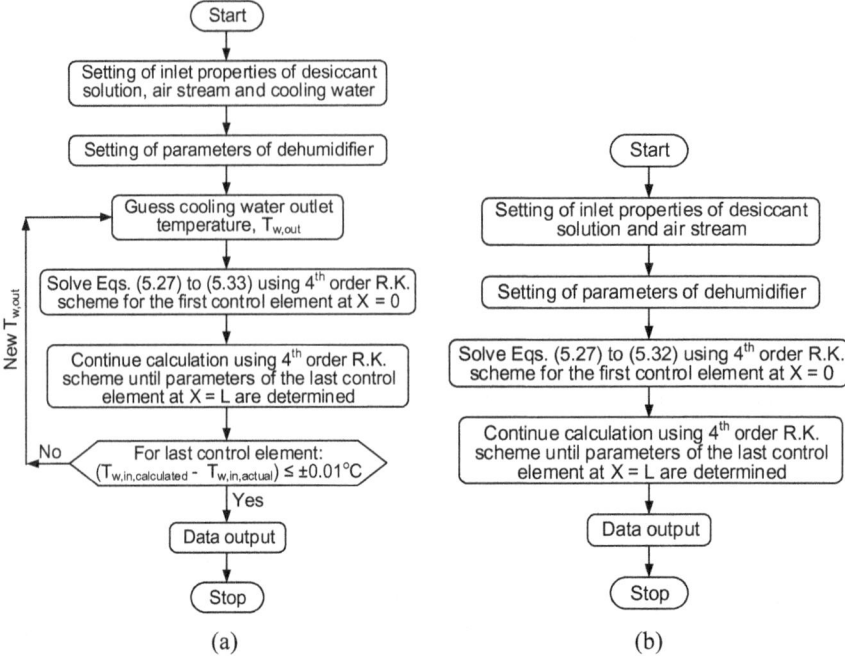

Fig. 5.6 Flow chart for calculation of parameters (**a**) internally cooled dehumidifier and (**b**) adiabatic dehumidifier

the solution flow direction. The NTU represents the number of airside sensible heat transfer unit. The simulated absolute moisture and temperatures of both the air and the desiccant solution are plotted in Fig. 5.7. It is observed that the developed theoretical model is able to predict the performance of the dehumidifier with good accuracy. The maximum discrepancies for the absolute moisture and the temperatures of the

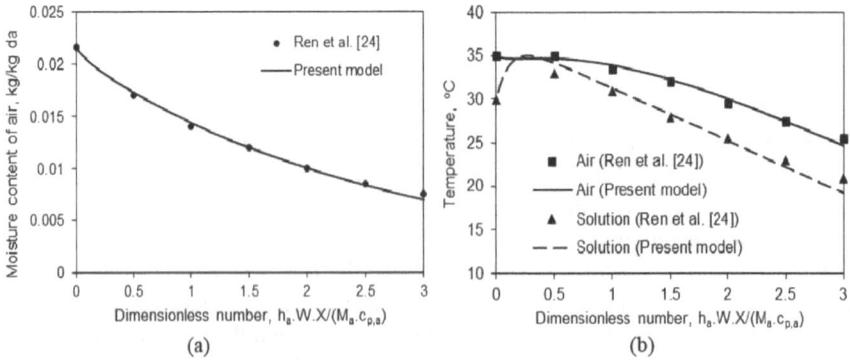

Fig. 5.7 Comparison between present simulation results and published data [24]: **a** Moisture content of air and **b** temperatures of air and desiccant solution

air and the desiccant solution are within ±6.0%. Therefore, upon conducting the validation exercise, the model is now capable of simulating the dehumidification and cooling performance of liquid desiccant dehumidifiers to ±6.0% accuracy.

5.7 Commonly Used Performance Indices

The dehumidification and cooling performance of the internally cooled and the adiabatic dehumidifiers are characterized using several performance indices as shown in Table 5.9 [2].

The flow rate of the liquid desiccant per unit width of the dehumidifier plate is the most sensitive parameter for the adiabatic dehumidifiers. The inlet temperature of the coolant, on the other hand, is the most sensitive parameter for the internally cooled dehumidifiers. The variation of the three key indices, namely specific moisture removal rate, latent heat ratio and cooling capacity, with the flow rate of liquid desiccant and the inlet temperature of the coolant for the adiabatic and the internally cooled dehumidifiers are shown in Fig. 5.8.

5.8 Performance of Dehumidifiers

The ensuing sections study the performance of liquid desiccant under different settings, specifically direct contact and membrane-based dehumidifiers.

5.8.1 Direct Contact Dehumidifiers

Researchers have examined the performance of different adiabatic and internally cooled liquid desiccant dehumidifiers via experiments and simulations. Liu et al. [44] experimentally investigated the dehumidification performance of lithium bromide aqueous solution using a celdek-structured cross-flow adiabatic dehumidifier. The effects of the dehumidifier inlet parameters, namely, air and desiccant flow rates, air inlet temperature and humidity ratio and desiccant inlet temperature and concentration on the moisture removal rate and dehumidifier effectiveness were studied. They varied the absolute moisture content of inlet air from 0.01 to 0.021 kg/kg, the temperature of inlet air from 24.7 to 33.9 °C and inlet concentration of desiccant from 42 to 49%, and reported the moisture removal rate of 1–3 g/s and the effectiveness of 40–70%. They found that the moisture removal rate increased with increasing air and desiccant flow rate, air inlet humidity ratio and desiccant inlet concentration and changed very little with air inlet temperature. The dehumidifier effectiveness increased with increasing desiccant flow rate and inlet temperature but was affected little by the desiccant inlet concentration, air inlet temperature, and humidity ratio.

Table 5.9 Commonly used performance indices to evaluate the performance of liquid desiccant air-conditioning systems [2]

Performance indices	Definitions	Expressions
Moisture removal rate (g/s)	The removal rate of moisture from the humid circulating air by the liquid desiccant	$\mathrm{MRR} = M_a \left(\omega_{a,in} - \omega_{a,out} \right)/1000$
Specific moisture removal rate (g/m^2 s)	The removal rate of moisture from the humid circulating air by the liquid desiccant of unit exposed area	$\mathrm{MRR_{sp}} = \frac{M_a \left(\omega_{a,in} - \omega_{a,out} \right)}{1000A}$
Moisture transfer effectiveness	The ratio between actual change in humidity ratio of the air across the dehumidifier to the maximum possible change in humidity ratio of air	$\varepsilon_m = \frac{\omega_{a,in} - \omega_{a,out}}{\omega_{a,in} - \omega_{s,in,eq}}$
Enthalpy effectiveness	The ratio between actual change in enthalpy of the air across the dehumidifier to the maximum possible change in enthalpy of air	$\varepsilon_i = \frac{i_{a,in} - i_{a,out}}{i_{a,in} - i_{s,in,eq}}$
Sensible heat ratio	The ratio between the sensible energy to the total energy removed from the air stream by the dehumidifier	$\mathrm{SHR} = \frac{c_{p,a} \left(T_{a,in} - T_{a,out} \right)}{i_{a,in} - i_{a,out}}$
Latent heat ratio	The ratio between the latent energy to the total energy removed from the air stream by the dehumidifier	$\mathrm{LHR} = \frac{\left(i_{a,in} - i_{a,out} \right) - c_{p,a} \left(T_{a,in} - T_{a,out} \right)}{i_{a,in} - i_{a,out}}$
Cooling capacity (kW)	Overall cooling effect produced per unit exposed area of desiccant solution. It is the difference between the enthalpies of inlet and outlet air	$\mathrm{CoolingCapacity} = \frac{M_a \left(i_{a,in} - i_{a,out} \right)}{1000A}$
Coefficient of performance	The ratio between the cooling capacity to the total consumed energy (electrical and thermal) by the liquid desiccant cooling system	$\mathrm{COP} = \frac{M_a \left(i_{a,in} - i_{a,out} \right)}{Q_{total,input}}$
Electrical coefficient of performance	The ratio between the system cooling capacity to the consumed electrical energy by the liquid desiccant cooling system	$\mathrm{COP_e} = \frac{M_a \left(i_{a,in} - i_{a,out} \right)}{Q_{electrical,input}}$
Thermal coefficient of performance	The ratio between the unit cooling capacity to the consumed thermal energy in the liquid desiccant cooling system	$\mathrm{COP_{th}} = \frac{M_a \left(i_{a,in} - i_{a,out} \right)}{Q_{heat,input}}$

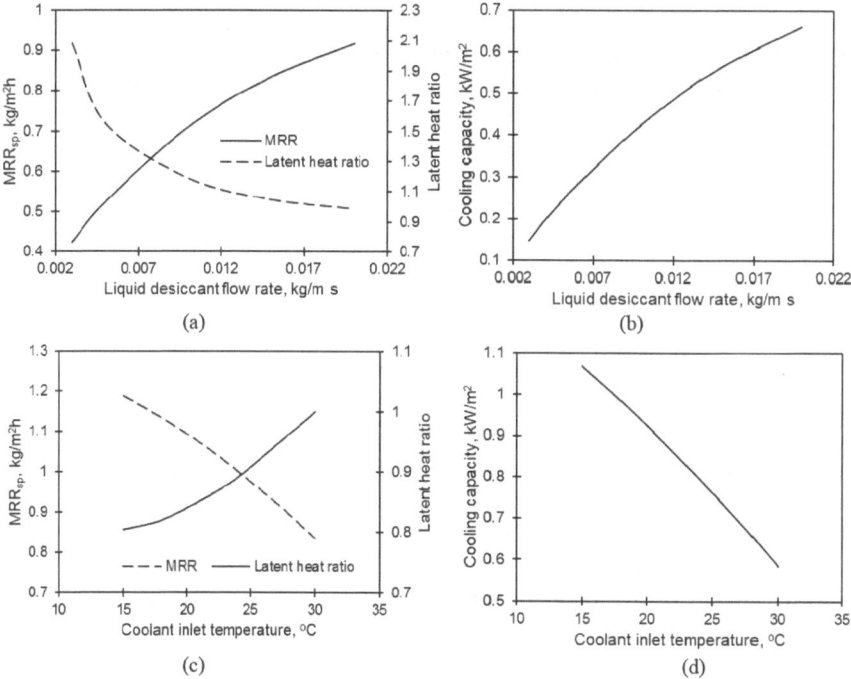

Fig. 5.8 Performance indices: **a** specific MRR and Latent heat ratio for the adiabatic dehumidifier, **b** cooling capacity for the adiabatic dehumidifier, **c** specific MRR and latent heat ratio for the water-cooled dehumidifier, **d** cooling capacity for the water-cooled dehumidifier

They proposed the correlations to predict the performance of the cross-flow dehumidifier. The dehumidifier effectiveness correlation can predict 99.4% of the total 179 cases within ±20%, 83.2% within ±10% and 56.4% within ±5% discrepancies. Their developed correlation of moisture removal rate can predict the experiment values with the absolute difference of 5%.

Bansal et al. [45] experimentally investigated the dehumidification performance of calcium chloride solution in an adiabatic and an internally cooled structured packed-bed dehumidifier. The internally cooled structured packed-bed absorber consisted of rigid media pads with cooling water flowing through the tubes embedded in the packing. The desiccant and the air were circulated in counter flow configuration. The moisture absorption rate, the effectiveness of the dehumidifier, and the mass transfer coefficients between air and solution were evaluated. The maximum moisture absorption rates for the adiabatic and the internally cooled dehumidifiers were 0.0034 kg/kg of dry air and 0.005 kg/kg of dry air, respectively. They reported that a small part of the moisture condensed on the surface of the cooling water tubes as the temperature of the cooling water was below the dew point temperature of the circulating air. The range of effectiveness for the internally cooled and the adiabatic dehumidifiers were 0.55–0.706 and 0.38–0.55, respectively, for the same range of operating parameters. The effectiveness of the dehumidifier reached to the maximum value for the desiccant flow rate of about 5 l/min and the liquid to air flow rate of about 1.0. The effectiveness of the internally cooled cross-flow dehumidifier was significantly less compared with that of the counter-flow configured dehumidifier.

Cho et al. [46] conducted a series of experiments to evaluate the dehumidification performance of counter-flow and cross-flow liquid desiccant dehumidifiers. Experimental results showed that the dehumidification effectiveness and enthalpy effectiveness of the counter flow dehumidifier increased from 51.2% to 82.4% and 40.2% to 72.5%, respectively, when the moisture content of the inlet air was increased from 10.1 g/kg to 22.7 g/kg. Under the same operating conditions, both the effectiveness values of the cross-flow dehumidifier increased from 62.3% to 63.4% and 46.3% to 55.0%, respectively. They also reported that the dehumidification effectiveness and the enthalpy effectiveness of the counter flow type dehumidifier depreciated from 71.7% to 45.4% and 64.4% to 47.3%, respectively, and those of the cross-flow type dehumidifier decreased from 65.3% to 54.8% and 55.8% to 45.2%, respectively, when the temperature of the inlet solution increased from 15.2 to 31.1 °C. The counter-flow dehumidifiers showed higher dehumidification performance in the highly humid inlet air conditions and at the lower desiccant solution temperatures. However, the cross-flow liquid desiccant dehumidifier provided a relatively stable dehumidification performance for a wide range of operating conditions. They also derived the empirical models for the dehumidification effectiveness and the enthalpy effectiveness for both flow configurations.

Liu et al. [47] numerically analysed the effect of flow pattern on the moisture dehumidification performance of the internally cooled plat plate dehumidifier as shown in Fig. 5.9. The cooling medium was circulated through the polypropylene double plates. A desiccant distributor was provided on top of each plate to develop a uniform downward flow of the desiccant solution over the polypropylene plates while the flow direction of air was changed to upwards or downwards to create different flow patterns. The results showed that the counter flow configuration of desiccant to air had better dehumidification performance due to the more uniform mass transfer driving force between the desiccant solution and the circulating air. The decrease

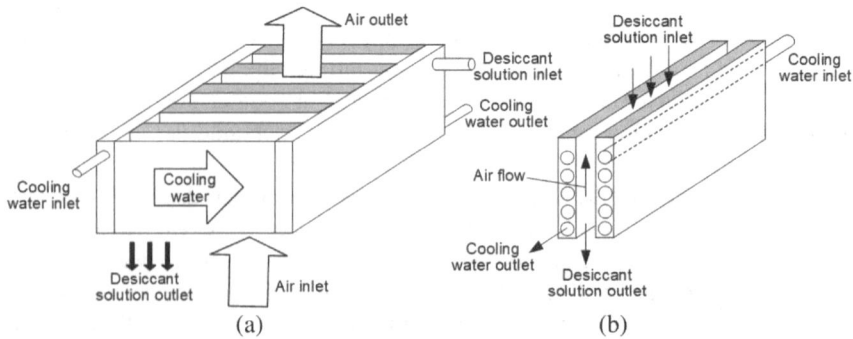

Fig. 5.9 Schematics of the internally cooled dehumidifier: **a** solid view and **b** inner view

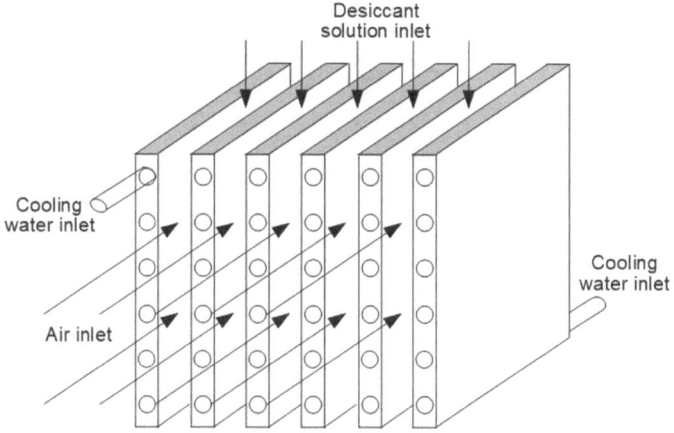

Fig. 5.10 Parallel plate packing material with embedded cooling coils in the plates

of concentration of the desiccant solution was identified as the main factor that influenced the dehumidification performance of the internally cooled dehumidifiers, while the increase of the desiccant temperature was reported as the key performance-restricting factor for the adiabatic dehumidifiers. Less desiccant flow rate could be adopted in the internally cooled dehumidifiers, which is about an order of magnitude less compared with that of the adiabatic dehumidifier to get the same dehumidification performance.

Yoon et al. [48] developed a theoretical model to study the simultaneous heat and mass transfer process in the absorption of vapour into a lithium bromide solution of internally cooled vertical plate absorber. The plate could be cooled by flowing the cooling water between two vertical plates or embedding the cooling coils in the plates of the packing materials as shown in Fig. 5.10 to remove the heat during the moisture absorption process. Simulation results further revealed that the absorption heat and mass fluxes increased rapidly at the inlet region and reached to the maximum values of 8.7336×10^3 W/m^2 and 3.1499×10^{-3} kg/m^2 s at the distance of 11.7×10^{-3} m from the entrance, and then decreased steadily along the solution flow direction. The heat and mass transfer coefficients developed to the highest values of 3.5386×10^3 W/m^2 K and 1.7527×10^{-4} m/s at a distance of 1.67×10^{-3} m and then decreased rapidly along the length of the plate. The absorption mass flux increased as the temperature of the inlet cooling water was lowered, while the inlet concentration of desiccant solution increased, and the inlet temperature of the solution decreased. In order to promote the contracting area of the desiccant and circulating air, the corrugated plates can be used instead of consecutive flat plates.

The detailed simulation and experimental study conducted by Islam [49] showed that the change in the desiccant solution concentration resulted from the absorption of moisture was confined to a small part of the solution film close to the interface as shown in Fig. 5.11. This was mainly due to the poor mass diffusivity of water in the solution. There was a sharp decrease in the interface concentration due to the

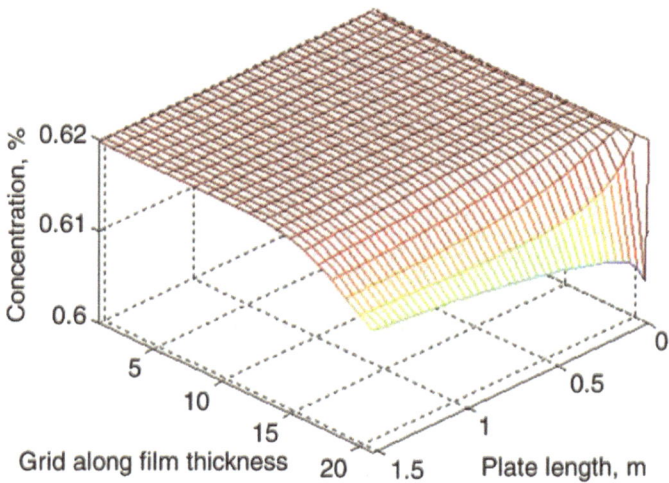

Fig. 5.11 Typical concentration distribution across the film of liquid desiccant

non-equilibrium between the moisture and the entering solution. This was followed by a more gradual decrease in the concentration because of moisture absorption at the interface which was also influenced by the change of temperature of the interface.

A novel absorber concept termed the film-inverting absorber was introduced by Islam et al. [50] with a view to developing a high-performance absorber. The performance of this new absorber was studied both experimentally and numerically and the results were compared with those of a conventional round tube absorber. The film-inverting absorber concept is based on the fluid flow characteristics of the falling film and the thermodynamic aspect of the vapour absorption process. The key features of the film-inverting absorber design include the interruption of the falling film at regular intervals, the cooling of the exposed surface of the falling film in a periodic manner, and the creation of a 'cross-flow' through the falling film to facilitate mixing. The round tube and vertical plate versions of the film-inverting absorbers are shown in Fig. 5.12. Due to the inversion of the desiccant film, both surfaces of the film are alternatively exposed to the moisture, resulting in moisture absorption in both surfaces of the film. The heat of absorption generated during the moisture absorption process is effectively transferred from the interfaces of the desiccant film to the cooling water. Moreover, the hydraulic and the thermal boundary layers are redeveloped at the entrance of each film inverting segment, which enhances the heat and mass transfer coefficient significantly. The maximum increase in the vapour mass flux for the film-inverting design was about 100% compared with that for the tubular absorber.

Lun et al. [51] presented a new method of controlling the rise of temperature of the desiccant solution during the dehumidification process. Anhydrous ethanol was added as a vaporizable coolant into the lithium chloride solution to prepare a self-cooled liquid desiccant solution to control the increase of temperature of the

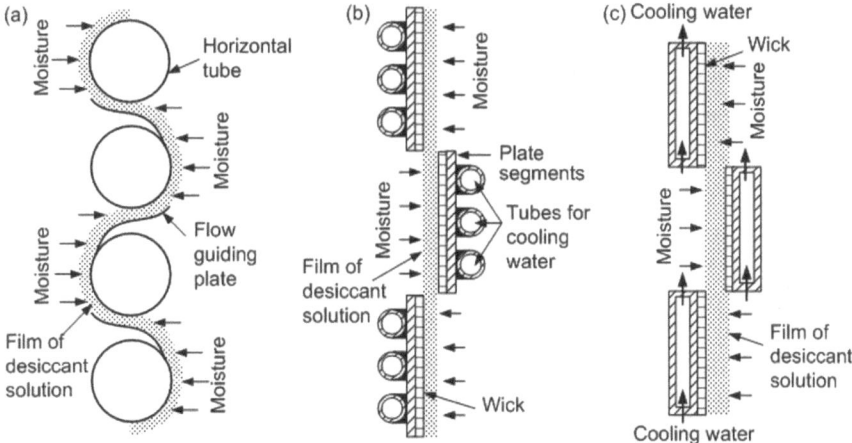

Fig. 5.12 Round tube and vertical plate film-inverting absorbers

desiccant solution. The dehumidification performance indices including moisture removal rate and dehumidification efficiency of the self-cooled solution were experimentally investigated and compared with the conventional lithium chloride desiccant solution. Experimental results indicated that the self-cooled desiccant solution was superior. The trends of the performance indices were identical to the self-cooled and conventional desiccant solutions. The dehumidification performance of the self-cooled desiccant solution increased with the increase in concentration and flow rate of the desiccant solution. The test results suggested that 25 °C and 40% were the optimum solution temperature and concentration. The dehumidification performance indices of the self-cooled desiccant were about 40% higher compared with that of the conventional liquid desiccant solutions.

The water-cooled metal tube banks are used in many industrial dehumidifiers as the heat and mass transfer contact area between the film of liquid desiccant and air because of its simplicity in design and fabrication. To increase the contact area, finned tube cooling coils are usually used. The horizontal and vertical distances between the tubes as well as the arrangement of the tubes are important for optimizing the heat and mass transfer coefficients while keeping the minimum pressure drop of the flowing air. The material of the tubes and fins should be non-corrosive or the metal tubes of high thermal conductivity could be coated with the anti-corrosion layer to develop the packing materials. Khan [52] developed a two-dimensional steady-state theoretical model and solved numerically to predict the dehumidification and cooling performance of an internally cooled horizontal tube bank as shown in Fig. 5.13. An easy-to-use NTU-effectiveness model was also developed to estimate the annual energy consumption using an hour-by-hour analysis. The total cooling load and the latent load removal performances predicted by the two-dimensional model and the easy-to-use model were compared with the performance data available in the catalogue. Study results demonstrated that the performance of the horizontal tube dehumidifier

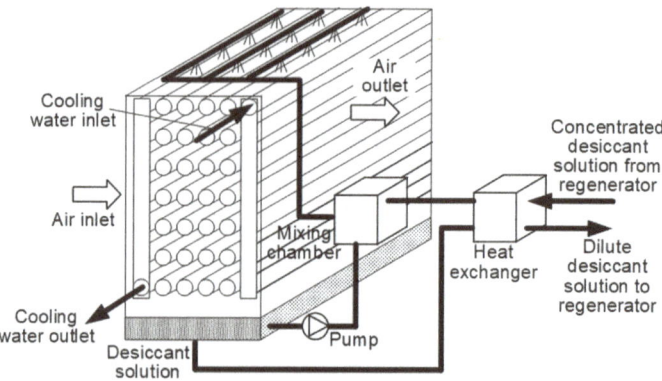

Fig. 5.13 Schematic diagram of the internally cooled horizontal tube bank liquid desiccant dehumidifier

is a strong function of carried-over regeneration heat, cooling water inlet temperature, water-side and solution side NUT, desiccant solution inlet concentration, and water-to-air mass flow ratio. The analysis of the desiccant flow behaviour suggested that a minimum flow rate of solution was required for the complete wetting of the coil surface and satisfactory operation of the system. Based on the local heat and moisture absorption rate of each tube, the cooling water flow rate for each tube could be modulated to optimize the overall performance of the horizontal tube dehumidifiers.

The desiccant solution is pre-cooled before spraying to the dehumidifier to increase its moisture absorption potential. The temperature of the solution increases as it absorbs moisture resulting in the decrease of the partial vapour pressure difference between the liquid desiccant and the circulating air. This problem leads to the reduction of moisture absorption rate and the overall performance of the conventional single-stage dehumidifier. Jiang et al. [4] presented the concept of multi-stage liquid desiccant dehumidification system to overcome the adverse temperature increase. Several single-stage dehumidifiers are connected in series as shown in Fig. 5.14 to

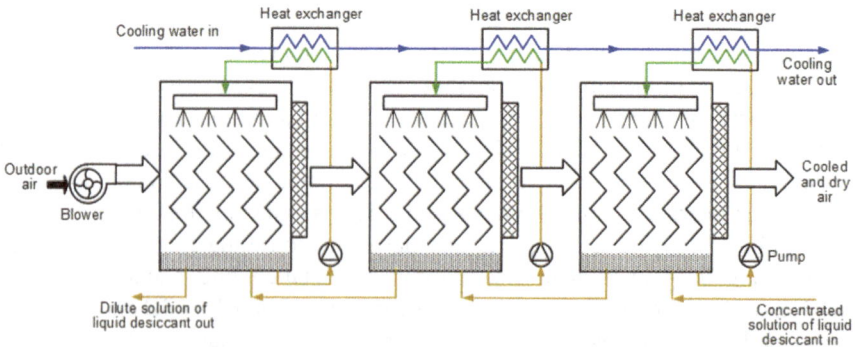

Fig. 5.14 Schematic diagram of the multistage liquid desiccant dehumidification system

develop the multi-stage liquid desiccant dehumidification system. Instead of cooling the liquid desiccant solution internally in each stage of the dehumidifier, the dehumidification performance of the multi-stage dehumidifier was enhanced by cooling the solution externally in stages at different temperatures using the external heat exchangers. The inlet humid air of high partial vapour pressure was first dehumidified using the weakest (most dilute) desiccant solution that had the highest partial vapour pressure. The difference of the partial vapour pressure of moist air and the weakest solution was still adequate for migrating the moisture from the air stream to the weakest desiccant solution. The moist air was partially dried and its partial vapour pressure was decreased. The air stream then moved to the next module of the dehumidifier where the concentration of the flowing desiccant solution was relatively high and the partial vapour pressure of the desiccant solution was relatively low. Hence, the difference of the partial vapour pressure of air and the solution remained enough for migrating the moisture from the air stream to the desiccant solution. Finally, the air of low partial vapour pressure flows through the last module of the dehumidifier where the strongest (most concentrated) desiccant solution with the lowest partial vapour pressure was used to dehumidify the air stream. An adequate vapour pressure difference between the air and the desiccant solution was maintained to realize effective dehumidification at each module of the dehumidifier. In the conventional single-stage cross-flow dehumidifier, the vapour pressure difference and the corresponding dehumidification rate gradually drop along the direction of air flow. The irreversible loss of the single-stage dehumidifier is greatly minimized in the multi-stage dehumidifier [53].

The mass flow rate of the desiccant solution in each module dehumidifier was small, thus the difference between the inlet and outlet concentration of the desiccant solution was increased, which facilitated the effective execution of the regeneration of weak solution. However, the relatively small mass flow rate of the desiccant solution could lead to the problem of partial wetting of the packing materials, which could reduce the exposed surface area of desiccant film and moisture absorption rate. The spray system and the packing material should be designed carefully to ensure the even distribution of the desiccant solution.

Cheng et al. [54] designed a multi-stage internal circulation liquid desiccant dehumidifier, where the packing bed and the cooling module were layered separately inside each stage as shown in Fig. 5.15. The solution concentration and flow rate could be evenly distributed using the internal circulation structure based on the moisture content distribution along the dehumidification process. They developed a steady-state heat and mass transfer model and studied the effects of the operating and the number of packing stages on the dehumidification performance. Study results showed that the effect of the water flow rate was less significant compared with the solution flow rate. They described that the flow rate of the solution affected both the driving force and the mass transfer coefficient between the air and the solution while the flow rate of the cooling water only indirectly affected the mass transfer driving force. The heat and mass transfer area increased with the higher packing series, which led to the increased dehumidification efficiencies. The dehumidification efficiency approached the limit value when the number of packing series reached 6 and 12 for

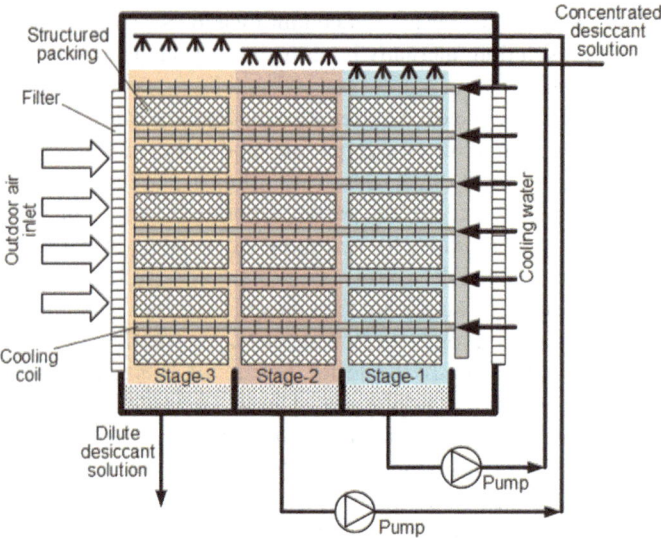

Fig. 5.15 Schematic diagram of the multistage internal circulation liquid desiccant dehumidifier

the mass flow rate of air of 1.63 kg/s and 2.44 kg/s, respectively. The volume heat transfer coefficient in the multi-stage internal circulation liquid desiccant dehumidifier varied from 15.7 to 31.4 kW/(m^3 K), which was significantly higher than that in the packed tower with cooling tubes of 2.73–4.28 kW/(m^3 K) [45]. The ultimate dehumidification efficiency of the unit was found to be 7.3% higher than that of the packed tower with cooling tubes.

Two-stage dehumidifiers with different desiccant solutions operated with high desiccant concentration difference as shown in Fig. 5.16 have several advantages such as low desiccant investment, high storage capacity while keeping good dehumidification performance. Xiong et al. [55] conducted an experimental study and exergy analysis based on the second thermodynamic law using a two-stage liquid desiccant dehumidification system to investigate the dehumidification and thermal coefficient of performance enhancement opportunities. Calcium chloride and lithium chloride were used as desiccants. In the system, the air was predehumidified in the first module using calcium chloride and then further dehumidified in the second module using the lithium chloride solution. Study results showed that the pre-dehumidification of calcium chloride solution reduced the irreversibilities in the dehumidification/regeneration process. Compared with the single-stage system, the thermal coefficient of performance and exergy efficiency of the two-stage dehumidification system increased from 0.24 to 0.73 and from 6.8% to 23.0%, respectively. The energy storage capacity of calcium chloride solution and lithium chloride solution at the concentration of 40% reached 237.8 MJ/m^3 and 395.1 MJ/m^3, respectively. The high desiccant concentration variance method and calcium chloride pre-dehumidification method contributed to the improvement. The performance of

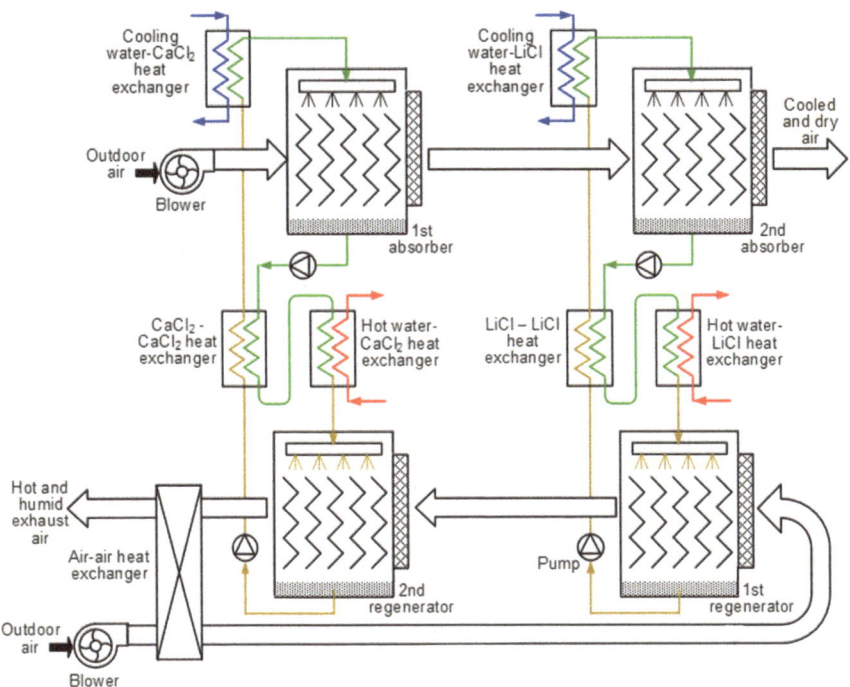

Fig. 5.16 Two-stage liquid desiccant dehumidification system

the system changed significantly with the distribution of the regenerators' volume between the lithium chloride and calcium chloride. The highest performance achieved when 55% of the volume was assigned to the lithium chloride regenerator. The calcium chloride solution was recommended to regenerate at 58 °C, 56 °C, 52 °C, and 47 °C when lithium chloride solution was regenerated at 72 °C, 68 °C, 64 °C, and 60 °C, respectively, to get the optimum performance. They reported that the investment cost could be reduced by about 53% in comparison with the single-stage lithium chloride dehumidification system as calcium chloride is cheaper than lithium chloride.

Xiong et al. [56] also evaluated the performance of the two-stage dehumidi-fier using calcium chloride and lithium bromide desiccant solutions. They installed an indirect evaporative cooling system between the calcium chloride dehumidifi-cation module and the lithium bromide dehumidification module to cool the pre-dehumidified air before entering to the lithium bromide dehumidification module. The system operated with a high-desiccant variance and the energy storage capacities of 481.2 MJ/m^3 for calcium chloride and 382.2 MJ/m^3 for lithium bromide solutions, which were 30 and 26 times higher than the conventional systems, respectively. The coefficient of performance of the two-stage system reached 2.13, which was 32% higher than those of the conventional lithium bromide alone system.

To address the carry-over problem at the high flow rate of the circulating air, Kumar et al. [57] proposed to add more absorber in parallel and analyzed the performance of two standalone multi-absorber liquid desiccant cycles. They used two parallel absorbers in Multi-Absorber Cycle-1 and three parallel absorbers in Multi-Absorber Cycle-2. The liquid desiccant was circulated in series through the absorber. The Multi-Absorber Cycle-1 allowed to double and the Multi-Absorber Cycle-2 allowed to triple the quantity of circulating air while maintaining the same desiccant to air flow ratio in each absorber. The Multi-Absorber Cycles operated on much higher concentration gradients of the desiccant solution and improved the coefficient of performance of the Multi-Absorber Cycle-1 and Multi-Absorber Cycle-2 by 67% and 116%, respectively, as compared with conventional single-stage liquid desiccant dehumidification system. A summary of the studies conducted by researchers to evaluate the performance of dehumidifiers under different flow configurations and operating conditions is presented in Table 5.10.

5.8.2   Membrane-Based Dehumidifiers

The membrane-based liquid desiccant dehumidification technology mostly developed in recent 10 years. The aspects of the development include dehumidification performance, membrane materials, feasibility, energy, and economic benefits. The researchers and design engineers mainly focused on the parallel-plate membrane and hollow fibre membrane dehumidifiers with the counter flow, cross-flow, and quasi-counter flow configurations. The membrane-based liquid desiccant dehumidifiers had been designed and tested as adiabatic and internally cooled. Single-stage and multi-stage systems had also been designed to meet specific requirements and improve dehumidification performance. Huang et al. [69] numerically and experimentally investigated the heat and mass transfer behaviour of a membrane-formed parallel-plate channel liquid desiccant dehumidifier as shown in Fig. 5.17. Lithium chloride solution was used as the liquid desiccant. The membrane was made up of a layer of PVDF porous membrane. Two layers of silica gel were applied to the porous membrane to prevent the liquid desiccant from leaking. The membrane only selectively permitted the transport of water vapour and heat. Two parallel rectangular channels are formed on both sides of the membrane. The dimensions of the channels were 2 mm height and 10 cm width. Air flows in the upper channel, while liquid desiccant flows in the lower channel in a cross-flow configuration. The naturally formed real-boundary conditions on membrane surfaces were numerically obtained by simultaneous solution of momentum, energy, and concentration equations for the two fluids. The local and mean Nusselt and Sherwood numbers along the channels were calculated using the boundary conditions. Simulation results showed that the heat transfer developed shortly after the entry, but the mass transfer was still developing even at the outlet. The mass transfer Sherwood numbers for the solution stream were much larger than the airside. The solution side Nusselt number was about 15%

Table 5.10 Summary of the performance of dehumidifiers of different flow configurations

Ref	Details of dehumidifier	Operating conditions	Performance
Patnaik et al. [58]	Desiccant: LiBr Configuration: Counter flow Packing: Tripack No. ½ polypropylene spheres (random); 0.81 m (D) × 0.4 m (H)	$T_{a,in}$ = 28.1–38.9 °C; $\omega_{a,in}$ = 12.1–22.7 g/kg; $M_{a,in}$ = 0.65–0.95 kg/s; $T_{s,in}$ = 23.1–32 °C; $\omega_{s,in}$ = 43.9–58.6%; $M_{s,in}$ = 0.3–0.48 kg/s	MRR = 0.96–5.59 g/s; ε_m = 13–41.2%
Liu et al. [34]	Desiccant: LiBr Configuration: Cross flow Packing: Celdek 7090 (structured. 396 m²/m³); 0.55 m × 0.4 m × 0.35 m	$T_{a,in}$ = 26.8–32.7 °C; $\omega_{a,in}$ = 12.0–17.3 g/kg; $M_{a,in}$ = 0.328–0.453 kg/s; $T_{s,in}$ = 21.2–27.8 °C; $\omega_{s,in}$ = 42.8–48.0%; $M_{s,in}$ = 0.31–0.64 kg/s	MRR = 1.08–2.31 g/s; ε_m = 41.3–68.0%
Chung et al. [59]	Desiccant: LiCl Configuration: Counter flow Packing: 1.6 cm polypropylene Flexi rings (random, 342m²/m³); 0.1525 m (D) × 42 m (H)	$T_{a,in}$ = 23.1–25.5 °C; $\omega_{a,in}$ = 10.6–13.9 g/kg; $M_{a,in}$ = 28–56 ft³/min; $T_{s,in}$ = 13.2–22.9 °C; $\omega_{s,in}$ = 30–40%; $M_{s,in}$ = 2–3.5 gal/min	MRR = 0.07–0.237 g/s; ε_m = 50–71.6%
Chung et al. [20]	Desiccant: LiCl Configuration: Counter flow Packing: Cross-corrugated cellulose (structured, 410 m²/m³); 0.1525 m (D) × 0.4 m (H)	$T_{a,in}$ = 24.4–29.6 °C; $\omega_{a,in}$ = 13.1–17.8 g/kg; $M_{a,in}$ = 27.5–44.0 ft³/min; $T_{s,in}$ = 19.1–21.0 °C; $\omega_{s,in}$ = 30–38%; $M_{s,in}$ = 2.3–3.5 gal/min	MRR = 0.09–0.2 g/s; ε_m = 62.–68. %
Fumo and Goswami, [60]	Desiccant: LiCl Configuration: Counter flow Packing: 1 in polypropylene RauschertHiflow rings (random, 210m²/m³); 0.24 m (D) × 0.6 m (H)	$T_{a,in}$ = 29.9–40.1 °C; $\omega_{a,in}$ = 14.2–21.5 g/kg; $M_{a,in}$ = 0.890–1.513 kg/(m²s); $T_{s,in}$ = 25.0–35.2 °C; $\omega_{s,in}$ = 33.1–34.9%; $M_{s,in}$ = 5.019–7.420 kg/(m²s)	MRR = 0.21–0.53 g/s; ε_m = 54.1–76.7%
Longo and Gasparella, [61]	Desiccant: LiCl Configuration: Counter flow Packing: 25 mm plastic Pall Rings (random); 0.4 m (D) × 0.725 m (H)	$T_{a,in}$ = 24.3–37.6 °C; $\omega_{a,in}$ = 7.3–23.3 g/kg; $M_{a,in}$ = 0.43–0.47 kg/(m²s); $T_{s,in}$ = 23.4–24.0 °C; $\omega_{s,in}$ = 39.2–40.6%; $M_{s,in}$ = 0.10–1.17 kg/(m²s)	MRR = 0.08–0.95 g/s; ε_m = 34–93%

(continued)

Table 5.10 (continued)

Ref	Details of dehumidifier	Operating conditions	Performance
Dong et al. [62]	Desiccant: LiCl Configuration: Counter flow Packing: S-shape PVC (structured, 81.8 m²/m³); 0.3 m × 0.3 m × 0.5 m	$T_{a,in} = 27$–34 °C; $\omega_{a,in} = 17$ g/kg; $M_{a,in} = 0.071$ kg/s; $T_{s,in} = 21.4$ °C; $\omega_{s,in} = 36\%$; $M_{s,in} = 0.051$–0.116 kg/s	MRR = 0.316–0.383 g/s; $\varepsilon_m = 34.5$–41.9%
Chen et al. [63]	Desiccant: LiCl Configuration: Cross flow Packing: Z-type gauze packing way 2 (structured, 160 m²/m³); 0.34 m(L) × 0.34 m(W) × 0.5 m(H)	$T_{a,in} = 34.9$–35.6 °C; $\omega_{a,in} = 21.5$–22.6 g/kg; $M_{a,in} = 0.13$–0.28 kg/s; $T_{s,in} = 14.5$–24.5 °C; $\omega_{s,in} = 39\%$; $M_{s,in} = 0.13$–0.53 kg/s	MRR = 0.039–2.007 g/s; $\varepsilon_m = 25.0$–41.7%
Kumar et al. [64]	Desiccant: CaCl₂ Configuration: Counter flow Packing: Wire mesh packing, 36 layers (structured); 0.3 m(W) × 0.3 m(L) × 0.18 m(H)	$T_{a,in} = 35.0$–44.8 °C; $\omega_{a,in} = 18.0$–27.1 g/kg; $M_{a,in} = 0.024$–0.048 kg/s; $T_{s,in} = 33.2$–38.8 °C; $\omega_{s,in} = 38.6$–41.3%; $M_{s,in} = 0.043$–0.071 kg/s	MRR = 0.05–0.20 g/s; $\varepsilon_m = 26.1$–55.2%
Moon et al. [65]	Desiccant: CaCl₂ Configuration: Cross flow Packing: Cross-corrugated cellulose paper sheets (structured. 608 m²/m³); 0.3 m × 0.3 m	$T_{a,in} = 26.8$–39.1 °C; $\omega_{a,in} = 16.4$–24.4 g/kg; $M_{a,in} = 0.92$–1.99 kg/(m²s); $T_{s,in} = 26.2$–38.2 °C; $\omega_{s,in} = 32.8$–43.0%; $M_{s,in} = 0.54$–3.18 kg/(m²s)	MRR = 0.48–1.24 g/s; $\varepsilon_m = 43.6$–77.8%
Oberg and Goswami, [66]	Desiccant: TEG Configuration: Counter flow Packing: 2.54 cm polypropylene RauschertHiflow rings (random, 210m²/m³); 0.24 m (D) × 0.6 m (H)	$T_{a,in} = 24.1$–36.1 °C; $\omega_{a,in} = 11.0$–23.0 g/kg; $M_{a,in} = 0.44$–1.56 kg/(m²s); $T_{s,in} = 24.0$–36.0 °C; $\omega_{s,in} = 94.0$–96.1%; $M_{s,in} = 4.48$–6.59 kg/(m²s)	MRR = 0.19–0.71 g/s; $\varepsilon_m = 73.8$–94.0%
Zurigat et al. [67]	Desiccant: TEG Configuration: Counter flow Packing: Arrays of aluminium plate stack (structured, 77 m²/m³); 0.48 m(H) × 0.0225 m²	$T_{a,in} = 25.6$–40.7 °C; $\omega_{a,in} = 16.0$–21.8 g/kg; $M_{a,in} = 1.50$–2.61 kg/(m²s); $T_{s,in} = 25.0$–43.2 °C; $\omega_{s,in} = 93.0$–98.0%; $M_{s,in} = 0.13$–0.82 kg/(m²s)	MRR = 0.101–0.25 g/s; $\varepsilon_m = 18.7$–46.3%

(continued)

Table 5.10 (continued)

Ref	Details of dehumidifier	Operating conditions	Performance
Longo and Gasparella [68]	Desiccant: KCOOH Configuration: Counter flow Packing: 25 mm plastic Pall Rings (random); 0.4 m (D) × 0.725 m (H)	$T_{a,in}$ = 22.6–35.8 °C; $\omega_{a,in}$ = 8.8–20.7 g/kg; $M_{a,in}$ = 0.48–0.52 kg/(m²s); $T_{s,in}$ = 21.9–24.8 °C; $\omega_{s,in}$ = 72.8–74.0%; $M_{s,in}$ = 0.09–1.23 kg/(m²s)	MRR = 0.12–0.85 g/s; ε_m = 34–93%

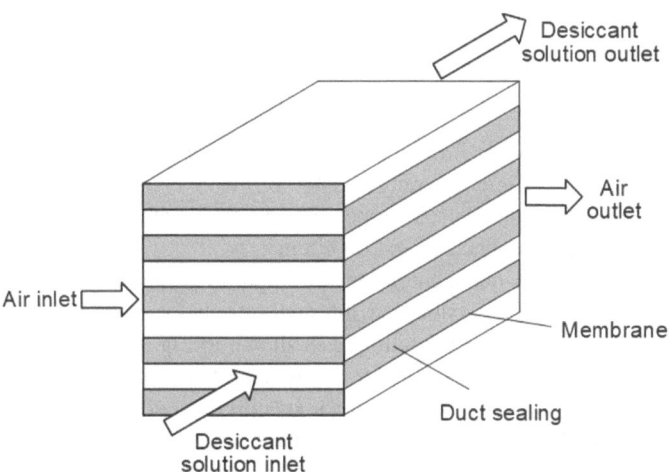

Fig. 5.17 Membrane-formed parallel-plate liquid desiccant dehumidifier

higher than the airside. The humidity values on the two surfaces of the membrane were different due to large resistance in mass transfer through membranes.

Xiao et al. [70] presented an internally cooled plate membrane-based liquid desiccant dehumidifier system. The schematic diagram of the dehumidifier system showing the solution channels, air channels, and cooling tubes with flow configurations of fluids and key operating parameters is illustrated in Fig. 5.18. The fresh air was dehumidified and cooled by the concentrated lithium chloride solution flowing through the adjacent channel. The desiccant solution was cooled by the cooling water flowing in the cooling tubes that were installed in the solution channel. The impacts of the key operating parameters on the specific dehumidification power, coefficient of performance, and dehumidification were investigated both experimentally and numerically. Heat and mass transfer process in the dehumidifier, regenerator, air-cooled chilled water, and other key components were modelled. Study results showed that both the specific dehumidification power and the coefficient of performance increased with the increase of air temperature, relative humidity, air flow rate,

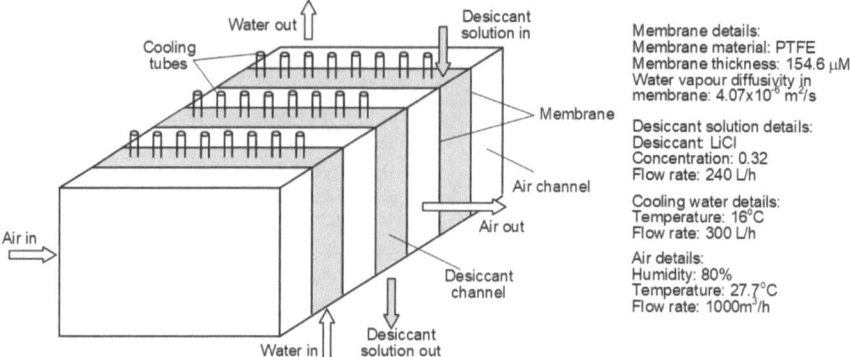

Fig. 5.18 Schematic of the internally cooled parallel-plate membrane liquid desiccant dehumidifier

solution flow rate, and water flow rate. The dehumidification efficiency (range 0.2–0.62) and the cooling efficiency (range 0.24–0.65) were changed significantly with the change of fresh air flow rates. The specific dehumidification power of the system at the design condition reached up to 263 g/h m^2.

Researchers had also made numerous attempts to develop quasi-counter flow hexagonal parallel-plate adiabatic and internally cooled membrane dehumidifiers and hollow fibre membrane dehumidifiers as shown in Fig. 5.19. Huang et al. [31] studied the fluid flow and heat transfer behaviour of the hexagonal parallel-plate membrane channels and reported that for a specific channel height and Reynolds number of the liquid desiccant, the Nusselt number for the desiccant solution stream became about 1.21–1.38 and 1.51–2.85 times of those for the water and the air streams, respectively. Qiu et al. [71 and Zhang et al. [72] focused on the influences of the channel structural parameters of the internally cooled hexagonal parallel-plate membrane with z-shaped cooling tubes on the Nusselt numbers and the product of mean friction factors and Reynolds number. These results would be useful for the performance optimization and structural design of the quasi-counter flow hexagonal membrane dehumidifiers.

The hollow fibre membrane dehumidifiers are developed based on the concept of shell and tube heat exchanger, where the desiccant solution flows in the tube side, while air flows in the shell side. The hollow fibre membrane dehumidifiers have inherent advantages of higher effectiveness, larger packing densities, and smaller airside pressure drop. However, the systems are more difficult to fabricate because of the sealing difficulties of the two ends [73]. Due to the smaller airside pressure drop, the hollow fibre membrane dehumidifiers could be installed in the multi-stage configuration as shown in Fig. 5.20 to dehumidify the circulating air to very low moisture content for specific industrial applications [2]. The desiccant solution is cooled before entering in each stage of the dehumidifier using external heat exchangers to reduce its partial vapour pressure and maintain sufficient partial vapour pressure difference between the desiccant and the flowing air throughout the dehumidification process.

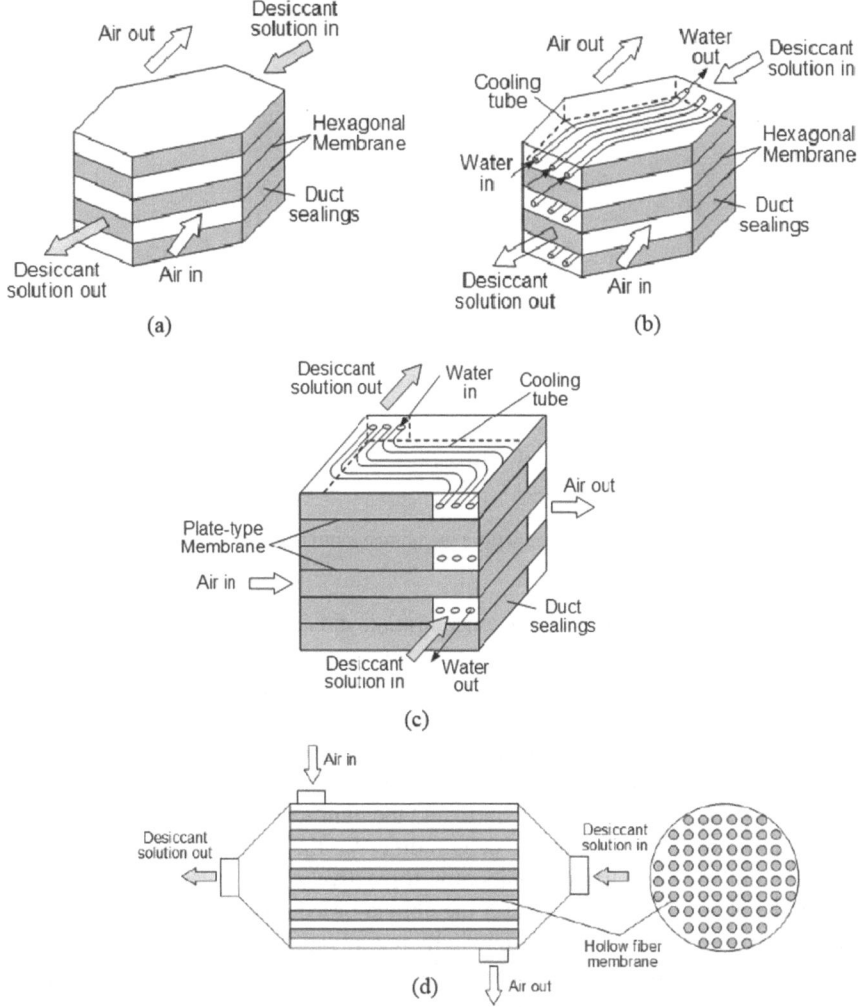

Fig. 5.19 Schematic diagrams of hexagonal, rectangular and hollow fibre membrane dehumidifiers **a** quasi-counter flow hexagonal adiabatic [31], **b** quasi-counter flow hexagonal with Z-shaped cooling tubes [71], **c** quasi-counter flow rectangular with Z-shaped cooling tubes [72], and **d** hollow fibre membrane [73]

The sensible, latent and total effectiveness as well as the moisture removal rate of a cross-flow corrugated membrane-based desiccant dehumidifier consisting of 22 air channels and 22 desiccant solution channels were experimentally investigated by Chu et al. [74]. The dimensions of the dehumidifier were 210 mm (H) × 230 mm (W) × 410 mm (L). The thickness and the thermal conductivity of the membrane were 0.5 mm and 0.3 W/m K, respectively. The gaps in the air channels and solution channels were 7.7 mm and 4.3 mm, respectively. The tested parameters include inlet

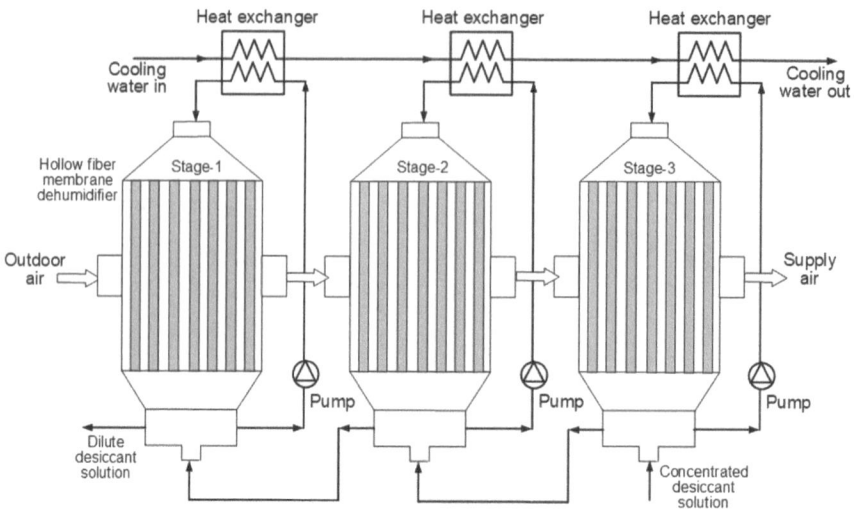

Fig. 5.20 Schematic diagram of hollow fibre membrane multistage dehumidifier

air relative humidity, inlet solution concentration and temperature, heat capacity rate ratio, and the number of heat transfer units. They reported that the moisture removal rate of the dehumidifier increased by 33.7% but total effectiveness decreased by 1.7% when inlet air relative humidity was changed from 62 to 74%. Desiccant solution concentration demonstrated a positive impact on the effectiveness. The highest sensible, latent, and total effectiveness of 0.823, 0.802, and 0.810, respectively, were achieved for the heat capacity rate ratio and the number of heat transfer units of 12. They recommended to operate the dehumidifier for the heat capacity rate ratio of 6 and the number of heat transfer units of 8, respectively.

Gurubalan et al. [75] proposed a membrane-based dehumidifier for the hybrid air-conditioning system and experimentally investigated its performance in terms of moisture removal rate, and sensible, latent and total effectiveness. The schematic diagram of the experimental setup and flow configurations of air, desiccant, and cooling water are shown in Fig. 5.21. The effect of flow configurations, namely, all counter flow, all parallel flow, cooling water and air in counter flow with air, cooling water and desiccant in counter flow with air as well as the operating parameters of air, cooling water, and desiccant on the performance of the dehumidifiers were thoroughly investigated. Experimental results revealed that the configuration with cooling water flow in counter and parallel to the desiccant and air, respectively, was optimum for the dehumidifier. The latent effectiveness was found to be independent of changes in the ambient specific humidity and temperature. The proposed dehumidifier is suitable for locations where the ambient temperature fluctuates over a wider range. The findings of the study are useful in optimizing the design of the membrane-based dehumidification system.

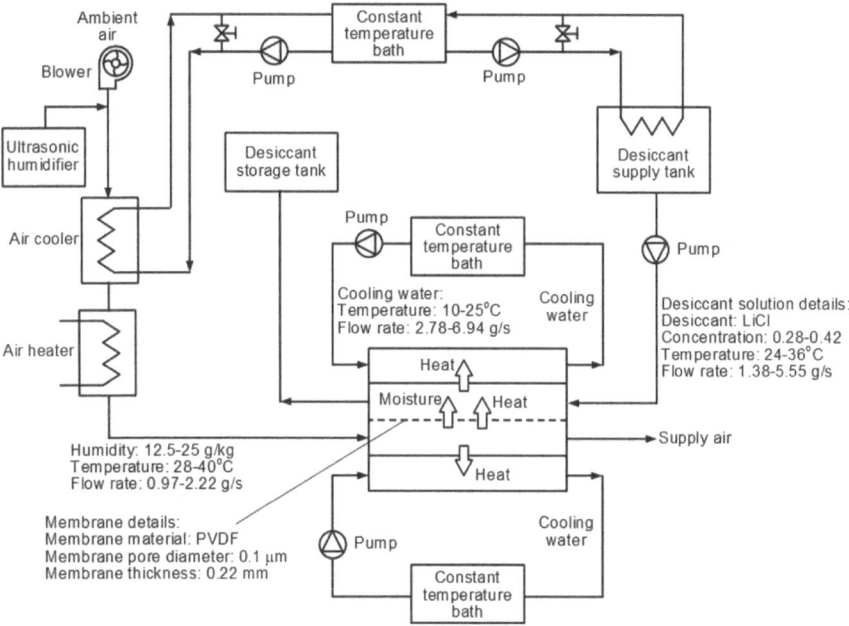

Fig. 5.21 Schematic diagram of the experimental setup and flow configurations of air, desiccant and cooling water

5.9   Liquid Desiccant Air-Conditioning Systems

The overall energy performance of the vapour compression air-conditioning systems and the vapour absorption air-conditioning systems can be improved by retrofitting them with the liquid desiccant dehumidification systems. Mansuriya et al. [76] designed and fabricated a hybrid air-conditioning system of 5 kW cooling capacity by combining the liquid desiccant dehumidifier and the vapour compression refrigeration system. A variable height cross-flow liquid desiccant dehumidifier of bed dimension 300 mm × 300 mm × 300 mm was retrofitted with a R143A driven vapour compression refrigeration unit of 1 refrigeration ton cooling capacity. The Polyvinyl Chloride (PVC) sheets were selected as packing material and the aqueous calcium chloride solution was used as the liquid desiccant. Four different air temperatures (35 °C, 36 °C, 37 °C, and 38 °C) and five different relative humidity of air (55%, 58%, 60%, 63%, and 65%) were considered for the study. The air flow velocity of 0.04 kg/s, the desiccant inlet temperature of 30 °C, and the desiccant inlet concentration of 0.41 kg of desiccant/kg of solution were maintained during the tests. They evaluated the dehumidification performance and the heat load of the dehumidifier and compared the energy performance of the hybrid system with the standalone vapour compression refrigeration system. Experimental results revealed that 54.93% of the latent heat load was supported by the liquid desiccant system and the coefficient of performance of the hybrid system was improved up to 27.54%. The payback period for the incremental cost to retrofit the liquid desiccant system was 4 years.

Mohan et al. [77] integrated two columns of lithium bromide solution with a conventional room air conditioner to enhance the dehumidification and energy performance of the unit. The absorber and the regeneration columns were installed after the evaporator and the condenser, respectively, as shown in Fig. 5.22. They studied the effect of varying the room temperature and humidity on the performance of the hybrid system. The liquid desiccant loop was found to enhance the dehumidification up to 2 g/kg by absorbing moisture in the absorber and then transferring to condenser air in the regenerator.

To provide a cost-effective thermal comfort condition particularly in humid climates, Cuce [78] designed, fabricated, and tested the performance of a liquid desiccant-based evaporative cooling system. Liquid potassium formate solution of concentration 74% was used as the desiccant owing to its favourable thermophysical properties such as less corrosive, lower density, and lower viscosity compared with the conventional absorbents such as lithium bromide solution [79]. Moreover, potassium formate is environmentally friendly and cost-effective compared with the alternatives. A specially designed desiccant filled fibre cloth dehumidifier was used to dehumidify the inlet fresh air. The dehumidifier air was then exposed to an evaporative cooling system to reduce its temperature to the thermal comfort level. The drained desiccant solution was collected from the bottom of the dehumidification channel and generated using a regeneration unit. Experimental results revealed that the system was able to dehumidify and cool the circulating air of average temperature 38.6 °C and relative humidity 94.7% to 33.3 °C and 65.5%, respectively. An average reduction of 5.3 °C in the supply air temperature as well as average dehumidification

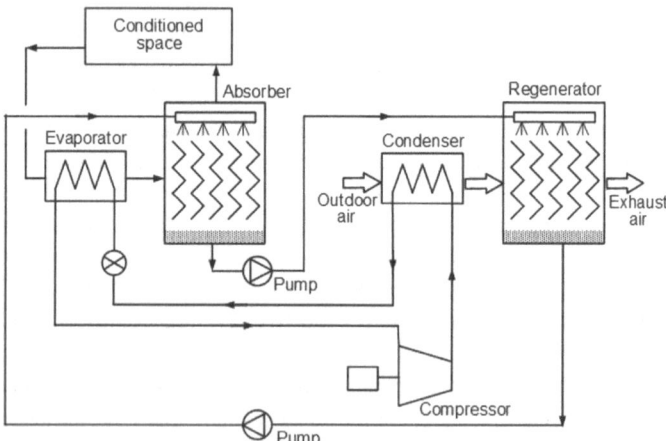

Fig. 5.22 Schematic diagram of the liquid desiccant dehumidifier retrofitted with the conventional vapour compression air conditioner

effectiveness of 63.7% were achieved. The average coefficient of performance of the system was 5.5 and 4.8, respectively, for the inlet air velocities of 0.3 and 0.5 m/s.

Mucke et al. [80] studied three different hybrid liquid desiccant air-conditioning systems as shown in Fig. 5.23 to evaluate their energy-saving potentials compared with a conventional vapour compression air-conditioning system for three different climatic design conditions. All considered systems consist of a moisture absorption system in combination with a vapour compression cooling system. Three outdoor climates were considered: the typical design conditions in Germany (temperature 32 °C, absolute humidity 12 g/kg), a medium summer day in Germany (temperature 27 °C, absolute humidity 10.5 g/kg), and a subtropical climate with average data of Kuala Lumpur, Malaysia (temperature 27 °C, absolute humidity 19 g/kg). The typical supply air condition (temperature 18 °C, absolute humidity 8 g/kg) and the return air condition (temperature 24 °C, absolute humidity 11 g/kg) were used in the study. The results showed that the achievable electric energy savings were 30%–60% depending on the design of the hybrid liquid desiccant air-conditioning systems and climatic conditions.

The multi-stage liquid desiccant dehumidification system could be retrofitted with the vapour compression and the vapour absorption cooling systems as shown in Fig. 5.24 and Fig. 5.25, respectively, to realize the better energy performance of the systems and achieve the desired humidity of the air-conditioned spaces using low-grade cost-effect desiccants [81]. Available roof space could be utilized to collect the solar energy using a flat plate or concentric solar collector for regenerating the dilute desiccant solution. A small cooling tower or an evaporative cooling system could be installed to precool the concentrated desiccant solution before spraying to the beds of absorbers. The vapour compression system could be operated at a higher pressure and temperature of the refrigerant to reduce the compression lift and energy consumption of the compressor. The vapour compression system would support only the sensible cooling load of the air-conditioned spaces. The temperature and moisture content of the exhaust air of air-conditioned are relatively low. The exhaust air could be used to cool the condenser coil of the vapour compression system, which would increase the energy efficiency of the vapour compression system. The hot exhaust air of low relative humidity could be circulated through the bed of regenerator to enhance the moisture evaporation rate of dilute desiccant solution. The energy consumption of a convention vapour compression cooling system could be saved by about 35% by retrofitting a liquid desiccant dehumidification system to a vapour compression cooling cycle [82].

The hybrid liquid desiccant dehumidification and vapour absorption cooling systems require more liquid desiccant solution compared with the hybrid liquid desiccant dehumidification and vapour compression cooling systems to meet the demand of air dehumidification as well as the sensible cooling load of the vapour absorption cycle. As a result, the solar collectors and auxiliary heaters of higher capacity are required to realize the better performance of the regenerator. Unlike the hybrid liquid desiccant dehumidification and vapour compression cooling systems, the exhaust air from the air-conditioned spaces of the hybrid liquid desiccant dehumidification and vapour absorption cooling systems could not be preheated before

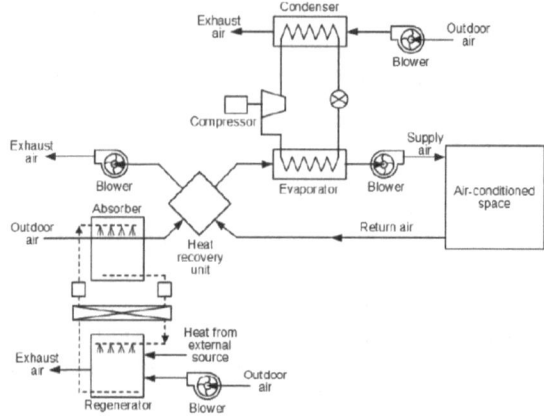

(a) Liquid desiccant system and vapour compression system are connected in series

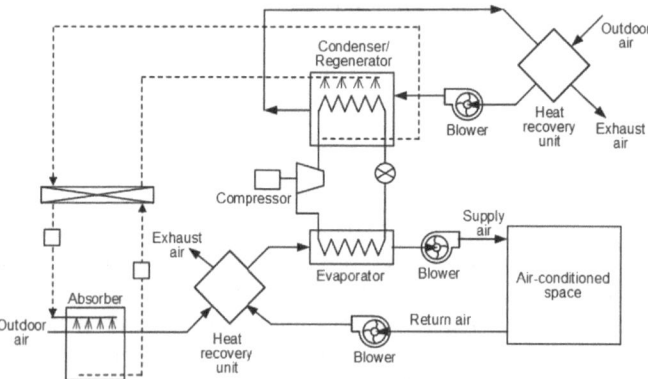

(b) Liquid desiccant system and vapour compression system are partly integrated

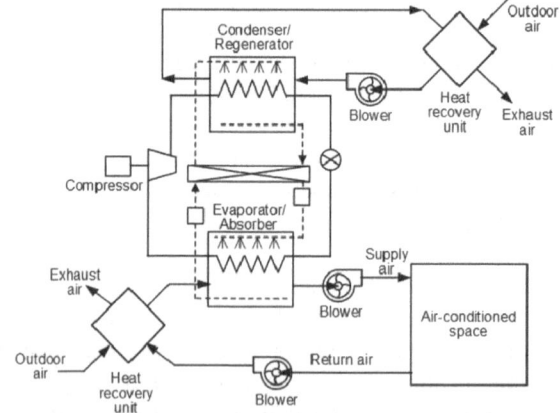

(c) Liquid desiccant system and vapour compression system are integrated

Fig. 5.23 Hybrid liquid desiccant air-conditioning systems **a** liquid desiccant system and vapour compression system are connected in series, **b** liquid desiccant system and vapour compression system are partly integrated since the condenser and regenerator are integrated into one unit, and **c** liquid desiccant system and vapour compression system are integrated since the condenser and regenerator as well as the evaporator and absorber are integrated into one unit

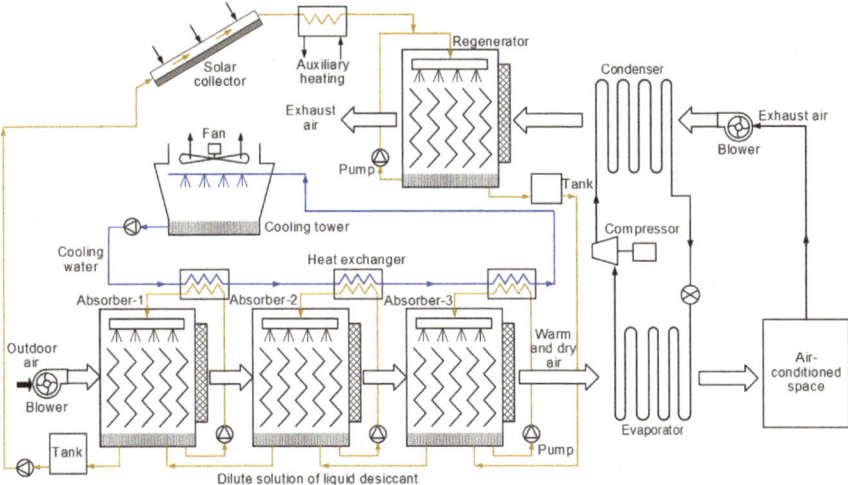

Fig. 5.24 Schematic diagram of the multistage liquid desiccant dehumidification system retrofitted with vapour compression cooling system

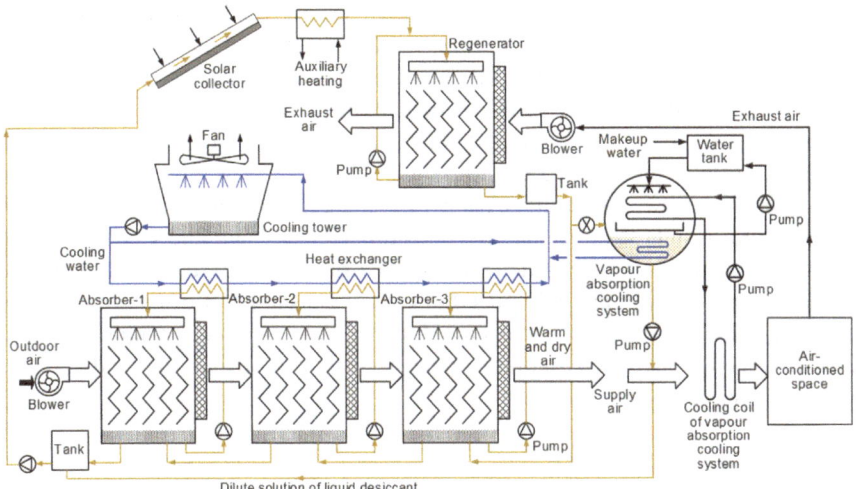

Fig. 5.25 Schematic diagram of the multistage liquid desiccant dehumidification system retrofitted with vapour absorption cooling system

supplying to the regenerator. As the exhaust air of low moisture content has the high driving potential of moisture absorption from the dilute desiccant solution, the circulation of exhaust air through the bed of regenerator would contribute remarkably in enhancing its regeneration performance. The overall coefficient of performance of a hybrid liquid desiccant dehumidification and vapour absorption cooling system is about 50% higher than that of a conventional standalone vapour absorption cooling system [83]. The renewable and low-grade energy could be fully utilized, with minimum consumption of electricity, in the hybrid liquid desiccant dehumidification, and vapour absorption cooling systems. The vapour compression and the vapour absorption cooling systems coupled with the multi-stage liquid desiccant dehumidification system are less affected by the ambient climate conditions as the vapour compression and the vapour absorption cooling systems handle only the sensible cooling load while multi-stage liquid desiccant dehumidification systems support the latent cooling load.

Zhang et al. [35, 72, 84] conducted a detailed study on the opportunities of retrofitting the hollow fibre membrane-based liquid desiccant dehumidification systems with the vapour compression air-conditioning system. They retrofitted first a two-stage membrane-based liquid desiccant dehumidifier system with a vapour compression refrigeration cycle. The system consisted of two membrane-based desiccant absorbers, two membrane-based desiccant regenerators, and a conventional vapour compression air-conditioning system. The concentrated desiccant solution was cooled by the evaporator of the vapour compression system before entering the membrane-based desiccant absorbers to form the internally cooled system. Similarly, the dilute desiccant solution was heated by the condenser of the vapour compression system before entering the membrane-based desiccant regenerator to form the internally heated system. The two-stage system generated a lower concentration of desiccant solution at the exit of absorbers. The COP of the system was about 4.5–5.5, which was about 20% higher than the standalone conventional vapour compression air conditioning systems. To match the dehumidification and cooling capacity, and increase the overall capacity of the system, they also developed another system, which consisted of two membrane-based desiccant absorbers, two membrane-based desiccant regenerators, two evaporators, and two condensers as shown in Fig. 5.26. Their experimental results showed that both dehumidification rate and COP decreased with higher inlet air temperature. Finally, they developed a genetic algorithm to optimize the operation and minimize the energy consumption of the system. The hourly optimal regulation of the dehumidification and cooling operations successfully reduced the energy consumption of the system by more than 20% under the hot and humid weather conditions.

5.10   Future Research Opportunities

Although the dehumidification capacity and the energy performance of the liquid desiccant dehumidification systems and the liquid desiccant air-conditioning systems are improved remarkably, the researchers and design engineers have identified the following challenges and limitations of the technology that should be addressed for further improvement of the overall performance of the systems:

Development of new generation mixed desiccant solution: Extensive research works have been carried out on the characterization and testing of the moisture absorption and regeneration performances of different single desiccant materials.

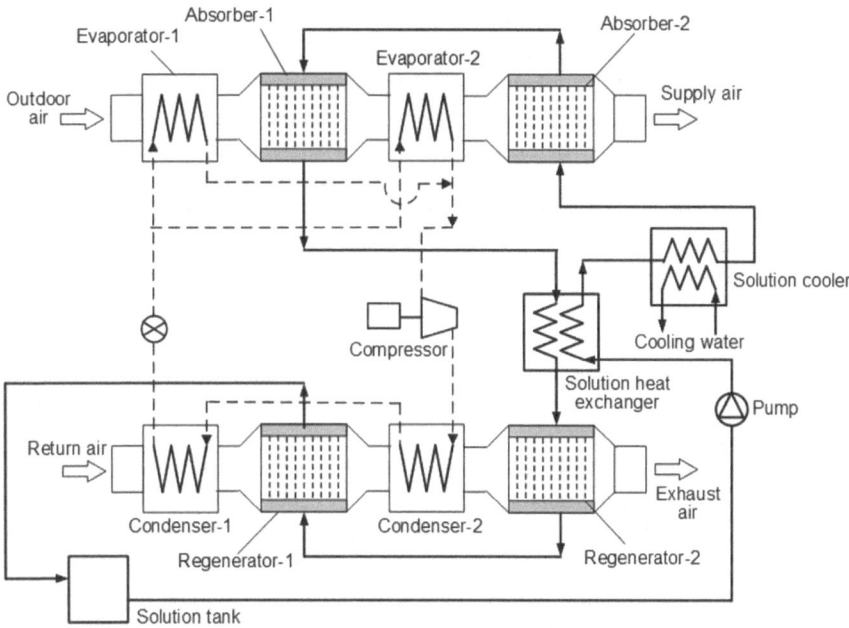

Fig. 5.26 Schematic diagram of hollow fibre membrane-based liquid desiccant dehumidification system retrofitted with the vapour compression air conditioning system

The desired properties of a good desiccant include low surface vapour pressure, high moisture absorption capacity, high moisture diffusivity, non-corrosive, low regeneration temperature, non-volatile, low viscosity, chemically stable, available, and cheap. The most widely tested used single desiccants are lithium chloride, lithium bromide, calcium chloride, and triethylene glycol, which do not have all the preferred properties of a good desiccant. Further research is needed to develop new desiccants to reduce the partial vapour pressure and increase the thermal performance of desiccant solutions. Several attempts have been made by the researchers to generate the new generation liquid desiccant solutions by mixing different desiccants and chemicals. Opportunities for the development of new generation mixed desiccant solutions possessing the preferred properties and long-term stability should be investigated further.

Packing materials: Moisture absorption and regeneration rate of liquid desiccants largely depend on the exposed surface generation performance of packing materials. Good wettability, non-reactive with desiccant materials, highly compact, high thermal conductivity, no carryover of liquid desiccant are other key characteristics of good packing materials. Plasma surface treatment with plasma nanolayer depositions, TiO_2 superhydrophilic self-cleaning coating, fibrous sheet attachment on the surfaces, grooving and coating with hydrophilic materials, etc., surface modification methods have been adopted to enhance the wettability of the packing materials. The development of low-cost and high durable coatings and micro/nanostructured hydrophilic surfaces are worthwhile to be investigated further. Maintaining good wettability particularly at a low desiccant flow rate remains a challenge for researchers and design engineers. The flow behaviour and the carryover potentials of liquid desiccants of different surface tensions under variable flow conditions should be analyzed to optimize the wettability of different structured and random packing materials. Further research is needed to develop the strategies of eliminating desiccant droplet carryover without compromising the capacity of the dehumidifiers. Maintenance guidelines for different treated and enhanced surface packing materials should be developed to minimize the degradation of performance throughout the lifespan. Limited design alternatives for the internally cooled packing materials have been developed. The influence of tube arrangement, design of fins with complex structures, and the wetting factor of fin surfaces should be further investigated. Research works should also be devoted to the development of high performance internally cooled packing materials.

Flow configuration of air and desiccant solution: Experimental and simulation work revealed that the dehumidification and regeneration performances of the counter-flow type dehumidifiers are generally better than that of the cross-flow type dehumidifiers under the similar practical operating conditions. In practice, the cross-flow type liquid dehumidifier tends to be preferred in commercial and residential buildings because of the limitations of floor height in most building applications. The moisture absorption rate gradually drops due to the decrease of the difference of partial vapour pressures of the air stream and desiccant solution along the direction of air flow for cross-flow dehumidifiers. Moreover, research results show that moisture absorption flux increases rapidly at the entry region of the desiccant solution and then drops progressively along the flow direction of the desiccant solution. The externally cooled multi-stage dehumidification systems [4, 53], multi-stage internal circulation liquid desiccant dehumidifiers [54], multi-stage dehumidifier with different desiccant solutions in different columns [55, 56, 57], and the concepts of localized cooling and heating of the packing materials [52] have the potential of developing the high-performance cross-flow type dehumidifiers. Further study on the optimization of local heat and mass fluxes is necessary to improve the dehumidification performance of the cross-flow liquid desiccant dehumidifier to achieve the performance comparable to that of the counter flow dehumidifier.

Detailed theoretical model: Several theoretical models of different levels of complexities have been developed to analyze the influences of different design and operating variables on the dehumidification performance of liquid desiccant dehumidifiers. However, the exact heat and moisture diffusion processes involved in the film of liquid desiccant, local distribution of the heat of absorption between the desiccant film, coolant and the air stream as well as the dynamic wetting behaviour of desiccant solution over the packing materials are still uncertain. It is necessary to develop the detailed theoretical models considering the local variations of the thermophysical properties of different desiccant solutions, surface characteristics of packing materials, and the features of enhancing the local heat and mass transfer coefficients. Design optimization of inserts is necessary by considering the trade-off among the enhancement of the local heat and mass transfer coefficients and pressure drop. The findings of the simulation will contribute to developing and optimizing new generation high-performance liquid desiccant dehumidifier systems.

Hybrid liquid desiccant air-conditioning system: Unlike solid desiccant systems, it is not necessary to switch the operation of the absorbers and regenerators of liquid desiccant dehumidification systems. As a result, a liquid desiccant dehumidifier can continuously supply dehumidified air of constant temperature and moisture content. Moreover, the absorbers and regenerators can be installed at strategic locations far from each other to avoid any short-circuiting issues such as the flow of warm and humid exhaust air of the regenerator to the absorber. The regenerators can be installed near the available heat sources and the generated concentrated solution can be circulated using a simple pumping system to the pre-cooling heat exchangers and absorbers located at various levels of the building. A hybrid liquid desiccant air-conditioning system can easily realize the independent control of the humidity and temperature of the circulating air and optimize the overall air-conditioning systems. However, the liquid desiccant air-conditioning system has the drawback of big dimension, which hinder its widespread applications. The researchers and design engineers should emphasis on the design of the compact hybrid system to widen the use of the energy smart liquid desiccant air-conditioning techniques.

Membrane-based liquid desiccant dehumidification systems: Membrane-based liquid desiccant dehumidification systems have received considerable attention in recent years due to the inherent selectivity and permeability properties of membranes that allow the moisture to flow from the stream of air to the desiccant solution while preventing the liquid desiccant to flow with the supply air. Membrane-based systems have been studied on the aspects of membrane materials, heat and mass transfer behaviour, energy and economic feasibility, environmental impacts, etc. The opportunities of reducing the resistance of moisture flow through the membrane and the effects of the membrane deformations on the thermal, momentum and mass transports both in the hollow fibre membrane dehumidifiers and flat plate membrane dehumidifiers should be studied. In-depth analysis of these studies and the thermodynamic aspects of the membrane systems are important for the design and optimization of the membrane-based dehumidification systems.

5.11 Conclusions

This chapter presented an overview of the opportunities for developing the liquid desiccant air-conditioning system as an energy-efficient alternative of the conventional vapour compression systems. An extensive literature review on the flow configurations and characteristics of the adiabatic and internally cooled dehumidifiers, thermophysical properties of widely adopted single and mixed desiccant solutions, wettability and effectiveness of different packing materials, findings of simulation and experimental studies, and the potential benefits of retrofitting the liquid desiccant dehumidification system with the vapour compression and vapour absorption systems have been conducted and coherently summarized.

The development of a theoretical model that involves coupled heat and mass transfer processes among the desiccant solution, cooling medium, and air stream of the dehumidifiers are systematically presented for easy reference. The usefulness of the theoretical model to evaluate the local heat and mass transfer rates as well as the variation of performance indices with the change of design and operating parameters are discussed. Researchers have used the findings from the simulations as their guides to design high-performance dehumidifiers.

The construction details, flow configurations, optimization opportunities, and the key findings of several single-stage and multi-stage innovative dehumidifier prototypes are discussed. Under the specific operating conditions, the maximum achievable moisture transfer effectiveness of about 94% and the moisture removal rate of about 5.6 g/s were reported for the polypropylene Rauschert Hi-flow rings and the Tripack No. ½ polypropylene spheres (random) packing materials, respectively. The performances of different design alternatives of the membrane-based dehumidifiers, that can eliminate the carryover issues of liquid desiccant, are elaborated. The design of multi-stage dehumidifiers using different desiccant solutions is discussed for the development of cost-effective high storage density dehumidifiers. With the view of developing energy-efficient liquid desiccant air-conditioning systems, the opportunities of retrofitting the single-stage and multi-stage dehumidifiers with the vapour compression and vapour absorption chiller systems are presented. Considering the recent technological development, the potential research directions including the development of new generation mixed desiccant solution, high-performance packing materials, optimization of the air stream flow configuration, and desiccant solution are also documented.

References

1. Isaac M, Vuuren DPV (2009) Modeling global residential sector energy demand for heating and air conditioning in the context of climate change. Energy Policy 37:507–521
2. Rafique MM, Gandhidasan P, Bahaidarah HMS (2016) Liquid desiccant materials and dehumidifiers—a review. Renew Sustain Energy Rev 56:179–195
3. Enteria N, Mizutani K (2011) The role of the thermally activated desiccant cooling technologies in the issue of energy and environment. Renew Sustain Energy Rev 15(4):2095–2122
4. Jiang Y, Li Z, Chen XL, Liu XH (2004) Liquid desiccant air conditioning system and its applications. HV&AC 34:88–98
5. Kessling W, Laevemann E, Kapfhammer C (1998) Energy storage for desiccant cooling systems component development. Sol Energy 64:209–221
6. Gomez-Castro FM, Schneider D, Pabler T, Eicker U (2018) Review of indirect and direct solar thermal regeneration for liquid desiccant systems. Renew Sustain Energy Rev 82:545–575
7. Conde MR (2004) Properties of aqueous solutions of lithium and calcium chlorides: formulations for use in air conditioning equipment design. Int J Therm Sci 43:367–382
8. Chaudhari SK, Patil KR (2002) Thermodynamic properties of aqueous solutions of lithium chloride. Phys Chem Liq 40–3:317–325
9. McNelly L (1979) Thermodynamic properties of aqueous solutions of lithium bromide. ASHRAE Transf 85:412–434
10. Kaita Y (2001) Thermodynamic properties of lithium bromide-water solutions at high temperatures. Int J Refrig 24:371–390

11. Wang G, Zeng M, Zhang Q (2020) Experimental investigation of saturated pressure and mass transfer characteristics of nano-lithium bromide solution. Int Commun Heat Mass Trans 115, Article 104605
12. Ertas A, Anderson EE, Kiris I (1992) Properties of a new liquid desiccant solution—lithium chloride and calcium chloride mixture. Sol Energy 49–3:205–212
13. Krolikowska M, Romanska K, Paduszynski K, Skonieczny M (2020) The study on the influence of glycols on the vapor pressure, density and dynamic viscosity of lithium bromide aqueous solution. Fluid Phase Equilib 519, Article 112640
14. Tsai C-Y, Soriano AN, Li M-H (2009) Vapour pressures, densities, and viscosities of the aqueous solutions containing (triethylene glycol or propylene glycol) and (LiCl or LiBr). J Chem Thermodyn 41–5, 623–631
15. Chen L-F, Soriano AN, Li M-H (2009) Vapour pressures and densities of the mixed solvent desiccants (glycol + water + salts). J Chem Thermodyn 41–6, 724–730
16. Chen S-Y, Soriano AN, Li M-H (2010) Densities and vapor pressures of mixed-solvent desiccant systems containing {glycol (diethylene, or triethylene, or tetraethylene glycol) + salt (magnesium chloride) + water}. J Chem Thermodyn 42–9, 1163–1167
17. Wen T, Lu L, Li M, Zhong H (2018) Comparative study of the regeneration characteristics of LiCl and a new mixed liquid desiccant solution. Energy 163:992–1005
18. Donate M, Rodriguez L, Lucas AD, Rodríguez JF (2006) Thermodynamic evaluation of new absorbent mixtures of lithium bromide and organic salts for absorption refrigeration machines. Int J Refrig 29:30–35
19. Gandhidasan P (2002) Prediction of pressure drop in a packed bed dehumidifier operating with liquid desiccant. Appl Therm Eng 22:1117–1127
20. Chung TW, Ghosh TK, Hines AL (1996) Comparison between random and structured packings for dehumidification of air by lithium chloride solutions in a packed column and their heat and mass transfer correlations. Ind Eng Chem Res 35–1:192–1988
21. Bravo JL, Rocha JA, Fair JR (1985) Mass transfer in gauze packings. Hydrocarbon Process 64–1:91–95
22. Ren H, Ma Z, Liu J, Gong X, Li W (2019) A review of heat and mass transfer improvement techniques for dehumidifiers and regenerators of liquid desiccant cooling systems. Appl Therm Eng 162:114271
23. Gandhidasan P (2003) Estimation of the effective interfacial area in packed-bed liquid desiccant contactors. Ind Eng Chem Res 42:3420–3425
24. Ren CQ, Tu M, Wang HH (2007) An analytical model for heat and mass transfer processes in internally cooled or heated liquid desiccant–air contact units. Int J Heat Mass Transf 50:3545–3555
25. Ali A, Vafai K, Khaled ARA (2004) Analysis of heat and mass transfer between air and falling film in a cross flow configuration. Int J Heat Mass Transf 47:743–755
26. Dai YJ, Zhang HF (2004) Numerical simulation and theoretical analysis of heat and mass transfer in a cross flow liquid desiccant air dehumidifier packed with honeycomb paper. Energy Convers Manage 45:1343–1356
27. Emhofer J, Beladi B, Dudzinski P, Fleckl T, Kuhlmann HC (2017) Analysis of a cross-flow liquid-desiccant falling-film. Appl Therm Eng 124:91–102
28. Nada SA (2017) Air cooling-dehumidification/desiccant regeneration processes by a falling liquid desiccant film on finned-tubes for different flow arrangements. Int J Therm Sci 113:10–19
29. Tu M, Ren C, Zhang L, Shao J (2009) Simulation and analysis of a novel liquid desiccant air-conditioning system. Appl Therm Eng 29:2417–2425
30. Luo Y, Chen Y, Yang H, Wang Y (2017) Study on an internally-cooled liquid desiccant dehumidifier with CFD model. Appl Energy 194:399–409
31. Huang SM, Zhong Z, Yang M (2016) Conjugate heat and mass transfer in an internally-cooled membrane-based liquid desiccant dehumidifier (IMLDD). J Membr Sci 508:73–83
32. Luo YM, Yang HX, Lu L, Qi RH (2014) A review of the mathematical models for predicting the heat and mass transfer process in the liquid desiccant dehumidifier. Renew Sustain Energy Rev 31:587–599

33. Shen S, Cai W, Wang X, Wu Q, Yon H (2016) Hybrid model for heat recovery heat pipe system in liquid desiccant dehumidification system. Appl Energy 182:383–393
34. Liu XH, Qu KY, Jiang Y (2006) Empirical correlations to predict the performance of the dehumidifier using liquid desiccant in heat and mass transfer. Renew Energy 31:1627–1639
35. Zhang N, Yin SY, Zhang LZ (2016) Performance study of a heat pump driven and hollow fiber membrane based two-stage liquid desiccant air dehumidification system. Appl Energy 179:727–737
36. Abdel-Salam AH, Ge G, Simonson CJ (2013) Performance analysis of a membrane liquid desiccant air-conditioning system. Energy Build 62:559–569
37. Islam MR, Wijeysundera NE, Ho JC (2006a) Heat and Mass effectiveness and correlations for counter-flow absorbers. Int J Heat Mass Transf 49:4171–4182
38. Islam MR, Alan SWL, Chua KJ (2018) Studying the heat and mass transfer process of liquid desiccant for dehumidification and cooling. Appl Energy 221:334–347
39. Patil KR, Tripathi AD, Pathak G, Katti SS (1990) Thermodynamic properties of aqueous electrolyte solutions. 1. Vapor pressure of aqueous solutions of lithium chloride, lithium bromide, and lithium iodide. J Chem Eng Data 35(2), 166–168
40. American Society of Heating, Refrigerating and Air-Conditioning Engineers (2009) ASHRAE handbook—fundamentals. 1791 Tullie Circle, N.E., Atlanta, GA 30329
41. Chaudahari SK, Patil KR (2002) Thermodynamic properties of aqueous solutions of lithium chloride. Phys. Chem. Liq 40(3):317–325
42. Incropera FP, DeWitt DP (2002) Fundamentals of heat and mass transfer. Wiley
43. Islam MR, Wijeysundera NE, Ho JC (2006b) Heat and mass effectiveness and correlations for counter-flow absorbers. Int J Heat Mass Transf 49:4171–4182
44. Liu XH, Zhang Y, Qu KY, Jiang Y (2006) Experimental study on mass transfer performances of cross flow dehumidifier using liquid desiccant. Energy Convers Manage 47:2682–2692
45. Bansal P, Jain S, Moon C (2011) Performance comparison of an adiabatic and an internally cooled structured packed-bed dehumidifier. Appl Therm Eng 31:14–19
46. Cho H-J, Cheon S-Y, Jeong J-W (2019) Experimental analysis of dehumidification performance of counter and cross-flow liquid desiccant dehumidifiers. Appl Therm Eng 150:210–223
47. Liu XH, Chang XM, Xia JJ, Jiang Y (2009) Performance analysis on the internally cooled dehumidifier using liquid desiccant. Build Environ 44:299–308
48. Yoon JI, Phan TT, Moon CG, Bansal IP (2005) Numerical study on heat and mass transfer characteristic of plate absorber. Appl Therm Eng 25:2219–2235
49. Islam MR (2003) Performance evaluation of absorbers for vapour absorption cooling systems, PhD Thesis, National University of Singapore
50. Islam MR, Wijeysundera NE, Ho JC (2003) Performance study of a falling-film absorber with a film-inverting configuration. Int J Refrig 26–8:909–917
51. Lun W, Li K, Liu B, Zhang H, Yang Y, Yang C (2018) Experimental analysis of a novel internally cooled dehumidifier with self-cooled liquid desiccant. Build Environ 141:117–126
52. Khan AY (1998) Cooling and dehumidification performance analysis of internally-cooled liquid desiccant absorbers. Appl Therm Eng 18–5:265–281
53. Li Z (2003) Liquid desiccant air conditioning and independent humidity control air conditioning systems. HV&AC 33:26–31
54. Cheng X, Rong Y, Zhou X, Gu C, Zhi X, Qiu L, Yuan Y, Wang K (2020) Performance analysis of a multi-stage internal circulation liquid desiccant dehumidifier. Appl Therm Eng 172:115163
55. Xiong ZQ, Dai YJ, Wang RZ (2010) Development of a novel two-stage liquid desiccant dehumidification system assisted by $CaCl_2$ solution using exergy analysis method. Appl Energy 87:1495–1504
56. Xiong ZQ, Dai YJ, Wang RZ (2009) Investigation on a two-stage solar liquid-desiccant (LiBr) dehumidification system assisted by $CaCl_2$ solution. Appl Therm Eng 29:1209–1215
57. Kumar R, Dhar PL, Jain S, Asati AK (2009) Multi absorber stand alone liquid desiccant air-conditioning systems for higher performance. Sol Energy 83:761–772
58. Patnaik S, Lenz TG, Lof GOG (1990) Performance studies for an experimental solar open-cycle liquid desiccant air dehumidification system. Sol Energy 44(3):123–135

59. Chung TW, Ghosh TK, Hines AL (1993) Dehumidification of air by aqueous lithium chloride in a packed column. Sep Sci Technol 28(1–3):533–550
60. Fumo N, Goswami DY (2002) Study of an aqueous lithium chloride desiccant system: air dehumidification and desiccant regeneration. Sol Energy 72(4):351–361
61. Longo GA, Gasparella A (2005a) Experimental and theoretical analysis of heat and mass transfer in a packed column dehumidifier/regenerator with liquid desiccant. Int J Heat Mass Transf 48(25–26):5240–5254
62. Dong C, Qi R, Lu L, Wang Y, Wang L (2017) Comparative performance study on liquid desiccant dehumidification with different packing types for built environment. Sci Technol Built Environ 23(1):116–126
63. Chen T, Dai Z, Yin Y, Zhang X (2017) Experimental investigation on the mass transfer performance of a novel packing used for liquid desiccant systems. Sci Technol Built Environ 23(1):46–59
64. Kumar R, Dhar PL, Jain S (2011) Development of new wire mesh packings for improving the performance of zero carryover spray tower. Energy 36(2):1362–1374
65. Moon CG, Bansal PK, Jain S (2009) New mass transfer performance data of a crossflow liquid desiccant dehumidification system. Int J Refrig 32(3):524–533
66. Oberg V, Goswami DY (1998) Experimental study of the heat and mass transfer in a packed bed liquid desiccant air dehumidifier. J Sol Energy Eng 120(4):289–297
67. Zurigat YH, Abu-Arabi MK, Abdul-Wahab SA (2004) Air dehumidification by triethylene glycol desiccant in a packed column. Energy Convers Manage. 45(1):141–155
68. Longo GA, Gasparella A (2005b) Experimental and theoretical analysis of heat and mass transfer in a packed column dehumidifier/regenerator with liquid desiccant. Int J Heat Mass Transf 48(25–26):5240–5254
69. Huang SM, Zhang LZ, Tang K, Peil X (2012) Fluid flow and heat mass transfer in membrane parallel-plates channels used for liquid desiccant air dehumidification. Int J Heat Mass Transfer 55(9–10):2571–2580
70. Xiao L, Yang M, Yuan W-Z, Huang S-M (2020) Performance characteristics of a novel internally-cooled plate membrane liquid desiccant air dehumidification system. Appl Therm Eng 172:115193
71. Qiu D, Wu Z, Huang SM (2017) Laminar flow and heat transfer in an internally-cooled hexagonal parallel-plate membrane channel (IHPMC). Appl. Therm. Eng. 124:767–780
72. Zhang N, Yin SY, Li M (2018) Model-based optimization for a heat pump driven and hollow fiber membrane hybrid two-stage liquid desiccant air dehumidification system. Appl Energy 228:12–20
73. Huang SM, Zhang LZ (2013) Researches and trends in membrane-based liquid desiccant air dehumidification. Renew. Sustain. Energy Rev. 28:425–440
74. Chu J, Zhu J, Bai H, Cui Y (2019) Experimental study of a membrane-based liquid desiccant dehumidifier based on internal air temperature variation. Appl Therm Eng 159:113936
75. Gurubalan A, Maiya MP, Tiwari S (2019) Experiments on a novel membrane-based liquid desiccant dehumidifier for hybrid air conditioner. Int J Refrig 108:271–282
76. Mansuriya K, Rajab BD, Patela VK (2020) Experimental assessment of a small scale hybrid liquid desiccant dehumidification incorporated vapor compression refrigeration system: an energy saving approach. Appl Therm Eng 174:115288
77. Mohan BS, Tiwari S, Maiya MP (2015) Experimental investigations on performance of liquid desiccant-vapor compression hybrid air conditioner. Appl Ther Eng 77:153–162
78. Cuce PM (2017) Thermal performance assessment of a novel liquid desiccant-based evaporative cooling system: an experimental investigation. Energy Build 138:88–95
79. Cuce PM, Riffat S (2016) A state of the art review of evaporative cooling systems for building applications. Renew Sustain Energy Rev. 54:1240–1249
80. Mucke L, Fleig D, Vajen K, Jordan U (2016) Hybrid liquid desiccant air-conditioning systems: A conceptual study with respect to energy saving potentials. Int J Refrig 69:64–73
81. Mei L, Dai YJ (2008) A technical review on use of liquid-desiccant dehumidification for air-conditioning application. Renew Sustain Energy Rev 12:662–689

82. Yadav YK, Kaushik SC (1991) Psychometric techno-economics assessment and parametric study of vapour compression and solid/liquid hybrid air-conditioning system. Heat Recov CHP 991(11–6):563–572
83. Ahmed CSK, Gandhidasan P, AL-Farayedhi AA (1997) Simulation of a hybrid liquid desiccant based air conditioning, system. Appl Therm Eng 17–2, 125–134
84. Zhang N, Yin SY (2017) Investigation on capacity matching in a heat pump and hollow fiber membrane-based two-stage liquid desiccant hybrid air dehumidification system. Int J Refrig 84:128–138

Chapter 6
Membrane Air Dehumidification

6.1 Introduction

Heating, Ventilation and Air-Conditioning (HVAC) has been widely employed to control both supply air temperature and humidity in order to provide indoor thermal comfort as well as sustain a hospitable environment for storing goods and equipment in residential, industrial, and commercial buildings. In tropical countries, the total energy consumed by the HVAC system is mainly due to the direct cooling process carried by vapour compression chillers. Outdoor air is passed over the cooling coils of the Air Handling Unit (AHU) to remove both sensible and latent heats. The cooled and dehumidified air is then reheated or mixed with the return air to raise its temperature to the human thermal comfort level. Often than not, due to the air's high humidity, chillers need to work harder to ensure heat exchangers operate below the air's dew point temperature for moisture condensation to take place. Excessive energy consumption during these deep cooling steps makes the present coupled cooling and dehumidification process an inefficient one [1].

It is, therefore, apparent that decoupling the latent load from the chiller's duty would significantly improve chiller efficiency where chillers are applied only for the sole purpose of sensible cooling. Thus far, attempts have been conducted to handle the latent load using solids and liquid desiccant dehumidifiers [2–5]. The overall efficiencies of these processes are low because these systems require excessive energy to regenerate the desiccant at high temperatures. Additionally, the air can be contaminated with undesired desiccating particles that are potentially entrained in the air stream.

Of recent years, vacuum-based membrane dehumidification (VMD) has gained significant research attention and traction [6–14]. In this process, the air passes over a membrane surface at normal pressure. A vacuum pressure is applied on the opposite side of the membrane to create a driving force for water to permeate through the membrane. Humid air is dehumidified without undergoing any temperature change. To achieve thermal comfort conditions, the dried air is then cooled down to thermal comfort levels with minimal energy consumption via a conventional vapour

© Springer Nature Singapore Pte Ltd. 2021

C. Kian Jon et al., *Advances in Air Conditioning Technologies*, Green Energy and Technology, https://doi.org/10.1007/978-981-15-8477-0_6

compression chiller. This isothermal dehumidification process is considered to be 'green' since no heat source is needed for thermal regeneration, resulting in minimal environmental emission [8].

Over the last two decades, there have been many attempts to develop highly perm selective membranes for air or gas dehumidification [9, 10, 14, 15]. Many types of polymer [10–14, 16–26], inorganic [8, 10, 27], liquid [6, 28–30], and mixed matrix [31, 32] membranes have been explored. Among them, dense polymeric membranes have received a significant level of attention due to their low cost, lightness, physical robustness, and ease of fabrication and modification. Mass transport in dense polymer materials is based on the solution diffusion mechanism [9, 10, 12, 16, 33–35]. Water molecules selectively permeate through the membranes due to both its smaller kinetic diameter as well as greater condensability compared with other gases [13]. For low temperature and working conditions in ventilation and air conditioning, stable performance of the polymeric membrane can be achieved [9, 10, 15].

The objective of the chapter is to succinctly summarize the current state of knowledge on the MAD for the researchers to create a complete understanding of MAD by discussion the findings presented in recent research on the MAD. To reach the goal, first, state of the art of the MAD process and its working principle is first presented, followed by the characteristics and configurations of the membranes used in the MAD systems. Then, the chapter further provides a detailed assessment of system design, modelling, and performance. It is concluded by a discussion of the research needs in the future.

6.2   Fundamental Principles of Membrane Dehumidification

Recently, vacuum-based membrane dehumidification (VMD) is being closely studied as an efficient alternative to remove moisture from humid air. In this process, the air is passed over a membrane surface at normal pressure. A vacuum pressure is applied on the opposite side of the membrane to create a driving force for water to permeate through the membrane. Such air conditioning process is illustrated as A-B-C pathway on psychrometric chart in Fig. 6.1. First, humid air is dehumidified without any temperature change from A to B. Then, the dried air is cooled down to the thermal comfort level with minimal energy consumption from B to C. In contrast to desiccant dehumidification, the membrane's isothermal dehumidification is considered to be an energy-efficient process because no heat source is needed for thermal regeneration, resulting in minimal environmental emission [4]. Figure 6.2 shows, in general, the membrane dehumidification process comprising the feed, permeate and retentate flows. The vapour of the humid air from the feed stream is selectively sieved out to form the permeate side and the rest of the air is released as the retentate stream for space dehumidification.

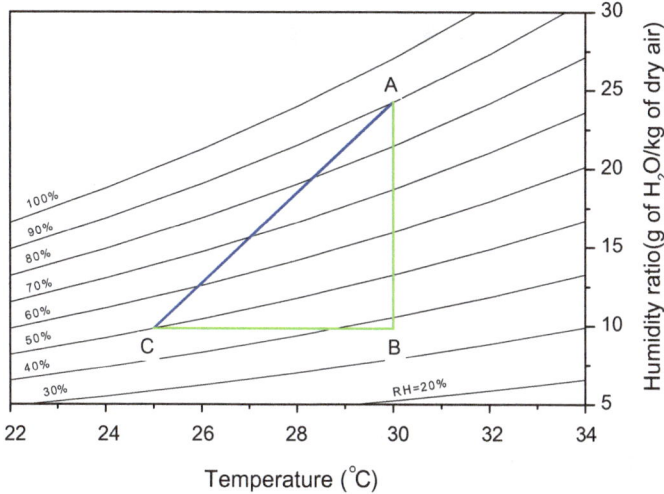

Fig. 6.1 The air conditioning processes from input condition A (at temperature $T_A = 30\ °C$ and relative humidity $RH_A = 90\%$) to desired condition C (at temperature $T_B = 25\ °C$ and relative humidity $RH_C = 50\%$)

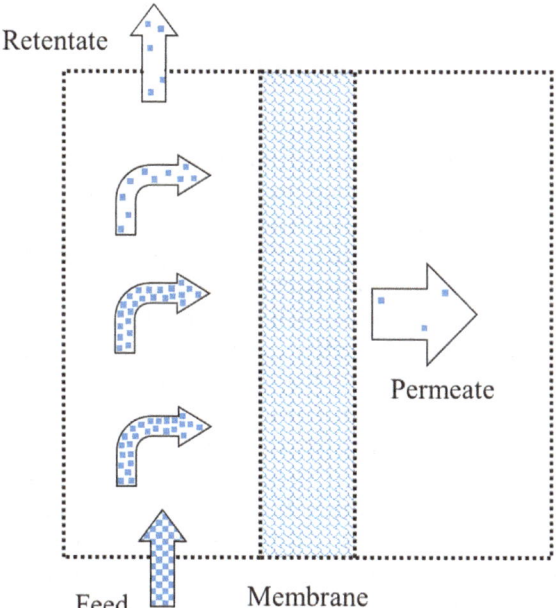

Fig. 6.2 Schematic diagram of the membrane dehumidification process

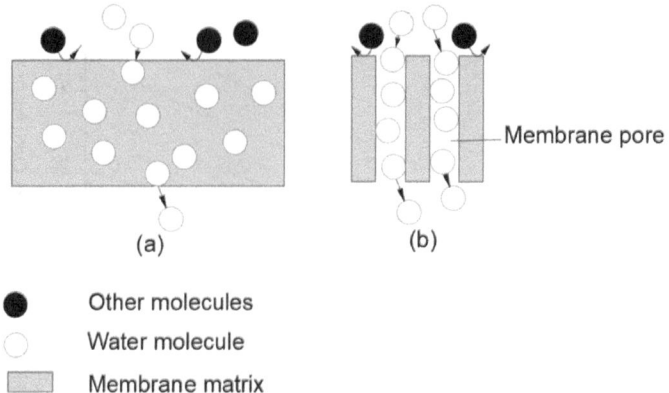

Fig. 6.3 Separation mechanisms of dehumidification membrane. **a** Solute diffusion in dense membrane. **b** Molecular sieving in microporous membrane

One salient feature of the MAD process is molecular separation functions of the membrane. The membrane sheet allows permeation of water vapour molecules while blocking other molecules as shown in Fig. 6.3. The vapour-phase membrane separation is a steady-state process operating under constant temperature and pressure (isothermal and isobaric process) in comparison to periodic switches between adsorption (or absorption) and regeneration required for the solid (or liquid) desiccating process.

Conventionally, MAD can be classified into two groups as portrayed in Fig. 6.3—dense and microporous. The dense membrane provides selective absorption of water over other molecules. The separation mechanism can be described by the solute diffusion model. Most polymeric membrane materials fall into this category, which include polyvinyl alcohol (PVA), chitosan, alginate, polysulfone (PSF), polyimides, polyamides, polyaniline, and polyelectrolyte membranes such as cation-exchanged Nafion. Separation in a microporous membrane is mostly achieved through molecular sieving based on molecular sizes and/or selective adsorption of water at entrance of the membrane pore. Microporous zeolite, silica, and carbon are common membrane materials. Water vapour molecule is often the smallest one in a mixture, and selective separation can be realized by making the membrane pore size small enough. However, it is worthy to note that the small pore size reduces molecule diffusion rates and may have a negative impact on vapour flux. Molecular sieving based on selective water adsorption enables the achievement of both high flux and selectivity.

A typical cross-flow membrane dehumidification setup is shown in Fig. 6.4(I). In this setup, the humid air is passed over the feed side of a working membrane module at ambient pressure. A vacuum pressure is applied on the permeate side of the membrane to create a driving force for the water vapour permeation to take place through the membrane. Water vapour is selectively sieved out of the air stream, pumped and discharged to the ambient by a mechanical means such as a vacuum

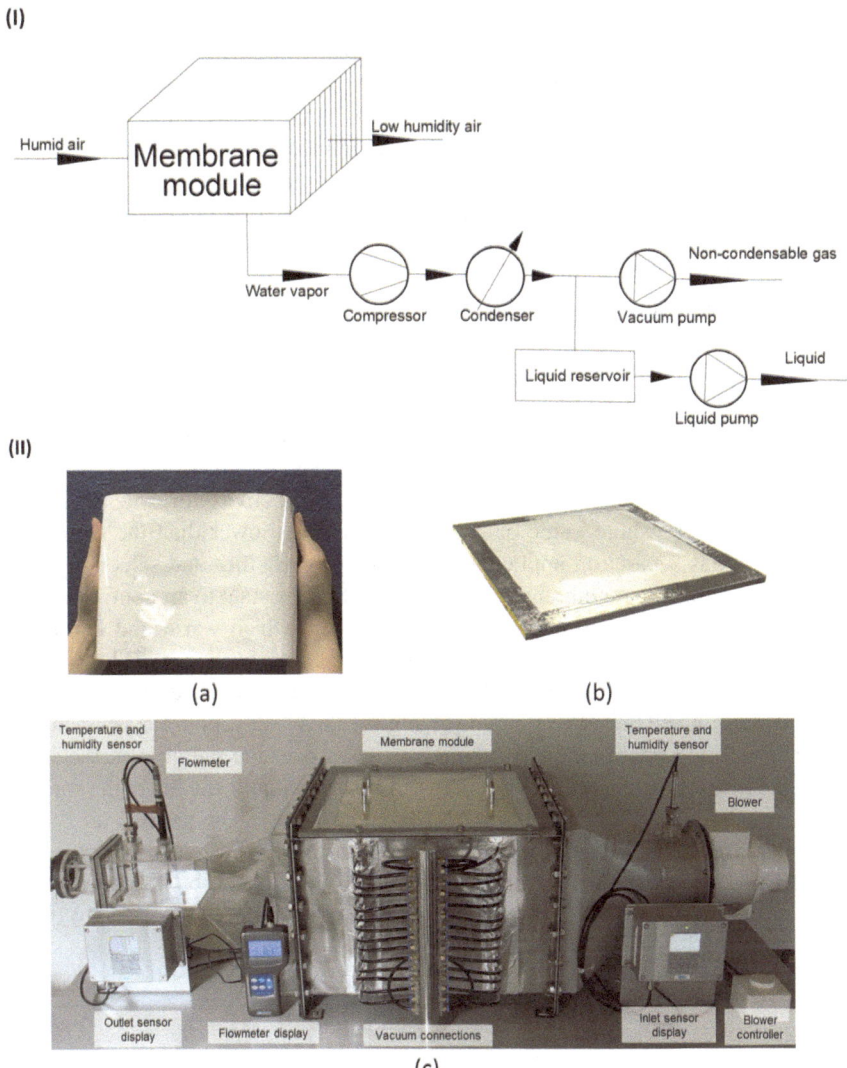

Fig. 6.4 (I) Schematic of a detailed dehumidification process of atmospheric air using membrane technology; and (II) (**a**) laboratory synthesized coated flat-sheet membrane; (**b**) water vapour pressure profile along the working membrane

pump. Figure 6.4 (II) further shows a pictorial perspective of the setup with laboratory a synthesized membrane module ready to be tested for its performance.

In simpler terms, the membrane employed for membrane in the MAD (Membrane Air Dehumidification) is a selective layer allowing some certain components to pass through but stops others, resulting in separating the components from the mixture. The membranes used for MAD separate vapour gas from the humid air. The driving force for vapour transfer is the chemical potential gradient between the feed side and the permeate side of the selective membrane [8–10].

Membrane permeability and selectivity are two critical characteristics that can be employed to determine the performance of the MAD in separating water vapour from the dried air [8, 12]. Permeability is often defined as the rate at which vapour permeates through a membrane per unit area or per unit driving force. A membrane with a high permeability to the vapour allows the vapour to pass more rapidly than the other gases. It facilitates same separation with less surface area. Selectivity is defined as the ratio of the permeability coefficient of vapour and the others in the humid air through the membrane. A membrane with a high selectivity allows vapour only to pass through the membrane, resulting in pure water vapour in the permeate side [8]. For air dehumidification, the purity of permeated vapour is not highly required. Based on the findings from Bui et al., the MAD can result in energy savings by using a relative low selective membrane with no requirement of a high purity [12]. Xing et al. [8] also predicted that a 50% or higher energy efficiency gain over a conventional vapour compression system was attainable when the selectivity for water vapour to air was above a threshold value of 200, which is much lower than the selectivity level used in gas separation application. Besides permeability and selectivity, it is worthy to note that the membrane for MAD should be resistant to air contamination, mechanical erosion, and bacteria attachment and growth in warm and humid air environment, as well as cost-effective to achieve better performance [36].

Figure 6.5a, b further illustrates how a membrane can be set up to conduct air dehumidification and the corresponding change of the water vapour pressure change along the working membrane. As humid air passes over the membrane, the water vapour pressure of the air stream gradually lowers. Essentially, the lower the vacuum pressure is, the more water vapour is removed and the drier the product air is. The work required in this dehumidification process is mainly the electrical power to compress the water vapour from the vacuum pressure to the ambient pressure. The energy efficiency (normally expressed as coefficient of performance (COP)) of the

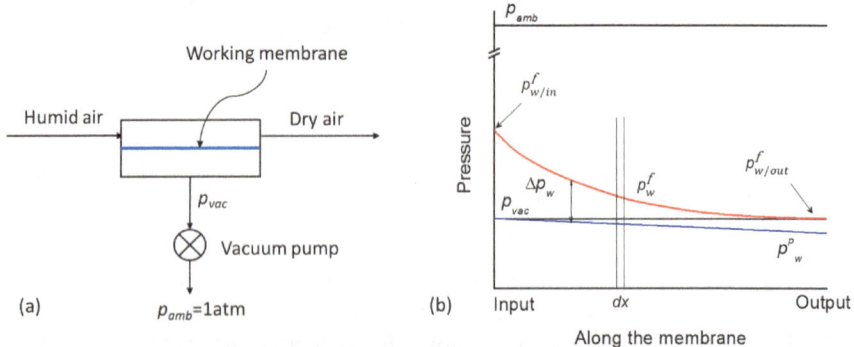

Fig. 6.5 The working mechanism of membrane air dehumidification: **a** typical component setup; and **b** water vapour pressure profile along the working membrane

dehumidification process is appropriately ratio of latent heat removed to the pump's work. It is apparent that without involving any phase change, VMD (Vacuum-assisted Membrane Dehumidification) potentially has higher COP than condensing water vapour by conventional processes. During VMD, dehumidification COPs spanning 2–3.5 has been reported for certain membranes and operation conditions [1, 4–6].

Beside energy efficiency (COP), the membrane's dehumidification performance needs to be considered in evaluating the overall system efficiency. The dehumidification performance is expressed as the ratio of water vapour removed to the water vapour in inlet air. Recently, Bui et al. have reported that the performance of a membrane system depends strongly on a few key parameters, namely, permeate pressure, velocity and RH of the feed air, permeability, and selectivity of the working membrane [6].

Water vapour permeability and selectivity have earlier been mentioned as two important transfer characteristics that can be employed to compare the performance of dehumidification membranes. Water permeability (P) is defined as the amount of water going through a unit of thickness and a unit of area of the membrane, under a unit transmembrane pressure and in a unit of time. It is usually expressed in Barrer, (1 Barrer $= 10^{-10}$ cm^3(STP) cm/cm^2 s cmHg $= 3.348.10^{-16}$ mol m/m^2 s Pa). The selectivity of a water towards N$_2$ is well defined as the ratio of their permeabilities.

The following sections provide the fundamental basis on how dehumidification takes place across a membrane sheet. First, a humid air stream is introduced to the feed side of the membrane at ambient pressure, p_{amb}. A vacuum pressure, p_{vac}, is applied on the permeate side of the membrane to create a driving force for water vapour to be selectively sieved out of the air stream, compressed, and discharged to the ambient by a vacuum pump. The water vapour pressure of the feed air stream (p_w^f) is gradually lowered as it passes over the membrane from its input value ($p_{w/in}^f$) to its output value ($p_{w/out}^f$) as portrayed in Fig. 6.5b. Lowering p_{vac} leads to more water vapour being removed and hence a drier product air.

Water vapour permeance and selectivity are two important mass transport properties of a membrane. For a membrane with water vapour and air permeances of P_w and P_a, respectively, the membrane's selectivity, S, is

$$S = \frac{P_w}{P_a} \tag{1}$$

Assuming that the contents of water and air are constant on both sides of the membrane along a small length increment of dx, the respective water vapour and air fluxes are [39]:

$$df_w = Pw(p_w^f - p_w^p)dx \tag{2a}$$

$$df_a = Pa(p_a^f - p_a^p) \tag{2b}$$

where p_w^f and p_w^p are water vapour partial pressures in feed and permeate streams, respectively, and p_a^f and p_a^p are air partial pressures in feed and permeate streams, respectively. Conducting a simple mass balance, the ratio of water vapour partial pressure to air partial pressure on the permeate side is equal to the ratio of water vapour flux to air flux shown as

$$\frac{p_w^p}{p_a^p} = \frac{df_w}{df_a} = S \frac{p_w^f - p_w^p}{p_a^f - p_a^p} \tag{3}$$

From Eq. (6.3) and the boundary conditions for water vapour partial pressures at the entrance and exit, a 1D model was developed. Water vapour and air fluxes through the entire membrane are:

$$F_w = \int df_w dx = P_w \int (p_w^f - p_w^p)dx \tag{4a}$$

$$F_a = \int df_a dx = P_a \int (p_a^f - p_a^p)dx \tag{4b}$$

6.3 Membrane Material Types

Membranes used in air dehumidification are closely related to the membranes used in other dehydration applications such as dehydration of compressed air, natural gas, flue gas and per-vapouration. Many types of organic and inorganic membranes have been explored for these industrial processes. In general, all the membranes have porous in spite of different sizes. According to the size of porous, the membranes used in MAD can be classified into two different types: dense and porous. Typically, the dense membranes have pore sizes in the order of 0.1 nm, while the porous membranes have pore sizes in the order of 0.1 μm [37]. Besides the size of pores in the membrane, the internal structure and arrangement of the pores are also employed to category the type of membranes. Specifically, the membranes can be grouped in two categories, namely, symmetrical or asymmetrical types [38]. A symmetrical membrane is uniform in composition, structure, and pore sizes. In contrast, an asymmetrical membrane is chemically or physically heterogeneous asymmetric. To optimize membranes' production productivity, many membranes are synthesized with the goal of minimizing their thickness and maximizing their area. Accordingly, membranes are usually either asymmetric or composite to minimize the thickness. In the ensuing sections, the common materials used to synthesize membranes for air dehumidification will be discussed. They include four main categories, namely, polymer, composite, zeolitic, and supported liquid.

6.3.1 Polymer Membranes

Polymeric membranes have received a significant amount of attention due to their low cost, lightness, physical robustness and ease of fabrication and modification. Among the different polymers, the hydrophilic organic polymers such as cellulose acetate (CA), ionic polymers, polyvinyl alcohol (PVA), polysulfone (PSF), and polyacrylonitrile [39, 40], are generally selected because the hydrophilic function in these polymer chains enhances the water solubility in the membrane. The diffusion of water molecules, therefore, is faster than the diffusion of other gases. Table 6.1 portrays the water vapour and nitrogen's permeability and selectivity of several polymer materials [41]. It provides a comprehensive perspective of the different polymers that are can be synthesized into membranes for air dehumidification. For low temperature and gentle working conditions in ventilation and air conditioning, stable performance of the polymeric membrane can be achieved. Mass transport in polymer materials is based on the solution diffusion mechanism. Gas molecules absorb on the membrane surface, diffuse through the membrane and then desorb at the opposite surface. Gas permeability through a membrane is a result of the gas dissolution (or sorption) and diffusion in the membrane material. The molecular interaction between the membrane material and permeant determines the permeation rate. Separation is achieved by the different degrees to which components are dissolved in and diffuse through the polymer.

As depicted in Table 6.1, all polymers having high water vapour permeability and selectivity are hydrophilic materials. Water molecule preferentially permeates through the membranes due to both its smaller kinetic diameter and higher condensability than the other gases. Most of the highly permeable polymers also have a very high selectivity. Additionally, the diffusivity of water greatly depends on its activity in air and its content polymer matrix. Higher water content in air can result in higher amount of adsorbed water in polymer matric and higher water permeability due to the plasticization. This effect of water vapour activity has been observed in almost all the materials. Highest increase of 42 times was reported in PVA membrane when water vapour activity increases from 0.5 to 0.9.

Polymer structure has a great effect on water vapour permeability. Non-polar membranes such as natural rubber, polystyrene, polypropylene and polyethylene have low permeabilities. The presence of strong polar functional groups such as ether or sulphonate groups markedly improves both water permeability and selectivity. The highest water vapour permeability and selectivity are reported in Nafion membranes. The unique properties of Nafion are a result of incorporating highly polar perfluorovinyl ether groups with sulfonate groups onto a hydrophobic tetrafluoroethylene backbone. The infusion of the sulphonate group into polyether ether keton, polysulfone, and polyimide structures produces similar result. Sulfonated polymers constitute a class of material that has high permeability and selectivity properties. Block polymers containing hydrophilic soft segments of poly (ethylene oxide) also exhibit high water vapour permeability and selectivity.

Table 6.1 Water vapour and gases' permeability of several polymer materials [41]

Polymers	T (°C)	RH (%)	P (barrer)		Selectivity (P_{H_2O}/P_{N_2})
			Water vapour	Nitrogen	
Nafion 115	30	90	1,905,000	0.27	7,055,556
Nafion 117	50	30–90	410,000–1,800,000	0.27	1,518,500–6,666,666
Sulfonated polyimides (NTDA–DMBDSA/BAPF (9/1))	50	50–85	300,000–610,000	–	–
Sulfonated polyimides (NTDA–ODADS/ODA (3/1))	50	50–85	190,000–330,000	–	–
Sodium sulfonated poly(ether ether ketone) (sulfonation degree of 120 to 200%)	30–52	–	40,000–150,000	0.005–0.038	1,052,632–30,000,000
Sodium-Sulfonated Poly(ether ether ketone)s (sulfonation degree of 120 and 160%)	30	–	39,000 (SD = 120) 60,100 (SD = 160)	–	3,430,000 (SD = 120) 10,500,000 (SD = 160)
Sulfonated poly(ether ether ketone) (ion exchange capacity 1.9 and 1.6 meq/g)	30–70	15–100	12,000–450,000 (1.9 meq/g) 5000–55,000 (1.6 meq/g)	0.029 (1.9 meq/g) 0.018(1.6 meq/g)	1,034,000–8,621,000(1.9 meq/g) 500,000–2,222,000(1.6 meq/g)
Sulfonated poly(ether ether ketone) (sulfonation degree of 60%)	30	32–93	20,000–490,000	0.01–0.18	300,000–2,000,000
Sulfonated polyetherketone with Cardo (SPEK-C)	–	–	2800–10,000	0.04–0.08	26,000–200,000
Sulfonated polyetherethersulfone (SPES)	30	–	4000–15,000	0.03–0.2	15,000–200,000

(continued)

Table 6.1 (continued)

Polymers	T (°C)	RH (%)	P (barrer)		Selectivity (P_{H_2O} / P_{N_2})
			Water vapour	Nitrogen	
Sulfonated poly(amide-imide)	30	–	4966–9767	0.018–0.027	183,000–496,000
Blend of polyether block amide (nylon12) and poly (ethylene oxide) (PEBAX 1074)	30	32–93	75,000–200,000	2.5	20,000–100,000
poly(ethylene oxide) poly(butylene terephthalate) block copolymers (1000PEO56PBT44)	30	30–82	93,846–192,366	2	46,920–96,180
Poly(ethylene oxide)-ran-poly(propylene oxide)	50	43–100	70,000–400,000	4.5–22	5000 (SS10000)–45,000 (SS2500)
Ethyl Cellulose	25	23–100	22,198–28,681	3.3	6730–8690
Cellulose Acetate	25–30	0–100	6000–15,000	0.25	24,000–60,000
Polyimides with fluorine-containing 6FDA dianhydride	35	–	2450–4400	–	5000–30,000 (CH4)

(continued)

Table 6.1 (continued)

Polymers	T (°C)	RH (%)	P (barrer)		Selectivity (P_{H_2O}/P_{N_2})
			Water vapour	Nitrogen	
poly(phenylene oxide)	30	–	4060	3.8	1068
Polyether-polyurethanes	22	–	1760 (PU400)–2340 (PU2000)	–	30,000 (PU400)–3000 (PU2000) (CH$_4$)
Polysulfone	40	20–80	2000–2100	0.25	8000–8400
Nylon 6	25	5–96	400–1400	0.025	16,000–56,000
Polycarbonate	25–80	20–100	1040–1125	–	–
Polycarbonate	30	–	1400	0.29	4830
Polyvinyl alcohol	38	50–90	21–880	0.00057	36,800–1,544,000
Polyacrylonitrile	30	–	300	–	1,000,000 (O$_2$)
Polyimide (Kapton)	30	–	640	0.00012	5,333,333
Polymethacrylonitrile	30	–	410	–	341,670 (O$_2$)
Polyvinylidene chloride	30	–	340	–	64,150 (O$_2$)

6.3.2 Composite Membranes

Albeit polymers being able to render high permeability and selectivity to membranes, they are rarely used as a single layer membranes in vacuum-based membrane dehumidification because of its low mechanical strength. Many of the membranes employed for the purpose of air dehumidification, thus far, have been constructed based on a composite structure. Composite membranes are multi-layer membranes comprising of support layers and an active layer. The support layers provide the membrane the properties of high mechanical strength and chemical stability. It is usually made of mesoporous or/and microporous materials. The active layer is a thin permselective layer that characterizes both permeability and selectivity of the composite membrane. A desired composite membrane is comprised of a highly microporous substrate and a thin, dense selective layer. Composite membranes have two main forms: flat-sheet and hollow-fibre membranes. Flat-sheet membranes can be packed in modules similar to plate-and-frame heat exchanger. In these modules, vacuum and air channels are arranged alternately. Hollow fibre membranes usually packed in modules similar to shell-and-tube heat exchangers. In these modules, vacuum is applied on the lumen side and air flows on the shell side inside, or vice versa. Hollow structure enables fibre membrane to withstand a large transmembrane pressure difference. However, supplying the air to the lumen side of the hollow fibres can lead to large pressure losses, while supplying the air to the shell side may result in poor flow distribution.

In order to quantify the water vapour permeation through a composite membrane, water vapour permeance is normally used instead of permeability. Permeance is expressed in GPU (1 GPU = 10^{-6} cm^3 (STP)/cm^2 s cmHg = $3.348.10^{-10}$ mol/m^2 s Pa). The selectivity of a permeant towards another permeant is also defined as the ratio of their permeances. Table 6.2 shows structure and water vapour and nitrogen's permeance and selectivity of several composite membranes with high water vapour permeance and selectivity for air dehumidification.

In a typical composite membrane, both the polymer and the inorganic fillers are selective to water vapour in the humid air, but the inorganic fillers typically have significantly higher selectivity than the neat polymer. This hybrid method improves membrane selectivity significantly. A few researchers investigated the performance of composite membrane for vapour permeation. Cheng et al. have used gelatin-silica and PVA to fabricate a composite membrane for separating vapour from the mixture of propylene and the vapour. The silica-based MMM was reported to achieve a seventime higher vapour permeance and a 14-time higher separation factor simultaneously [42]. Zhou et al. fabricated the polyurethane/TiO2 or SiO$_2$ nanohybrid membranes and found the nanohybrid membrane materials demonstrated two times or more water vapour permeability than the pure polyurethane membranes, especially under thermal stimulation [43]. Bui et al. fabricated a MMM by dipping a twilled Dutch weave stainless steel mesh scaffold, fine and porous TiO$_2$, in a hydrophilic solution of polyvinyl alcohol and lithium chloride at various ratios. The measured data showed

Table 6.2 Several composite membranes with high water vapour permeance and selectivity for air dehumidification [41]

Active layer material	Membrane structure	Temperature (°C)	Pressure condition	Permeance (GPU)		Selectivity
				Water vapour	N$_2$	
Sulfonated poly(ether ether ketone) (sulfonation degree of 60%)	A 2–5 µm-thick top layer on polyethersulfone microfiltration hollow fibre (OD = 1.2 mm, ID = 0.8 mm)	25	Lumen side: 1 mbar Shell side: ambient pressure	1500–2500	0.013	115,000–192,000
Pebax®1657	A 2 mm-thick Pebax® 1657 layer on an intermediate microporous support on a flat-sheet polyester substrate	21	Sweeping gas at ambient pressure	1800 (with substrate) 6000 (without substrate)	1	1800 6000
sodium alginate–poly(vinyl alcohol) blend	The top layer is dip-coated on porous polysulfone hollow fibre substrate. Skin layer 2 µm	25	Lumen side: swept by N2, 1.8 bar Shell side feed: 3.5 bar	2390–5380	–	> 15,000(C$_3$H$_6$)
Poly(N,N-dimethylaminoethyl methacrylate)	1.5 µm top layer is interfacially crosslinked onto the 70 µm thick polyacrylonitrile substrate	25	Ambient pressure on feed side 17 mbar on permeate side	4000–5000	–	3500–4500(CH$_4$)

(continued)

Table 6.2 (continued)

Active layer material	Membrane structure	Temperature (°C)	Pressure condition	Permeance (GPU)		Selectivity
				Water vapour	N_2	
cellulose triacetate	Single layer hollow membrane with 225 μm O.D. and 70 μm I.D	35	Lumen side: normal pressure Shell side: feed air at 2.29 bar	9000 GPU	4.2 GPU	2140
poly(vinyl alcohol) doped with EDTMPA	the top layer is dip-coated on porous polysulfone hollow fibre substrate	27	Lumen side: sweeping N2 at 1.8 bar Shell side: 3.5 bar	200–1000	–	> 5400 (propylene)
PVA/LiCl blend	Intermediate TiO_2 powder layer and PVA/LiCl top layer are consecutively coated on a fine stainless steel wire mesh	24	Ambient pressure on feed side 1 mbar on permeate side	1790–4780	–	1000–5000

(continued)

Table 6.2 (continued)

Active layer material	Membrane structure	Temperature (°C)	Pressure condition	Permeance (GPU) Water vapour	Permeance (GPU) N_2	Selectivity
PVA/gelatin–silica hybrid	The top layer is dip-coated on porous polysulfone hollow fibre substrate. (ID = 150 μm; OD = 350 μm)	25	Lumen side: swept by N2, 1.8 bar Shell side feed: 2.5–4.5 bar	4.1–15	–	2980–67,511(C_3H_6)
Ionic liquids [emim][ESU], [emim][DCA], [emim][BF$_4$]	A hydrophobic flat substrate (thickness: 100 μm, pore size: 0.1 μm, porosity: 70%) is coated with: 10.3 μm thick [emim][ESU], 11.5 μm [emim][DCA] or 9.9 μm [emim][BF$_4$]	25	Feed at atmospheric pressure 45–60 mbar on permeate side, with sweep air	4524 4805 3620	–	> 1000 (air)

(continued)

Table 6.2 (continued)

Active layer material	Membrane structure	Temperature (°C)	Pressure condition	Permeance (GPU)		Selectivity
				Water vapour	N_2	
ionic liquids [emim][Tf2N] [N(4)111][Tf2N] [emim][BF4]	the ionic liquids are spread over hydrophilic polyethersulfone (PES) flat substrate with a thickness of 132 μm	31	Both sides at normal pressure. N2 sweeping on permeate side	635 570 1050	–	3843 3290 16,300
Triethylene glycol Polyethylene glycol	50 μm top layer is supported on a highly hydrophobic flat substrate (pore size = 0.1 μm, thickness = 125 μm; and porosity = 70%)	15–30	Ambient pressure on feed side A pressure < 2 mbar on permeate side	180–270 150	–	1700–2500 2000
Triethylene glycol	18 μm TEG liquid layer spreads, wets and then soaks on the top hydrophilicized surface of a flat hydrophobic substrate (0.1 μm micropores and 95 μm thickness)	25	Ambient pressure on feed side Sweeping air at 20–70 mbar on permeate side	706	–	2420

(continued)

Table 6.2 (continued)

Active layer material	Membrane structure	Temperature (°C)	Pressure condition	Permeance (GPU)		Selectivity
				Water vapour	N_2	
silica–alumina	The top layer is coated on the outer surfaces of coarse porous ceramic cylinders (ID = 10 mm, thickness = 1 mm, porosity = 50%, average poresize = 1 μm)	40–59	Lumen side: vacuum pressure Shell side: normal pressure	12,445–33,187	2.5–16.6	> 1000

Fig. 6.6 Schematic representation of mixed matrix membrane [45]

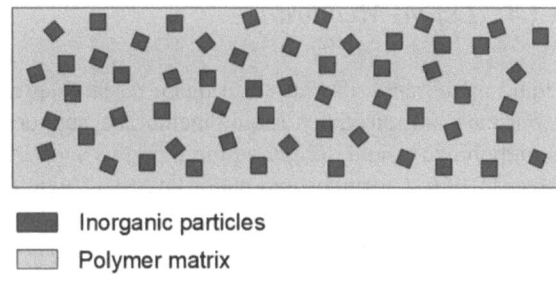

■ Inorganic particles
▢ Polymer matrix

that the membrane had a selective factor of 450 for vapour to air [44]. Figure 6.6 presents a schematic presentation of the mixed matrix membrane structure [45].

6.3.3 Zeolite Membranes

The zeolite membranes are one specific type of inorganic membranes often employed for vapour permeation. They are thermally and chemically stable and provide better separation performance because of the crystalline inorganic framework structures with uniform, molecular-sized pores. The pores are constructed from aluminium, oxygen, and silicon with alkali or alkaline-earth metals such as sodium, potassium, and magnesium, having water vapour molecules trapped in the gaps between them [46–48].

As the zeolitic materials are usually not self-supporting, they need to be nurtured and grow on porous support materials like aluminium or stainless steel. Zeolite can grow to form a continuous film either by employing in situ or seeded method. The seeded method is often preferred as it is able to offer greater flexibility in controlling the orientation of the zeolite crystals and the micro-structure of the zeolite membrane. However, it is an expensive procedure due to additional processing steps. For example, to reduce the production cost of zeolite membranes, Zhang and coworkers from Pacific Northwest National Laboratory, developed a low-cost, inorganic membrane module by using a hydrothermal membrane growth method to fabricate a membrane of a thin NaA zeolite on a flexible porous Ni substrate. They have successfully demonstrated its use for waste and ethanol separation [49]. Xing et al. [8] used a combination of the in situ and seeded methods to developing their membrane for air dehumidification by first coating a Ni sheet as the support with small seeding crystals of the targeted zeolite framework and then contacting it with a growth solution composed of NaOH, AlO_3, and sodium silicate to grow by the in situ method.

6.3.4 Liquid Membrane

Liquid membranes also attracted much research attention due to their high water permeance and selectivity. Liquid membrane, supported on a substrate, is designed to immobilize a liquid phase within a porous support membrane through capillary forces [50]. It constitutes two major layers: a single laminate of a supported liquid membrane of a hygroscopic liquid-like triethyleneglycol, polyethylene glycol 400, or ionic liquids, and a highly hydrophobic microporous membrane [51–53]. The liquid is usually stabilized in the support membrane by either direct immersion or a vacuum setting. The vacuum construction maintains a vacuum on the permeate side of the liquid membrane. Since the rate of vapour diffusion in liquids has been reported to be at least three to four times higher than those observed in polymeric membranes, the liquid membrane is capable of improving both vapour fluxes and selectivity [54]. Recently, Ito and co-workers have developed an ionic liquid-based liquid membranes demonstrated a vapour permeability of 26,000–46,000 Barrer and a separation factor over 1000 [55]. Despite its superior performance, liquid membranes suffer from low stability and material loss due to their high saturation vapour pressure.

6.4 Membrane Configurations

The MAD system uses chemical gradient, namely, the concentration or pressure gradient between two surfaces of the membrane, to drive the vapour out of the humid air across the membrane surface to the permeate side. Three different techniques can be employed to create the required partial vapour pressure difference: feed compression, vacuum pumping, and gas sweep [56]. As shown in Fig. 6.7a, the feed compression technique employs a compressor to increase the pressure in the feed side. In contrast, the vacuum pumping method, indicated in Fig. 6.7b, employs a vacuum pump to reduce the pressure at the permeate side. Both feed compression and vacuum pumping techniques require energy to produce either higher or lower pressure. Another viable method is the gas sweeping method, as portrayed in Fig. 6.7c, subject an inert gas to the permeate side for dilution, resulting in a lower partial vapour pressure. It is worthy to note that the gas sweeping process operates the permeate

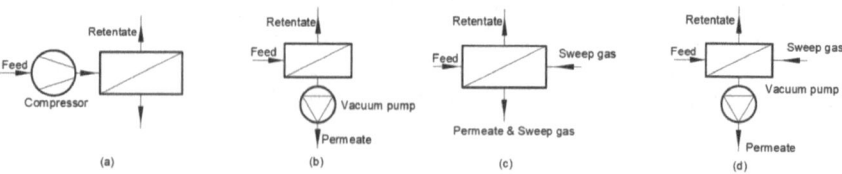

Fig. 6.7 Four different types of MAD configurations to generate the required driving force on the permeate side: **a** feed compression, **b** vacuum pumping, **c** gas sweeping, and **d** combination of vacuum pumping and gas sweeping

side at a higher total pressure and requires an additional compound like inert gas, nitrogen, or air; rendering it a costly method. Some other MAD systems combine and hybrid both vacuum pumping and gas sweeping methods, as shown in Fig. 6.7d, in order to promote better water vapour dehumidification. Despite improved vapour removal, such a combination raises capital and operating energy costs [57].

Comparatively, the vacuum pumping method is systematically preferable when a high purity of permeate is required [57]. Since MAD is often considered a new application for membrane processes, many of related literature found are laboratory work related, among which vacuum pumping accounts more than 50% [44, 58, 59]. There are a few studies that employed the gas sweeping system configuration as reported by Metz et al. [36], while the remaining studies employed a combination of vacuum pumping and gas sweeping method [7, 55].

6.5   Membrane Dehumidification Flow Organization

The organization of the flow configurations plays a critical role in the system efficiency. There are four ways the flows and directions of the air can be organized, namely, (i) perfect mixing with two fans in both feed and permeate sides, (ii) cross-plug flow with the feed stream perpendicular to the permeate stream, (iii) co-current flow with the feed and permeate streams at same direction, and (iv) counter current flow with the feed and permeate streams in opposite direction, as indicated in Fig. 6.8. The cross-plug flow with a reasonable precision is typically hollow-fibre membrane modules when the permeate pressure is low enough. On the other hand, a counter-current flow generally is used in flat-sheet modules [60].

MAD modules in the form of flat-sheet membranes and hollow-fibre membranes are commonly designed and employed for air dehumidification. Flat-sheet

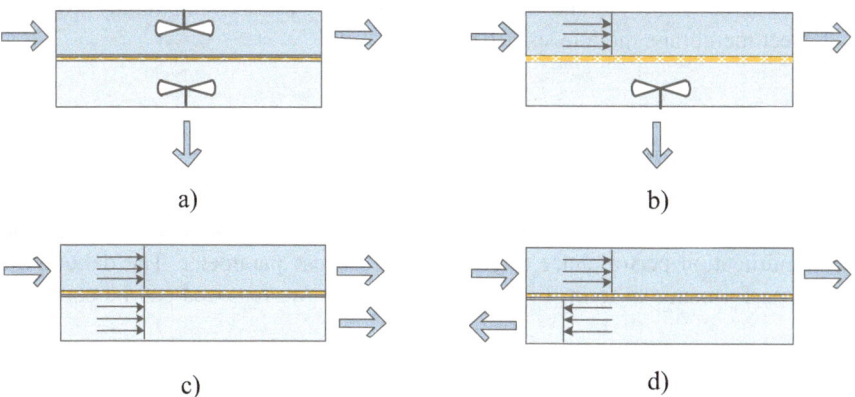

Fig. 6.8 Flow configurations for membrane-based dehumidification: **a** perfect mixing, **b** cross-plug flow, **c** co-current flow, and **d** counter-current flow [60]

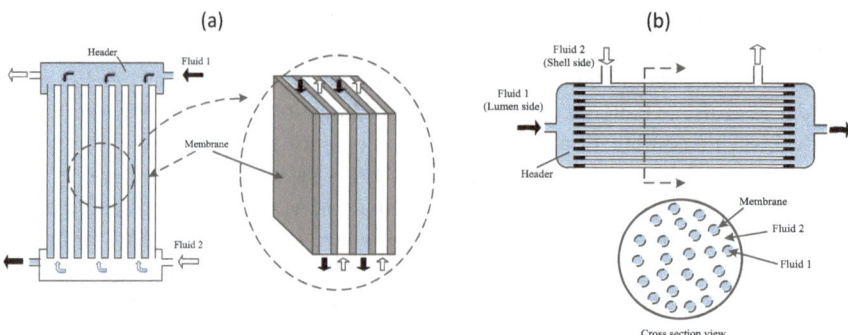

Fig. 6.9 Schematic of **a** flat-sheet module; and **b** hollow-fibre module

membranes are large sheets, usually on the order of 100 μm thick. They have module constructs similar to plate-and-frame heat exchangers (Fig. 6.9a). Hollow fibres are tubes, typically around 500 μm in diameter, where the wall of the tube is the membrane. They are used to construct modules similar to shell-and-tube heat exchangers (Fig. 6.9b). In these modules, one fluid flows inside the tubes (the lumen side) and the other flows around the tubes (the shell side). Both forms can be employed for the purpose of MAD to lower the humidity of the process air.

One key disadvantage of employing hollow fibres is their large pressure drop. Zhang [61] measured a pressure drop for their 1.2-mm i.d. fibre lumens of around 350 Pa. The pressure drop in the module employed by Kistler and Cussler [51], which had 0.6-mm i.d. fibres, were likely much higher, but they were not reported. However, it is worthy to note that pressure drops of above 350 Pa are unlikely to be used in these devices because fan initial cost generally increases with the supplied pressure, and because of increased parasitic energy use. Comparatively, Mardiana-Idayu and Riffat [62] reported that the incurred pressure drops for their flat-sheet modules is about 30 Pa, and the pressure drops for most commercially available flat-sheet membrane module span 100–300 Pa.

6.6 Performance Evaluation

In this section, several performance indicators are defined to facilitate the membrane dehumidification performance under varying process parameter. The dehumidification performance in terms of percentage of moisture removed can be computed as:

$$\text{Dehumidification performance} = \frac{\left(p_{(w/in)}^f - p_{(w/out)}^f\right)}{\left(p_{(w/in)}^f\right)} \cdot 100\% \tag{5}$$

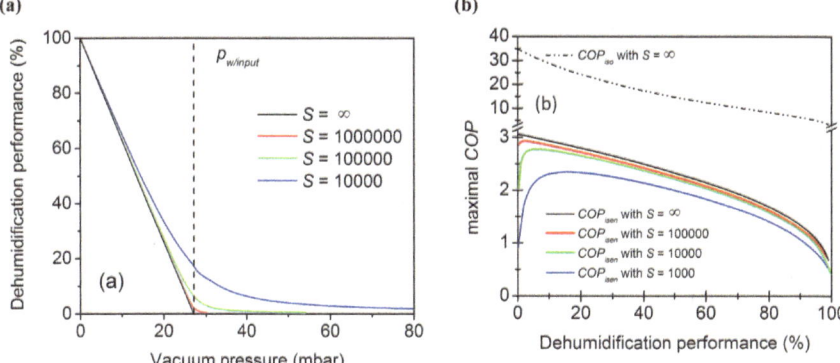

Fig. 6.10 **a** The effect of membrane selectivity (S) on dehumidification performance as a function of vacuum pressure, p_{vac}, with the vertical dashed line indicating the input water vapour pressure; **b** COP$_{iso}$ and COP$_{isen}$ as a function of dehumidification performance with different membrane selectivity. The inlet air's temperature and humidity are typical outdoor air condition in Singapore, at 31 °C and 60% (17 g/kg dry air), respectively

The effect of membrane selectivity (S) on dehumidification performance as a function of vacuum pressure is shown in Fig. 6.10a. At lower membrane selectivity, a larger fraction of air permeates through the membrane, reducing p_w^p. This results in a higher driving force for water vapour permeation. Therefore, a high dehumidification performance is obtained even at $p_{vac} > p_{w/in}^f$ and lower S results in higher dehumidification performance. Similarly, higher dehumidification performance is also achieved with the use of a sweep gas in permeate flow.

The work required in the separation process is the electricity to compress the water vapour from p_{vac} to p_{amb}. Theoretically, the pump works most efficiently in an isothermal process, which requires a maximal cooling. However, because water vapour is a condensable gas, during the isothermal compression, pressure increases and maintains at its saturation pressure (p_{sat}). Therefore, isothermal work equation is only applicable for the compression from p_{vac} to p_{sat}. Energy for pumping liquid water from p_{ws} to p_{amb} is negligible because the volume of liquid water is a few orders smaller than that of water vapour. Assuming that the pump's efficiency is 100%, the maximal obtainable VMD's COP in isothermal compression can be determined by Eq. 6.6:

$$COP_{iso} = \frac{F_w \bullet 45000\left(\frac{J}{mol}\right)}{RT(F_w + F_a)\ln\left(\frac{p_{sat}}{p_{vac}}\right)} \tag{6}$$

The vapour compression is least efficient in an isentropic process, which involves no cooling. With 100% pumping efficiency, the maximal obtainable VMD's COP in isentropic compression can be determined by Eq. 6.7:

$$\text{COP}_{\text{isen}} = \cfrac{F_{\text{w}} \bullet 45000\left(\frac{\text{J}}{\text{mol}}\right)}{\frac{F_{\text{w}} k_{\text{w}} RT}{k_{\text{w}}-1}\left[\left(\frac{P_{\text{amb}}}{P_{\text{vac}}}\right)^{\frac{k_w-1}{k_w}}-1\right] + \frac{F_{\text{a}} k_{\text{a}} RT}{k_{\text{a}}-1}\left[\left(\frac{P_{\text{amb}}}{P_{\text{vac}}}\right)^{\frac{k_a-1}{a}}-1\right]} \tag{7}$$

Practically, the pump operates in a polytropic process that involves an intermediate degree of cooling. In order to fully understand the COP limits of VMD, the two extremes of isothermal ($T = \text{constant}$ and $PV = \text{constant}$) and isentropic ($PV^{k_w} = \text{constant}$ and $PV^{k_a} = \text{constant}$, with k_w and k_a denoting the specific heat ratio, C_p/C_v, of water vapour and air, respectively) operations are considered. When $S = \infty$, COP_{iso} and COP_{isen} as the functions of dehumidification performance are plotted in Fig. 6.10b.

The energy consumption is much lower, and hence, COP is much higher for isothermal compression than for isentropic compression, as shown in Fig. 6.10b. With the assumption that pump's efficiency is 100% and membrane selectivity being infinite, COP_{iso} and COP_{isen} constitute the upper and lower thermodynamic COP limits of VMD systems. The upper limit is 10–35 while the lower one is 2–3. Both decrease when a higher percentage of the moisture is being removed. This indicates that the drier the product air becomes, the less efficient the VMD is. This is because a lower vacuum pressure, which incurs more pumping energy, is required to drive the separation in order to achieve a drier product air.

As shown in Eqs. (6.5)–(6.7), the dehumidification COP is independent of the membrane's water vapour permeance. Instead, it depends strongly on the membrane selectivity. The effect of membrane selectivity (S) on maximal COP via isentropic compression as a function of dehumidification performance is shown in Fig. 6.10b. The fact that a higher dehumidification performance is obtained with lower S or a permeate sweep gas [7, 44], shown in Fig. 6.10a, does not mean that a poorer selectivity implies a higher energy efficiency. It is because a greater amount of energy is required to pump the permeated air. Therefore, the dehumidification COP decreases with lower S. Additionally, COP peaks as a function of dehumidification performance; an observation consistent with reported results [44]. From Fig. 6.10b, it is apparent that when the membrane selectivity exceeds 1000 (attainable with current polymer membranes [9, 44]), COP that can be achieved is about 2–3 in isentropic conditions, far higher values than existing desiccant dehumidifiers.

It is noteworthy that the theoretical limits depicted as the curves in these graphs are the maximal COPs that a VMD system is able to attain when the efficiency of the pump is assumed to be 100%. In practice, a pump's efficiency is always lower than 100% due to energy losses arising from friction or heat loss. Therefore, the practical COP is always less than the theoretical limit. Nonetheless, the COP limits shown in these graphs demonstrate the potential of a VMD system that is able to realize improved performance.

Many of the existing works focus on making and evaluating water permeability and selectivity of new membranes [6, 7, 8, 10–32]. Only a few of them evaluate the efficiency of VMD by means of theoretical analysis [7, 8, 9]. It is worth noting that COP is the most suitable term to describe the energy efficiency of VMD because it can have a magnitude that is greater than 1 and is commonly used in the heating, venti-lation and air conditioning (HVAC) community. It is apparent that without involving any phase change, VMD potentially has higher *COP* than condensing water vapour by conventional processes. Theoretical input work was estimated from the permeate flow using the isothermal compression work equation [8, 9, 11]. VMD's COP of 2–3.5 has been reported with different membranes and operation conditions [7, 8, 9, 11]. This energy efficiency is significantly higher than that of dehumidification by desiccants, which have reported COPs of less than 1 [4, 5, 8].

Table 6.3 summarizes the performance of different system configurations under varying operational condition, and membrane selectivity. Generally, the supported liquid membranes are able to operate at a relatively high pressure at the permeate side compared with the other three types of membranes. Albeit the limited number of case studies, Table 6.3 shows that the zeolite membrane with a selectivity of 178 synthesized by Xing and co-workers [8] demonstrated the best dehumidifica-tion performance with an 83.6% moisture removal which translates to a latent COP of 3.0. The composite membrane tested by Bui and co-workers [44] on the other hand, was able to realize a dehumidification COP of 2.5. It is noteworthy that these available studies do provide some favourable indications that potentially MAD may be competitive to the conventional air dehumidification systems with the improved membrane materials and system design.

6.7 Future Challenges and Direction

The development of HVAC membrane processes comes hundreds of years after the invention of the vapour compression air conditioner. These processes continue to evolve as researchers and product developers work to make them competitive with conventional technology. The focus of the research is starting to transition from feasibility and proof of concept to cost and longevity. This is leading to design for manufacturing, accelerated life testing, and demonstrated performance in field installations. However, each technology is at a different stage, and each has its own set of research needs.

Vacuum membrane dehumidification is just past the proof of concept phase. While similar modules have been used for industrial drying applications, the largest module tested for this concept was around 10 cm^2 [40]. Tests on larger modules are needed, including hollow-fibre modules. Research is also needed to better understand how to size these devices in compact form factors, specifically the membrane module and the compressor. Other configurations can also improve efficiency, primarily ones that lower the compression ratio of the compressor.

One existing challenge for MAD is membrane durability that is particularly related to the issue of fouling. The success of industrial membrane processes shows that it is possible to operate membrane systems either without fouling or with controlled fouling and scheduled maintenance. The HVAC processes discussed here should have less fouling issues than common membrane filtration processes, where pressure

Table 6.3 Membrane materials, operational conditions, and system performance in different MAD systems with performance indicator COP

Membrane material	Permeation method (VP-Vacuum Pumping; GS—Gas Sweeping)	Condition T (°C)/RH	Permeate pressure (kPa)	Water permeance mol/(m^2 Pa s)	H$_2$O selectivity	%moisture removal/COP$_{latent}$	References
Porous Ni-supported NaA Zeolite	VP	32/90%	0.8	6.8×10^{-6}	178	83.6%/3.0	[8]
Ionic liquid [emim][BF4]	VP + GS	31/94%	6.6	3.5×10^{-7}	16,300	60%/0.75	[7]
Polysulfone hollow fibre	VP	32/100%	4.7	1.8×10^{-7}	529	–	[58]
MMM PVA: LiC, TiO	VP	24/90%	3	4.5×10^{-7}	450	20%/2.5	[44]
Cellulose Triacetate	VP	35/95%	0.28	1.0×10^{-6}	–	55%/–	–
Triethylene glycol (TEG)	VP	22/72%	0.13	7.5×10^{-8}	–	83.6%/–	[58]
Ionic liquid [emim][DCA]	VP + GS	25/94%	4.5	1.6×10^{-6}	1000	NA	[28]
Ionic liquid [emim][ESU]	VP + GS	25/94%	4.5	1.5×10^{-6}	1000	NA	[28]
Ionic liquid [emim][BF4]	VP + GS	25/94%	4.5	1.2×10^{-6}	1000	NA	[28]

forcibly pushes water through the membrane. However, more research is still needed to understand these potential fouling phenomena, and how environmental conditions and pollutants will affect the life of the membrane.

Also, improved understanding is necessary on the degradation of the material from exposure to temperature and humidity cycles, airside particulate fouling on the membrane surface, or degradation from chemical reactions or oxidation of the membrane material. The outlook for these membrane devices is uncertain. Advances in membrane technology over the previous few decades have enabled these unique devices for HVAC applications, which could potentially lead to more energy-efficient products. But more research and development are needed if these products are to compete economically against the inexpensive and established air-conditioning technologies.

As far as the future MAD is concerned, constant improvement of the existing materials and discovery of new materials of membrane with high water permeability and selectivity are needed to further enhance dehumidification performance. The materials must be inexpensive, strong enough to sustain the pressure, and no or little physical degradation. New membrane modules with novel configurations that promote better mass transport than the existing module designs need to be studied and developed. The new developed membrane module should address the challenges of the durability, the deformation, and the maldistribution of water vapour membrane sieving. In addition, an optimal balance must be found among the moisture removal, the dehumidification COP, and the cost during the selection of membrane and the design of the pressure difference based on desired supplied-air conditions.

As an initial first step, the advancement of system optimal design, operation, and control of the MAD process can be coupled with existing air-conditioning processes to deliver comfortable indoor conditions. This hybrid process is considered a low-hanging fruit upon harvesting will produce immediate impact on cutting energy consumption of existing air-conditioning systems. Further, the development of system energy performance models is highly needed for identification of the best system design, operation, and control to minimize energy consumption, the impacts from the operating parameters, and the maintenance and operational system cost, while to achieve high energy efficiency and overall system COP.

Thus far, almost all of the MAD studies in the literature were based on laboratory scale experimentation. The studies on large-scale MAD systems are still considered scarce. Membrane modules, system configuration, capacity, operation, and control for scale-up systems need to be investigated in order to reduce the size, pressure drop while enhancing heat and mass transfer from the feed stream to the permeate stream to maximize moisture dehumidification.

6.8 Conclusions

MAD is a recently emerged air dehumidification technology, separates the moisture from the humid air by using a selective membrane. During MAD, only water vapour molecules are transferred from the one side of the membrane at a high concentration to the other side at a low concentration. The MAD process has superior performance in energy and economic than other traditional dehumidification technologies.

This chapter provides a comprehensive discussion of the MAD technology including the membrane materials and characteristics, membrane forms and modules, system configuration and operation, and describes the existing challenges and future directions of MAD. The key discussions in the chapter include the description and discussion on (i) four types of membranes used in MAD include polymeric, zeolite, composite, and supported liquid, (ii) three distinct techniques used in MAD to create the required partial vapour pressure difference to water vapour migration to take place are feed compression, vacuum pumping, and gas sweep, (iii) four ways to organize the flows and directions of the air in the MAD, namely, perfect mixing, cross-plug flow, co-current flow, and counter-current flow; and (iv) various parameters to quantify the performance of the membranes and membrane modules under varying feed and desired supplied-air conditions.

Currently, MAD technology is new and is still at the research infancy phase. Apparently, it still lacks essential knowledge on system performance, performance consistency and reliability, optimal design and system analysis to achieve significant energy savings, reduced energy consumption, and low operational cost. The technology is still relatively new for direct deployment and implementation on industrial scale compared with other well-established air dehumidification technologies. Key advances on MAD technology must be made if it is to be widely employed air dehumidification applications. Research efforts have to step up to facilitate a deeper understanding of the MAD technology and to address its environmental, economic, and energy impacts.

References

1. Chua KJ, Chou SK, Yang WM, Yan J (2013) Achieving better energy-efficient air conditioning—A review of technologies and strategies. Appl Energy 104:87–104
2. Rafique MM, Gandhidasan P, Bahaidarah HMS (2016) Liquid desiccant materials and dehumidifiers—a review. Renew Sustain Energy Rev 56:179–195
3. Rambhad KS, Walke PV, Tidke DJ (2016) Solid desiccant dehumidification and regeneration methods—a review. Renew Sustain Energy Rev 59:73–83
4. Sahlot M, Riffat SB (2016) Desiccant cooling systems: a review. Int J Low-Carbon Technol. ctv032
5. Mina EM, Newell TA, Jacobi AM (2005) A generalized coefficient of performance for conditioning moist air. Int J Refrig 28:784–790
6. Li J, Ito A (2008) Dehumidification and humidification of air by surface-soaked liquid membrane module with triethylene glycol. J Membr Sci 325:1007–1012
7. Scovazzo P, Scovazzo AJ (2013) Isothermal dehumidification or gas drying using vacuum sweep dehumidification. Appl Ther Eng 50:225–233
8. Xing R, Rao Y, TeGrotenhuis W, Canfield N, Zheng F, Winiarski DW et al (2013) Advanced thin zeolite/metal flat sheet membrane for energy efficient air dehumidification and conditioning. Chem Eng Sci 104:596–609
9. Woods J (2014) Membrane processes for heating, ventilation, and air conditioning. Renew Sustain Energy Rev. 33:290–304

10. Bolto B, Hoang M, Xie Z (2012) A review of water recovery by vapour permeation through membranes. Water Res. 46:259–266

11. Bui DT, Ja MK, Gordon JM, Ng KC, Chua KJ (2017) A thermodynamic perspective to study energy performance of vacuum-based membrane dehumidification. Energy 132:106–115

12. Bui TD, Nida A, Chua KJ, Ng KC (2016) Water vapour permeation and dehumidification performance of poly(vinyl alcohol)/lithium chloride composite membranes. J Membr Sci 498:254–262

13. Zhao B, Peng N, Liang C, Yong WF, Chung T-S (2015) Hollow fiber membrane dehumidification device for air conditioning system. Membranes (Basel). 5:722–738

14. Yang B, Yuan W, Gao F, Guo B (2013) A review of membrane-based air dehumidification. Indoor Built Environ. 24:11–26

15. Wolińska-Grabczyk A, Jankowski A (2015a) Membranes for vapour permeation: preparation and characterization BT—pervapouration, vapour permeation and membrane distillation. In: Publ W (ed) Ser. Oxford, Energy, Woodhead Publishing, pp 145–175

16. Bui TD, Wong Y, Thu K, Oh SJ, KumJa M, Ng KC et al (2017) Effect of hygroscopic materials on water vapour permeation and dehumidification performance of polyvinyl alcohol membranes. J Appl Polym Sci 44765:1–9

17. Lin H, Thompson SM, Serbanescu-Martin A, Wijmans JG, Amo KD, Lokhandwala KA, et al (2012) Dehydration of natural gas using membranes. Part I: Composite membranes. J Membr Sci 413–414:70–81

18. Scovazzo P, Hoehn A, Todd P (2000) Membrane porosity and hydrophilic membrane-based dehumidification performance. J Membr Sci 167:217–225

19. Liu S, Wang F, Chen T (2001) Synthesis of poly(ether ether ketone)s with high content of sodium sulfonate groups as gas dehumidification membrane materials. Macromol Rapid Commun. 22:579–582

20. Sijbesma H, Nymeijer K, van Marwijk R, Heijboer R, Potreck J, Wessling M (2008) Flue gas dehydration using polymer membranes. J Membr Sci 313:263–276

21. Du JR, Liu L, Chakma A, Feng X (2010) Using poly(N, N-dimethylaminoethyl methacrylate)/polyacrylonitrile composite membranes for gas dehydration and humidification. Chem Eng Sci 65:4372–4381

22. Xie W, Geise GM, Freeman BD, Lee HS, Byun G, McGrath JE (2012) Polyamide interfacial composite membranes prepared from m-phenylene diamine, trimesoyl chloride and a new disulfonated diamine. J Membr Sc 403–404:152–161

23. Potreck J, Nijmeijer K, Kosinski T, Wessling M (2009) Mixed water vapour/gas transport through the rubbery polymer PEBAX® 1074. J Membr Sci 338:11–16

24. Metz SJ, Potreck J, Mulder MHV, Wessling M (2002) Water vapour and gas transport through a poly(butylene terephthalate) poly(ethylene oxide) block copolymer. Desalination 148:303–307

25. Li Y, Jia H, Pan F, Jiang Z, Cheng Q (2012) Enhanced anti-swelling property and dehumidification performance by sodium alginate-poly(vinyl alcohol)/polysulfone composite hollow fiber membranes. J Membr Sci 407–408:211–220

26. Pan F, Jia H, Jiang Z, Zheng X (2008) Enhanced dehumidification performance of PVA membranes by tuning the water state through incorporating organophosphorus acid. J Membr Sci 325:727–734

27. Shin Y, Liu W, Schwenzer B, Manandhar S, Chase-Woods D, Engelhard MH et al (2016) Graphene oxide membranes with high permeability and selectivity for dehumidification of air. Carbon N. Y. 106:164–170

28. Alina K, Kamiya T, Hirota Y, Ito A (2016) Dehumidification of air using liquid membranes with ionic liquids. J Membr Sci 499:379–385

29. Krull FF, Fritzmann C, Melin T (2008) Liquid membranes for gas/vapour separations. J Membr Sci 325:509–519

30. Scovazzo P (2010) Testing and evaluation of room temperature ionic liquid (RTIL) membranes for gas dehumidification. J. Membr. Sci. 355:7–17

31. Cheng Q, Pan F, Chen B, Jiang Z (2010a) Preparation and dehumidification performance of composite membrane with PVA/gelatin-silica hybrid skin layer. J Membr Sci 363:316–325

32. Akhtar FH, Kumar M, Peinemann K-V (2016) Pebax®1657/Graphene oxide composite membranes for improved water vapour separation. J Membr Sci
33. Wijmans JG, Baker RW (1995) The solution-diffusion model: a review. J Membr Sci 107:1–21
34. Metz SJ, Van Der Vegt NFA, Mulder MHV, Wessling M (2003) Thermodynamics of water vapour sorption in poly(ethylene oxide) poly(butylene terephthalate) block copolymers. J Phys Chem B 107:13629–13635
35. Verliefde ARD, Van der Meeren P, Van der Bruggen B, Hoek EMV, Tarabara VV (2013) Solution-diffusion processes. In: Encyclopedia of Membrane Science and Technology. Wiley
36. Metz SJ, Van de Ven WJC, Potreck J, Mulder MHV, Wessling M (2005) Transport of water vapour and inert gas mixtures through highly selective and highly permeable polymer membranes. J Membr Sci 251(1–2):29–41
37. Woods J (2014) Membrane processes for heating, ventilation, and air conditioning 33:290–304
38. Baker RW (2001) Membrane technology. In: Encyclopedia of polymer science and technology
39. Bolto B, Hoang M, Gray S, Xie Z (2015) Chapter 9—New generation vapour permeation membranes. In: Pervapouration, vapour permeation and membrane distillation. Woodhead Publishing, Oxford, pp 247–73
40. Murali RS, Sankarshana T, Sridhar S (2013) Air separation by polymer-based membrane technology 42, pp 130–86
41. Chua KJ, Bui DT, M KumJa., Islam MR, Oh SJ (2017) Air Conditioning systems: cooling and dehumidification (Wiley Publication). In: Seidel A (ed) Kirk-Othmer Encyclopaedia of chemical technology. Wiley, United States, pp 1–34
42. Cheng Q, Pan F, Chen B, Jiang Z (2010b) Preparation and dehumidification performance of composite membrane with PVA/gelatin–silica hybrid skin layer. J Membr Sci 363(1–2):316–325
43. Zhou H et al (2008) The polyurethane/SiO2 nano-hybrid membrane with temperature sensitivity for water vapour permeation. J Membr Sci 318(1–2):71–78
44. Bui TD, Chen F, Nida A, Chu KJ (2015) Experimental and modeling analysis of membrane based air dehumidification. Sep Purif Technol 144(15):114–122
45. Wolińska-Grabczyk A, Jankowski A (2015b) 6—Membranes for vapour permeation: preparation and characterization. Woodhead, Oxford, pp 145–175
46. Okamoto K, Kita H, Horii K, Tanaka K (2001) Zeolite NaA membrane: preparation, single gas permeation, and pervapouration and vapour permeation of water/organic liquid mixtures. Ind Eng Chem Res 40:163
47. Woodford C (2009) Zeolites. https://www.explainthatstuff.com/zeolites.html. Accessed 21 Feb 2016
48. Zhang Y et al (2012) Hydrogen-selective zeolite membrane reactor for low temperature water gas shift reaction. Chem Eng J 197:314–321
49. Zhang J, Liu W (2011) Thin porous metal sheet-supported NaA zeolite membrane for water/ethanol separation. J Membr Sci 371(1–2):197–210
50. Vandezande P. 5 (2015) Next-generation pervapouration membranes: recent trends, challenges and perspectives. In: Woodhead publishing series in energy. Woodhead Publishing, Oxford, pp 107–41
51. Lozano LJ, Godinez C, De los Rios AP (2011) Recent advances in supported ionic liquid membrane technology. J Membr Sci 376:1–14
52. Malik MA, Hashim MA, Nabi F (2011) Ionic liquids in supported liquid membrane technology. Chem Eng J 171:242–254
53. Grünauer J et al (2015) Ionic liquids supported by isoporous membranes for CO2/N2 gas separation applications. J Membr Sci 494(15):224–233
54. Ong YT, Yee KF, Cheng YK, Tan SH (2014) A review on the use and stability of supported liquid membranes in the pervapouration process. Sep Purif Re 43:62–88
55. Kudasheva A, Kamiya T, Hirota Y, Ito A (2016) Dehumidification of air using liquid membranes with ionic liquids. J Membr Sci 499, pp 379–85
56. Harlacher T, Wessling M (2015) Chapter Thirteen—Gas–gas separation by membranes. In: Progress in filtration and separation. Oxford, pp 557–84

57. Vallieres C, Favre E (2004) Vacuum versus sweeping gas operation for binary mixtures separation by dense membrane processes. J MembrSci 244(1–2):17–23
58. Ito A (2000) Dehumidification of air by a hygroscopic liquid membrane supported on surface of a hydrophobic microporous membrane. J Membr Sci 175:35–42
59. El-Dessouky H, Ettouney H, Bouhamra W (2000) A novel air conditioning system: membrane air drying and evapourative cooling. Chem Eng Res Des 78(7):999–1009
60. Favre E (2010) 2.08—Polymeric membranes for gas separation. In: Comprehensive membrane science and engineering. Elsevier, Oxford, pp 155–212
61. Zhang LZ (2010) An analytical solution for heat mass transfer in a hollow fiber membrane based air-to-air heat mass exchanger. J Membr Sci 360:217–225
62. Mardiana-Idayu A, Riffat SB (2011) An experimental study on the performance of enthalpy recovery system for building applications. Energy Build 43:2533–2538

Chapter 7
Dissipative Losses in Cooling Cycles

Nomenclature

S	Entropy generation rate, W/K
m	Mass flow rate, kg/s
Q	Heat transfer rate, W
s	Specific entropy, J/kg/K
t	Time, s
T	Temperature, K
W	Work, W

Greek letters

η	Efficiency,

Subscripts

II	Second
0	Ambient
ac	Actual
cool	Cooling
evap	Evaporation
g	Generation
in	Inlet
L	Loss
rev	Reversible work
sys	System

© Springer Nature Singapore Pte Ltd. 2021
C. Kian Jon et al., *Advances in Air Conditioning Technologies*, Green Energy
and Technology, https://doi.org/10.1007/978-981-15-8477-0_7

7.1 Introduction

With the growth of global cooling demand, the high energy consumption of cooling systems has become a great challenge, and improving the energy efficiencies of cooling cycles has attracted interests at both research and industrial levels. Substantial efforts have been devoted to analyse and optimize the design and operational parameters of cooling cycles to minimize their energy consumption.

All of the existing cooling systems involve heat and mass transfer and fluid flow, which result in internal dissipations due to finite size of the actual devices, and finite times and speeds of real processes [1, 2]. Therefore, a major research direction is to identify these internal dissipations in order to pinpoint the main sources of irreversibility within the system and figure out the inefficient components that require optimization. The conventional first-law analysis based on heat and mass balances is unable to capture the thermodynamic irreversibility. It is the second-law analysis that captures the essence and causes of internal dissipations in a thermal system. Second-law analysis has been conducted on different energy systems, such as fuel cells [3], solar collectors [4], latent heat storage systems [5], desalination systems [6], and internal combustion engines [7].

Second-law analysis has also been adopted for cooling systems. Klein and Reindi [8] analysed the effect of heat exchanger allocation on the entropy generation rate of different refrigeration cycles. It was found that the system performance was maximized when the evaporator and the condenser had equal heat transfer areas. Ng et al. [9] conducted a thermodynamic analysis on an absorption chiller to identify the internal dissipations. The entropy generation of the chiller was found to be dominated by the heat exchangers. Chua et al. [10] analysed the entropy generation rates of a silica gel–water adsorption chiller. The adsorption/desorption beds were found to be the main contributors of irreversibility. Li et al. [11] presented an entropy analysis on a zeolite–water adsorption cooling cycle. The specific entropy generation with respect to the cooling capacity was found to be reduced by increasing the cycle time. Caliskan et al. [12] presented a second-law analysis on an indirect evaporative cooler based on the Maisotsenko cycle. The irreversibility of the cycle was minimized when the reference temperature was 23.88 °C.

This chapter is intended to conduct a systematic second-law analysis on different cooling cycles, including a mechanical vapour compression cycle, an absorption chiller, an adsorption chiller, and an indirect evaporative cooler. Based on the thermodynamic states of these systems reported in the literature, entropy generation rates of different components as well as the whole system will be calculated and analysed to identify the major sources of internal dissipations in these systems. Parametric studies will also be conducted to optimize the design and operational variables. Finally, the system-level entropy generation rates will be normalized with respect to the cooling capacities in order to evaluate and compare the thermodynamic efficiencies of these systems.

7.2 Methodology

The entropy generation rate for an open system can be derived through black-box entropy conservation analysis. For an open system shown in Fig. 7.1, entropy conservation is expressed as

$$\frac{\mathrm{d}S_{\mathrm{sys}}}{\mathrm{d}t} = \sum (\dot{m}s)_{\mathrm{in}} - \sum (\dot{m}s)_{\mathrm{o}} + \dot{S}_{\mathrm{g}} + \sum \frac{\dot{Q}}{T} \tag{7.1}$$

where S_{sys} is the entropy of the system, $\dot{m}$ is the mass flowrate, s is the specific entropy of the stream, $\dot{Q}$ is the heat input, and $\dot{S}_{\mathrm{g}}$ is the entropy generation rate.

Rearranging Eq. 7.1, entropy generation rate for the system can be derived as

$$\dot{S}_{\mathrm{g}} = \sum (\dot{m}s)_{\mathrm{o}} - \sum (\dot{m}s)_{\mathrm{in}} + \frac{\mathrm{d}S_{\mathrm{sys}}}{\mathrm{d}t} - \sum \frac{\dot{Q}}{T} \tag{7.2}$$

When the system operates under steady state, which represents the situations of most cooling systems, the entropy change of the system is zero, and Eq. 7.2 evolves

$$\dot{S}_{\mathrm{g}} = \sum (\dot{m}s)_{\mathrm{o}} - \sum (\dot{m}s)_{\mathrm{in}} - \sum \frac{\dot{Q}}{T} \tag{7.3}$$

Equations 7.2 and 7.3 can be employed to calculate entropy generation rates for different cooling systems at both component and system levels.

In addition to entropy generation, exergy loss and second-law efficiency are also effective measurements of the system's irreversibility. The exergy loss can be calculated from the entropy generation rate

$$W_{\mathrm{L}} = T_0 \dot{S}_{\mathrm{g}} \tag{7.4}$$

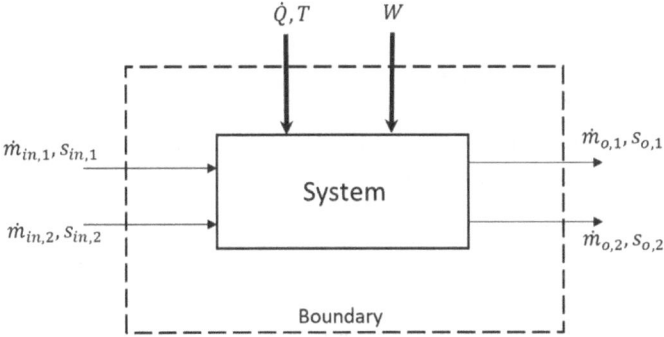

Fig. 7.1 Schematic of entropy balance for a black-box open system

where T_0 is the ambient temperature. Based on the exergy loss, the second-law efficiency can be derived, which is the ratio of the minimum work requirement to the actual work input

$$\eta_{\text{II}} = \frac{W_{\text{rev}}}{W_{\text{ac}}} = \frac{W_{\text{rev}}}{W_L + W_{\text{rev}}} \tag{7.5}$$

where W_{ac} and W_{rev} are the actual work input and the minimum work requirement, respectively. The minimum work requirement for a cooling cycle is determined from

$$W_{\text{rev}} = \dot{Q}_{\text{cool}}(\frac{T_0}{T_{\text{evap}}} - 1) \tag{7.6}$$

7.3 Accounting for Losses

Employing the equations derived previously, this section presents an entropy analysis on different cooling systems, including a mechanical vapour compression chiller, an absorption chiller, an adsorption chiller, and an indirect evaporative cooler. Experimental measurements of performance data will be obtained from the open literature. Based on these data, entropy generation rates of different components will be analysed to highlight the major sources of internal dissipations in these systems. Specific entropy generation with respect to the cooling capacity will also be calculated to compare the efficiencies of these systems.

7.3.1 Mechanical Vapour Compression Chiller

Figure 7.2 shows the schematic of a mechanical vapour compression chiller, and Table 7.1 summarizes the thermodynamic states of a typical chiller using R-134a refrigerant at a flow rate of 0.01 kg/s [13]. The cooling capacity provided by the chiller is 1.67 kW, and the compressor work input is 0.31 kW. The corresponding coefficient-of-performance (COP) is 5.4. Employing Eq. 7.3, the overall entropy generation rate is calculated as 0.54 W/K.

Figure 7.3 illustrates the percentage contributions of entropy generation from different components. The compressor accounts for the highest portion of entropy generation, followed by the evaporator and the expansion valve. The contribution of the condenser is less significant compared with the other components. The entropy generation rates, which represent irreversible losses within the system, can be attributed to frictional losses of fluid flow (e.g. compressor and expansion valve) and the finite size of devices (evaporators and condensers) for heat transfer.

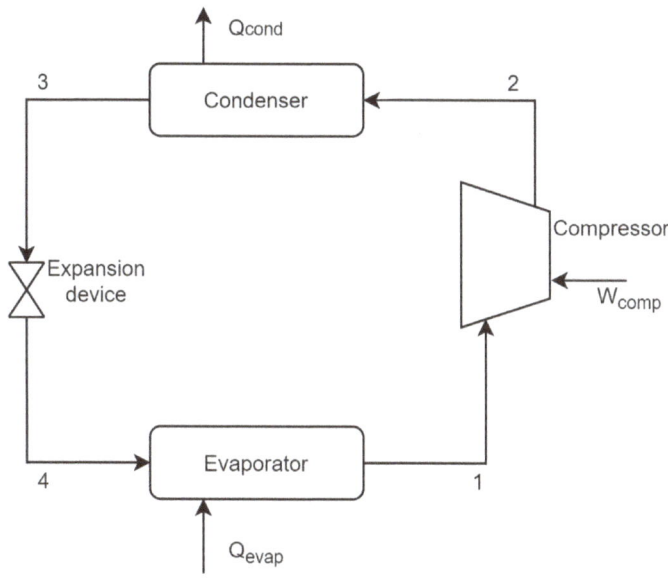

Fig. 7.2 Schematic of a mechanical vapour compression chiller [13]

Table 7.1 Thermodynamic states for a mechanical vapour compression chiller

Point	State	P (kPa)	T (°C)	h (kJ/kg)	s (kJ/kg K)	X
1	Low-pressure vapour	402.2	17	263	0.953	1
2	High-pressure vapour	815.9	57.1	293.6	1	1
3	High-pressure liquid	815.9	32	96.5	0.357	0
4	Two-phase mixture	402.2	9.1	96.5	0.363	0.169

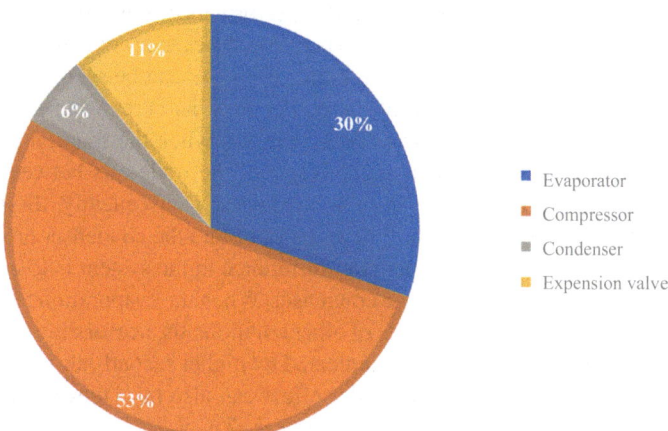

Fig. 7.3 Percentage contributions of entropy generation for different components of a mechanical vapour compression chiller

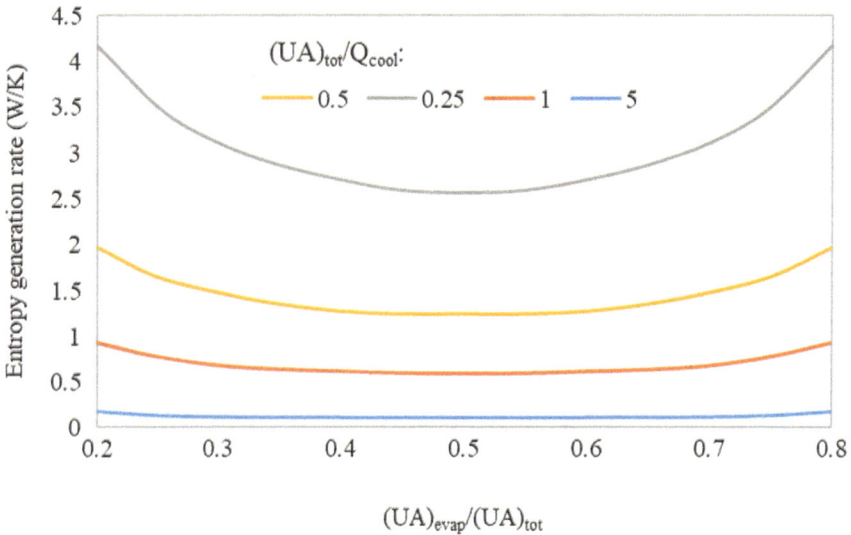

Fig. 7.4 Entropy generation rate under different overall thermal sizes and heat exchanger allocation [8]

One effective way to reduce the irreversibility is to reduce the temperature difference in the heat exchangers. In actual design, the temperature difference can be reduced by increasing the heat transfer area or promoting the heat transfer coefficient. Figure 7.4 portrays the entropy generation rate under different thermal sizes. The entropy generation of the system can be reduced by one order of magnitude when the normalized thermal size is increased from 0.5 to 5. Such trends are consistent with the common understanding that increasing the heat exchanger area or improving the heat transfer coefficient will promote chiller performance. For a fixed overall heat transfer area, optimal allocation of the area between the evaporator and the condenser also helps to minimize the system irreversibility [8]. As can be seen from Fig. 7.4, the overall entropy generation rate of the chiller first increases when more area is allocated to the evaporator, and the trend is reversed when the fraction of the area in the evaporator exceeds 0.5. This is the result of the trade-offs between entropy generation rates in the evaporator and the condenser. Consequently, dissipation is minimized when the thermal sizes of the evaporator and the condenser are equal.

Another way to reduce the temperature difference in the system is to control the temperatures in the evaporator and the condenser. When the evaporation temperature is higher, i.e. closer to the temperatures of other components, irreversible dissipations due to heat transfer can be reduced, as indicated by higher second-law efficiency and less exergy loss in Fig. 7.5 [14]. A similar effect can also be achieved by reducing the condensation temperature, as evidenced in Fig. 7.6.

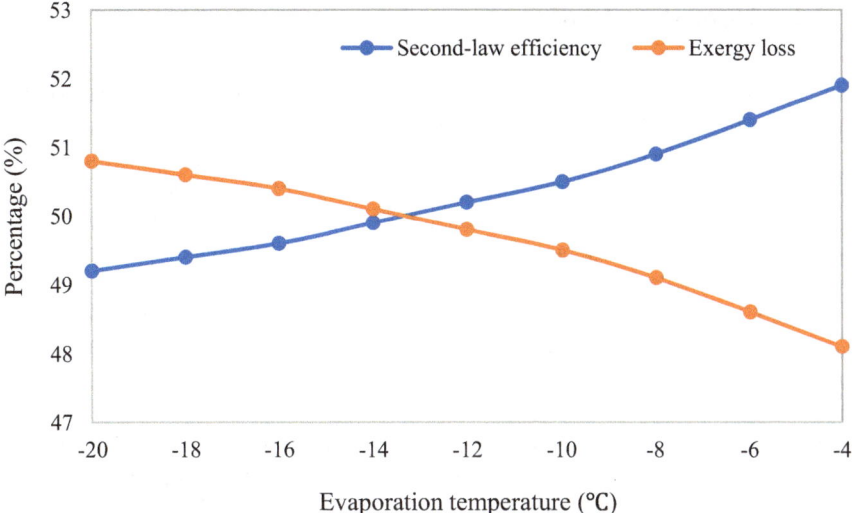

Fig. 7.5   Exergy loss and second-law efficiency of a mechanical chiller under different evaporation temperatures

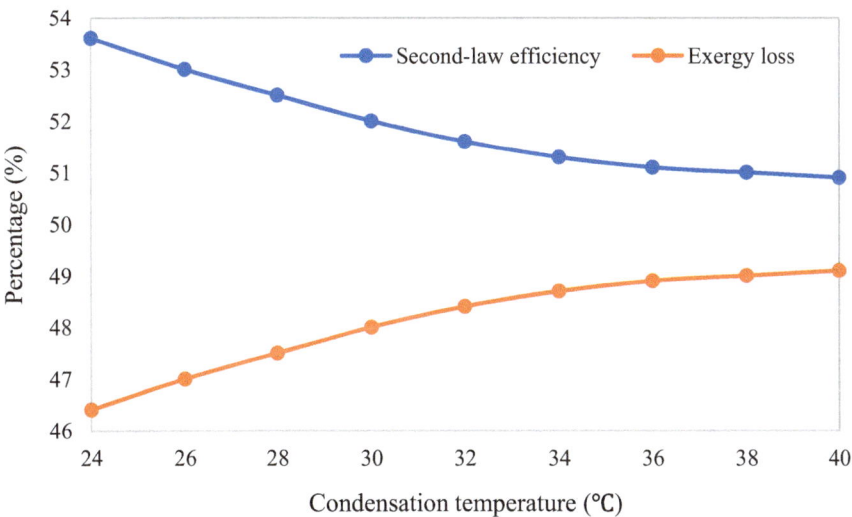

Fig. 7.6   Exergy loss and second-law efficiency of a mechanical chiller under different condensation temperatures

7.3.2   *Absorption Chiller*

Absorption chillers exploit thermal power instead of electricity to achieve cooling, and Second-law analysis of an absorption chiller is more complicated than mechanical chillers [9]. The reason is that there are four heat reservoirs (absorber, generator, evaporator, and condenser) in the absorption chiller, as compared with two (evaporator and condenser) in the mechanical chiller. In addition to evaporation and condensation, absorber and generator also contribute significant amounts of entropy generation during heat and mass transfer.

Figure 7.7 shows the schematic of an ammonia–water absorption chiller manufactured by the Daikin Corporation. Three additional regenerative heat exchangers are included in the system to allow internal heat recovery at the (a) generator (GHE), (b) absorber (AHE), and (3) generator-absorber (GAX). Table 7.2 summarizes the thermodynamic states of the chiller at its rated conditions [9]. The cooling capacity of the chiller is 6.89 kW, and the generation heat consumption is 8.28 kW, leading to a COP of 0.83. The entropy generation rate is calculated as 2.86 W/K.

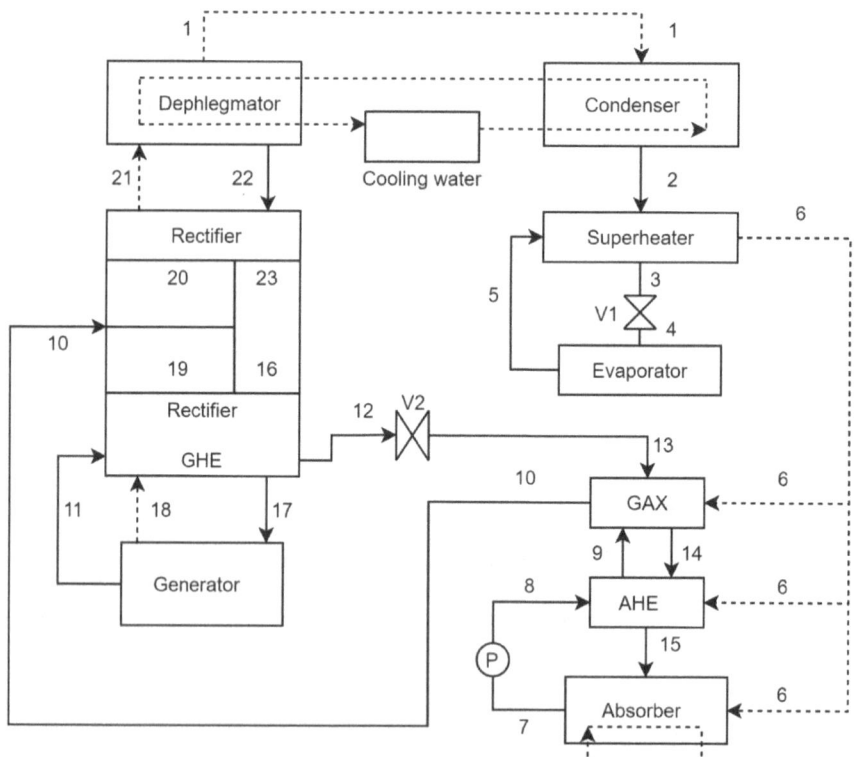

Fig. 7.7   Schematic diagram of an ammonia–water absorption chiller cycle

Table 7.2 Thermodynamic states for an absorption chiller [9]

State	T (°C)	p (bar)	X	h (kJ/kg)	s (kJ/kg K)	m (kg/s)
1	76.24	17.95	0.999	1394	4.429	0.00606
2	44	17.35	0.999	211	0.72	0.00606
3	25	17.35	0.999	121	0.426	0.00606
4	5.00	5.02	0.999	121	0.453	0.00606
5	5.00	5.02	0.999	1268	4.571	0.00606
6	38.36	5.02	0.999	1359	4.922	0.00606
7	41	4.72	0.476	−60	0.42	0.01347
8	41	18.75	0.476	−60	0.42	0.01347
9	90.66	18.75	0.476	171	1.098	0.01347
10	106.99	18.55	0.398	253	1.312	0.01347
11	194.00	17.95	0.047	801	2.325	0.00742
12	134.14	17.95	0.047	535	1.72	0.00742
13	134.14	4.72	0.047	535	1.72	0.00742
14	95.66	4.72	0.197	275	1.251	0.0088
15	72.19	4.72	0.306	116	0.9	0.01096
16	106.99	17.95	0.398	253	1.312	0.01202
17	145.89	17.95	0.226	489	1.818	0.01096
18	169.95	17.95	0.6	2064	5.816	0.00354
19	106.99	17.95	0.963	1512	4.754	0.0046
20	106.99	17.95	0.963	1512	4.754	0.00645
21	100.48	17.95	0.974	1483	4.678	0.00655
22	76.24	17.95	0.571	115	0.922	0.00049
23	106.99	17.95	0.398	253	1.312	0.00039

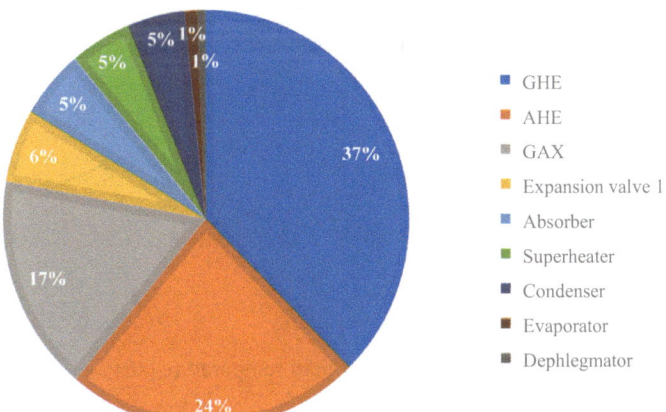

Fig. 7.8 Percentage contributions of entropy generation for different components of an absorption chiller

Figure 7.8 shows the relative contributions of each component on the entropy generation. Most of the entropy is generated in the regenerative heat exchangers, i.e. GHE, AHE, and GAX, which account for ~80% of the overall entropy generation. The reason is that these heat exchangers span the largest differences of temperature and chemical potential. It is natural to consider removing these heat exchangers to eliminating the corresponding entropy generation. However, excluding these heat exchangers will induce additional dissipations in other components, resulting in higher system irreversibility. For instance, entropy generation of the absorption chiller without regenerative heat exchangers was observed to be several orders higher [15], while the cooling capacity is only two times of the system illustrated in Fig. 7.8.

Similar to the mechanical chiller, the dissipations of the absorption chiller can also be reduced by increasing heat transfer areas, improving heat transfer coefficients, and regulating the temperatures of the evaporator and the condenser. Figure 7.9 shows the second-law efficiency of an absorption chiller under different chilled water inlet temperatures [16]. When increasing the chilled water inlet temperature, the second-law efficiency shows an increasing trend. The main reason is that chilled water with a lower temperature has a larger potential to create the cooling effect, which compensates the dissipation losses due to larger temperature differences.

As a heat-driven system, the second-law efficiency of the absorption chiller is also impacted by the heat source temperature. Figure 7.10 shows the second-law efficiency of an absorption chiller under different hot water inlet temperatures [16]. Second-law efficiency decreases when increasing the hot water temperature. Under a higher hot water temperature, more exergy is supplied to the system. Meanwhile,

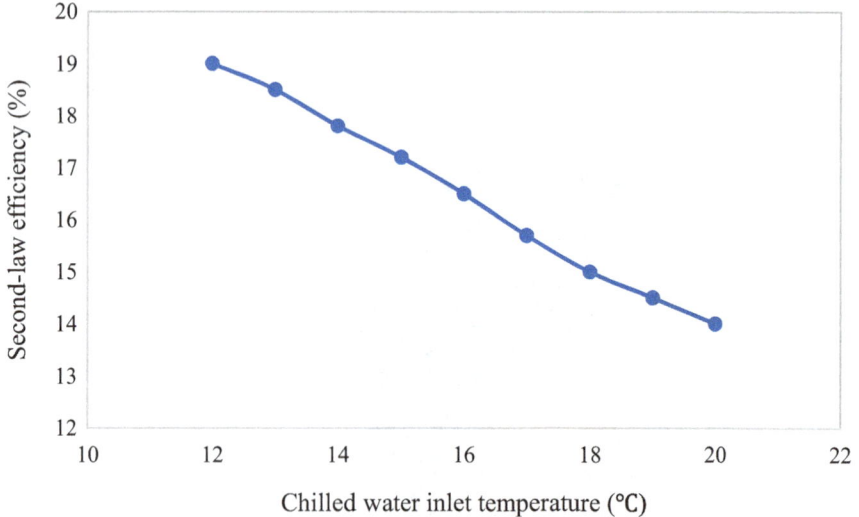

Fig. 7.9 Second-law efficiency of an absorption chiller under different chilled water inlet temperatures [16]

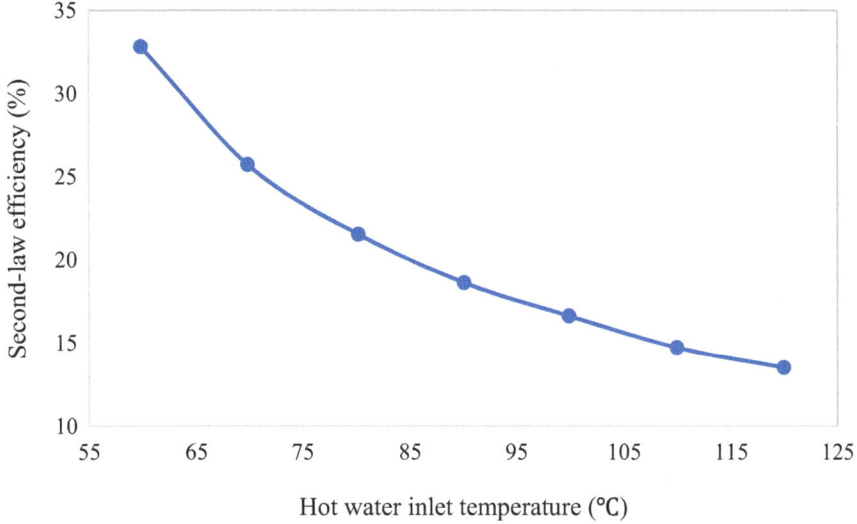

Fig. 7.10 Second-law efficiency of an absorption chiller under different hot water inlet temperatures [16]

dissipation losses are more significant when the operating temperature differences are larger.

In addition to the aforementioned factors, the load condition during operation also impacts the system's irreversibility. As shown in Fig. 7.11, the second-law efficiency of the chiller decreases with increasing the load factor, while the exergy loss increases [15]. Such an observation differs from the common expectation that a chiller working on partial load has lower energy efficiencies. The main reason is that the heat transfer rate in each component is increased when the load factor is higher, leading to higher irreversibility of different components.

7.3.3 Adsorption Chiller

Similar to absorption chillers, adsorption chillers are also driven by thermal energy, while the required heat source temperature is lower than that of the absorption chiller. Another difference is that adsorption chillers have to be operated in the batch mode, e.g. adsorption/desorption beds have to be switched periodically, and states of the system vary with time in each cycle.

Figure 7.12 shows the schematic of an adsorption chiller, while Fig. 7.13 portrays the profiles of water temperatures leaving each component of a zeolite–water adsorption chiller [11]. The inlet temperatures for the hot water, cooling water and chilled water are 65 °C, 27 °C, and 12 °C, respectively, while the cycle time and switching

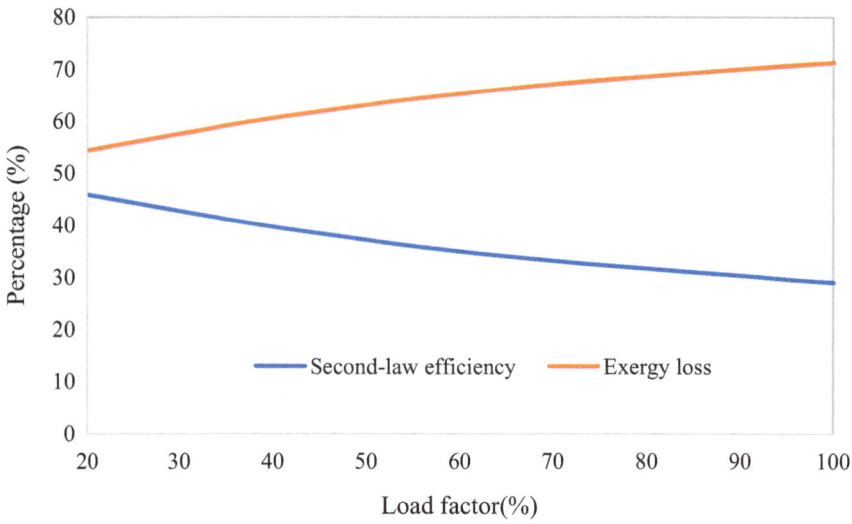

Fig. 7.11 Second-law efficiency and exergy loss of an absorption chiller under different load factors
[15]

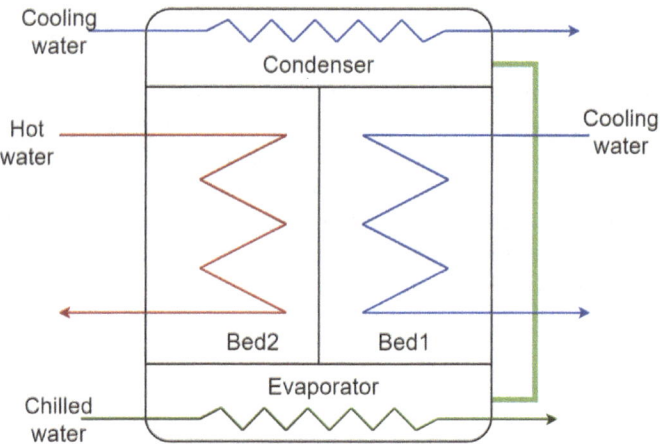

Fig. 7.12 Schematic of an adsorption chiller

period are 250 s and 22 s, respectively. The water temperatures leaving the adsorption/desorption beds change significantly at the beginning of each cycle and become steady afterwards. The reason is that during the switching period, hot water is circulating in the bed that is still cold while cooling water is supplied to the bed that is at a high temperature. Due to a large potential for adsorption/desorption, evaporation and condensation rates are also promoted at the beginning of each cycle, and, as a

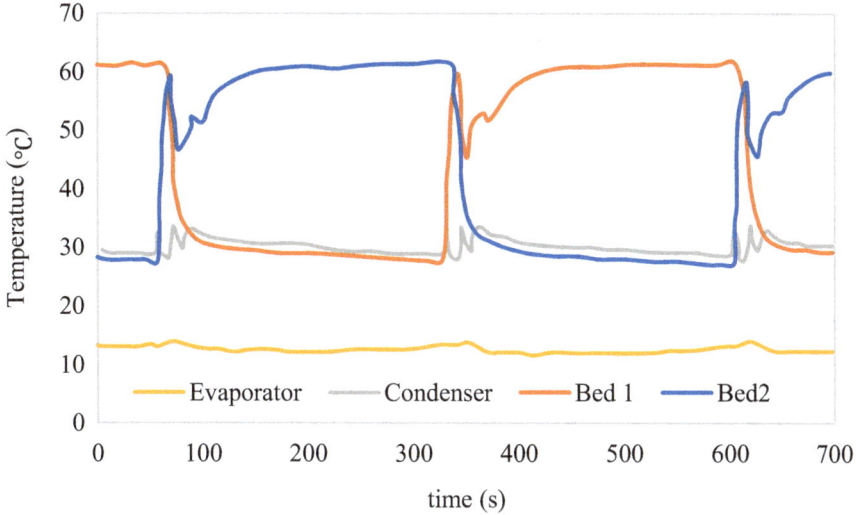

Fig. 7.13 Water temperature profile for a zeolite–water adsorption chiller at hot water temperature of 65 °C and cycle time of 250 s [11]

result, the temperatures of chilled water and cooling water also fluctuate markedly at the beginning of each cycle.

Figure 7.14 shows the instantaneous entropy generation rates of different components as well as the whole system for the zeolite–water system with a cycle time of 200 s [11]. Inlet temperatures for the hot water, the cooling water, and the chilled

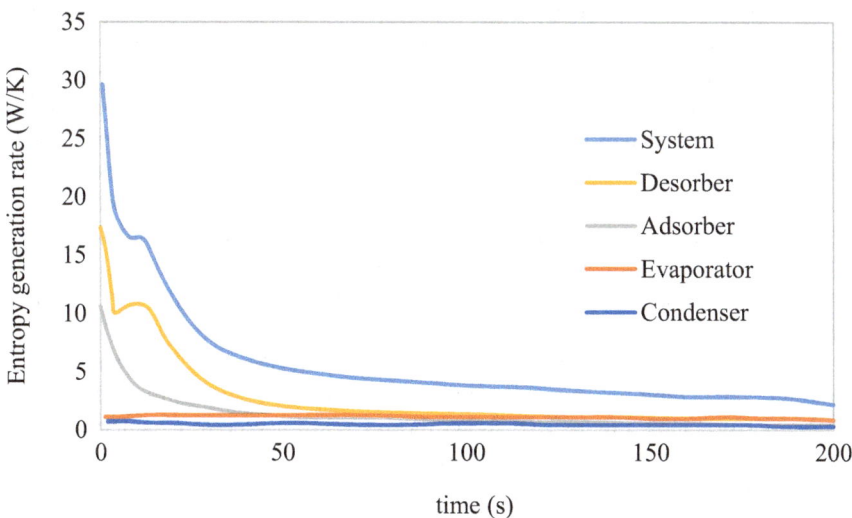

Fig. 7.14 Temporal entropy generation rate of different components for a half cycle [11]

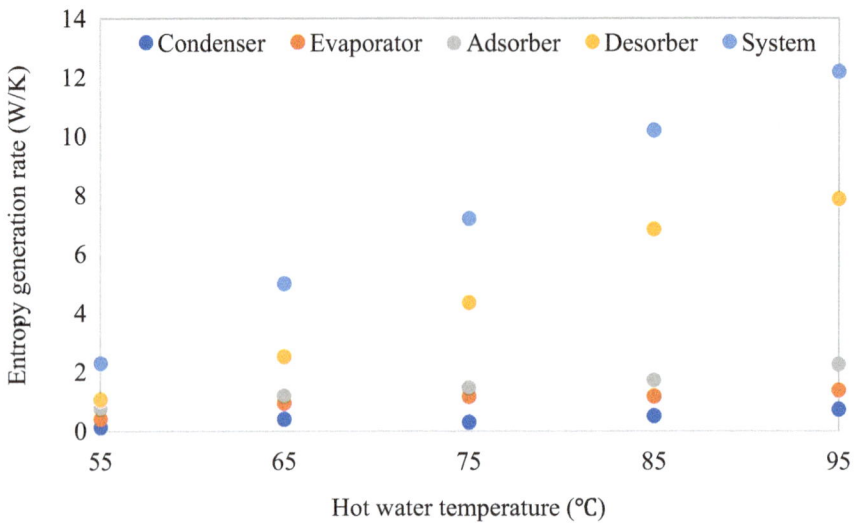

Fig. 7.15 Cycle-average entropy generation rates for different components of a zeolite–water adsorption chiller under different regeneration temperatures [11]

water are 65 °C, 26.5 °C, and 11 °C, respectively. As shown in the figure, entropy generation rates for the adsorption and desorption beds are several orders of magnitude higher than those of the evaporator and the condenser. This is explained by the high temperature/vapour pressure differences in these two beds, which provide large driving forces for heat and mass transfer. After approximately 70 s, adsorption and desorption approach the saturation states, and the irreversibility diminishes, as revealed by smaller entropy generation rates. The evaporator and the condenser are subjected to less irreversibility due to smaller temperature differences, which is a result of the higher heat transfer coefficients.

Similar to the absorption chiller, the entropy generation rate of the adsorption chiller is also impacted by the hot water temperature. Figure 7.15 illustrates the cycle-average entropy generation under different hot water temperatures. The desorption bed has the highest entropy generation rate under all situations, followed by the adsorption bed and the evaporator. The contribution of the condenser is the smallest. Also, entropy generation rates of different components as well as the whole system increase almost linearly with the regeneration temperature, which is a result of larger temperature differences.

A unique feature of the adsorption chiller is that it operates in the batch mode. Therefore, switching between adsorption and desorption also significantly impacts system performance. Figure 7.16 shows the cycle-average entropy generation rates for different components of a zeolite–water adsorption chiller under different cycle times. As clearly shown in the figure, the entropy generation rates of the system show a descending trend when increasing the cycle time. This is because a longer cycle

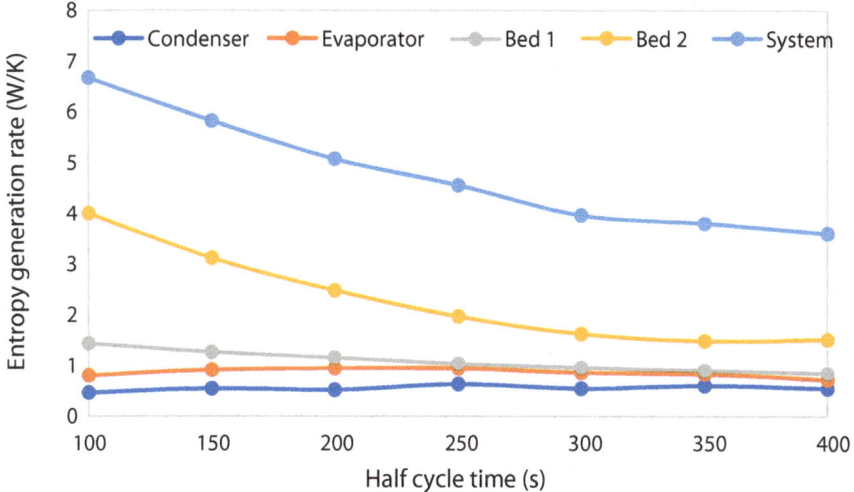

Fig. 7.16 Cycle-average entropy generation rates for different components of a zeolite–water adsorption chiller under different cycle times [11]

time enables more complete heat and mass transfer in different components, thus reducing irreversible dissipations.

7.3.4 Indirect Evaporative Cooler

The indirect evaporative cooler uses the evaporative potential of unsaturated air as the driving force for cooling. The only source of energy input is the power consumption of the fan. Therefore, the COP of the indirect evaporative cooler is much higher than other cooling systems.

Figure 7.17 is a schematic of the indirect evaporative cooler, and Table 7.3 summarizes its thermodynamic states under typical operation conditions (air temperature 30 °C, air humidity 11 g/kg, and 55% purge ratio). The cooling capacity is 4.76 kW, and the electricity consumption can be <10% of the cooling power when the cooler

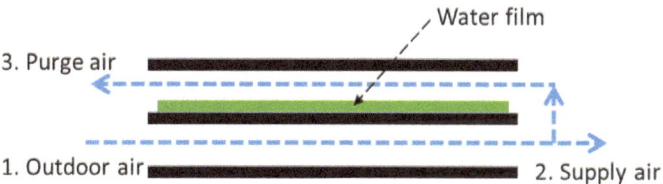

Fig. 7.17 Schematic diagram of an indirect evaporative cooler

Table 7.3 Thermodynamic states for an indirect evaporative cooler

Point	State	T (°C)	w (g/kg)	h (kJ/kg)	s (kJ/kg-K)	m (kg/s)
1	Outdoor air	30.00	11.00	58.31	0.202	1
2	Supply air	19.70	11.00	47.73	0.167	1
3	Purge air	26.78	15.78	67.19	0.230	0.55

is properly designed. The corresponding entropy generation rate is marginal at 0.085 W/K.

Figure 7.18 shows the diagram of exergy loss and flow for an indirect evaporative cooler [12]. Maximum exergy loss is observed from the exergy destruction inside the system, while exergy loss to the ambient also accounts for a significant portion. The useful exergy output, which equals to the second-law efficiency, is calculated as 16%.

Figure 7.19 highlights the entropy generation rate of the indirect evaporative cooler under different inlet air temperatures. The outlet temperatures of different air steams are adopted from the experimental results reported in [17], based on which the entropy generation rates are calculated using the developed model. As expected, the entropy generation rate increases exponentially with the increase of the inlet air temperature, which is a result of a higher temperature difference between the air streams. The entropy generation rate also increases when the inlet air humidity is smaller. The reason is that a drier airstream has a larger driving force for heat and mass transfer and leads to more dissipations and irreversibility.

Fig. 7.18 Exergy loss and flow diagram of an indirect evaporative cooler [12]

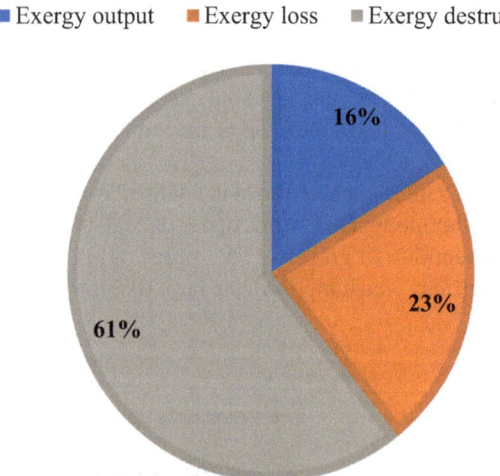

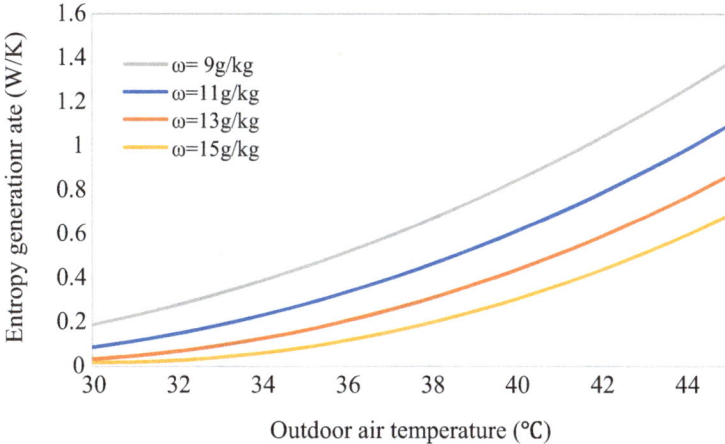

Fig. 7.19 Entropy generation rates of the indirect evaporative cooler under different outdoor air temperatures and humidity

7.4 Specific Entropy Generation

From the above results, it is observed that the entropy generation rates are quite different for different cooling systems. For example, the value for an adsorption chiller can be greater than 10 W/K, while that of the indirect cooler is less than 1 W/K. However, these values cannot reveal the actual irreversibility of these systems, because they are derived for systems with different cooling capacity.

To allow a direct comparison of different systems, the entropy generation rates are normalized with respect to the cooling capacity, as shown in Fig. 7.20. It can be seen from the figure that the specific entropy generation rates of the different cooling systems have the same order of magnitude. The absorption chiller has the highest specific entropy generation because heat transfer is poor in the reactor beds filled with porous adsorbents. On the other hand, the indirect evaporative cooler has the lowest specific entropy generation, indicating the least irreversibility. The observation is consistent with the common impression that the indirect evaporative cooler is most energy-efficient, while the adsorption chiller has the lowest energy efficiency.

7.5 Conclusions

This chapter quantifies the internal dissipative losses in cooling cycles via second-law analysis. The entropy generation rates of different cooling systems are calculated based on their thermodynamic states published in the literature. Results have revealed that the compressor is the main source of energy dissipation in the mechanical vapour compression chiller, while regenerative heat exchangers contribute the most amount

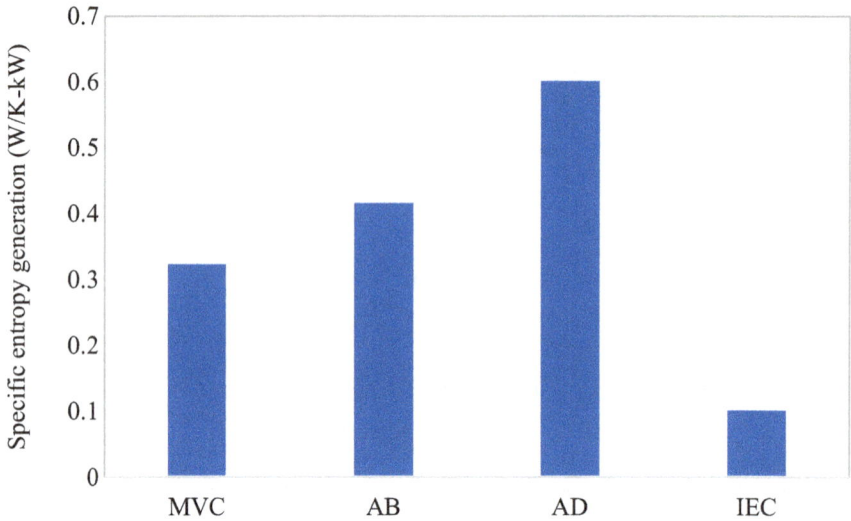

Fig. 7.20 Comparison of specific entropy generation rates for different cooling systems (MVC: Mechanical Vapour Compression; AB: Absorption; AD: Adsorption; and IEC: Indirect Evaporative Cooler)

of entropy generation in the absorption chiller. The adsorption chiller is the most inefficient due to poor heat transfer in the reaction beds filled with porous adsorbents, while the indirect evaporative cooler has the lowest specific entropy generation. Further improvement of the thermodynamic efficiency can be achieved by promoting heat and mass transfer in both evaporators and condensers of the mechanical chiller, increasing the heat source temperature of the absorption chiller, and optimizing the cycle time of the adsorption chiller. The derived results will be valuable towards enhancing the performance of different cooling cycles.

References

1. Bejan A (2013) Entropy generation minimization: the method of thermodynamic optimization of finite-size systems and finite-time processes. CRC press
2. Bejan A (1996) Entropy generation minimization: the new thermodynamics of finite-size devices and finite-time processes. J Appl Phys 79(3):1191–1218
3. Li X, Faghri A (2011) Local entropy generation analysis on passive high-concentration DMFCs (direct methanol fuel cell) with different cell structures. Energy 36(1):403–414
4. Makhanlall D, Munda JL, Jiang P (2013) Entropy generation in a solar collector filled with a radiative participating gas. Energy 60:511–516
5. Guelpa E, Sciacovelli A, Verda V (2013) Entropy generation analysis for the design improvement of a latent heat storage system. Energy 53:128–138
6. Chen Q, Ja MK, Li Y, Chua K (2017) On the second law analysis of a multi-stage spray-assisted low-temperature desalination system. Energy Convers Manage 148:1306–1316

7. Caton JA (2000) A review of investigations using the second law of thermodynamics to study internal-combustion engines. SAE Trans, 1252–1266
8. Klein S, Reindl D (1998) The relationship of optimum heat exchanger allocation and minimum entropy generation rate for refrigeration cycles
9. Ng K, Tu K, Chua H, Gordon J, Kashiwagi T, Akisawa A, Saha BB (1998) Thermodynamic analysis of absorption chillers: internal dissipation and process average temperature. Appl Therm Eng 18(8):671–682
10. Chua H, Ng K, Malek A, Kashiwagi T, Akisawa A, Saha BB (1998) Entropy generation analysis of two-bed, silica gel-water, non-regenerative adsorption chillers. J Phys D Appl Phys 31(12):1471
11. Li A, Ismail AB, Thu K, Ng KC, Loh WS (2014) Performance evaluation of a zeolite–water adsorption chiller with entropy analysis of thermodynamic insight. Appl Energy 130:702–711
12. Caliskan H, Hepbasli A, Dincer I, Maisotsenko V (2011) Thermodynamic performance assessment of a novel air cooling cycle: Maisotsenko cycle. Int J Refrig 34(4):980–990
13. Database JSE (2020) Introduction to refrigeration. Mech Eng
14. Yumrutaş R, Kunduz M, Kanoğlu M (2002) Exergy analysis of vapor compression refrigeration systems. Exergy, An Int J 2(4):266–272
15. Ahmed S, Mohammed M, Gilani SU-H (2012) Exergy analysis of a double-effect parallelflow commercial steam absorption chiller. J Appl Sci 12(24):2580–2585
16. Şencan A, Yakut KA, Kalogirou SA (2005) Exergy analysis of lithium bromide/water absorption systems. Renew Energy 30(5):645–657
17. Anisimov S, Pandelidis D, Jedlikowski A, Polushkin V (2014) Performance investigation of a M (Maisotsenko)-cycle cross-flow heat exchanger used for indirect evaporative cooling. Energy 76:593–606

Chapter 8
Efficacy Comparison for Cooling Cycles

8.1 Introduction

The air conditioners can be classified in four categories namely, (i) air-conditioner v/s heat pump, (ii) non-ducted v/s ducted, (iii) constant speed v/s variable speed, and (iv) single unit v/s split system as presented in Table 8.1. Some of these equipment are under board categories whereas others refer to different configurations of the same equipment. These different configurations and control are to cover the test procedures of various countries [1].

The air-conditioning systems test procedure varies with locations and they are tailored to meet the local requirements. With various equipment and different climate conditions, the minimum and highest temperature test conditions each test procedure are slightly different in each country. The most common test standards are:

i.   *ISO Standards*
     The ISO 5151 applies to small ducted air conditioners and heat pumps (rated capacity less than 8 kW), non-ducted air-cooled unit and air to air heat pumps. This standard has been adopted by many countries either by reference or in full. The scope of the standard covers both split and packaged systems but limits the split systems to single thermostat-controlled multi-split systems (however, ISO 5151 references ISO 15,042 test method for these products) [1, 2].

ii. *US Standards*
     The US standard document was updated in 2017 and it will be effective by 2023. The department of energy published the updated test procedure for residential ACs and heat pumps (US Code of Federal Regulation 10 CFR 430 Subpart B Appendix M) and a new test procedure is established listed as 10 CFR 430 Subpart B Appendix M1 [1, 3].
     The current test procedure applies to both air conditioners and heat pumps as a split and single unit. It also specifies that the split system units can be designed as multi-head mini-split, multi-split (including VRF) and multi-circuit systems. The definitions include specification of air source heat pumps [1, 3].

iii. *European Union Standards*

© Springer Nature Singapore Pte Ltd. 2021
C. Kian Jon et al., *Advances in Air Conditioning Technologies*, Green Energy
and Technology, https://doi.org/10.1007/978-981-15-8477-0_8

Table 8.1 Classification of air-conditioners

Description	Unit
(i) Air-conditioner v/s heat pump Air conditioners utilize refrigerants lower space temperature. On the other hand, heat pump operates on same cycle but in reverse order, add heat to space	
(ii) Non-ducted v/s ducted In non-ducted unit, the air is directly passed over the evaporator coil usually mounted in space. In ducted unit, a distribution fan supply conditioned air through duct work to the desired spaces	
(iii) Constant speed v/s variable speed Traditionally, the compressor operates at constant speed and once set temperature met, they turn-off completely Presently, the improved compressor operates at variable speed to meet demand. It helps to reduce the fluctuations in temperature, flow velocity and power consumption Visually, both compressors look same so they cannot be distinguished	
(iv) Single unit v/s split system In single or packaged unit, all components such as evaporator, condenser, compressor and expansion valve are mounted in same housing. It is always ducted because unit is located out of building In split system, the condenser, compressor and expansion valves are mounted in one unit and evaporator housed separately. The refrigerant circulates through pipe between both units. It can be ducted or non-ducted	

The European Union test procedure applies to both split and packaged air-conditioning units. These systems can be fixed speed, variable capacity, non-ducted, ducted, multi-split and single split [1, 4–8].

In 2018, the European Union test procedure (EU 2007) was replaced with BS 14,511 that covers all the scope of the earlier version. In addition, new scope of CO_2 refrigerant unit testing was included. In new test procedure, methodologies are also improved to reduce the test burden. An additional term was also added to the total cooling capacity calculation equation using Calorimeter Room method to accommodate the heat removed from the indoor compartment that compensates for heat leakage and allows smaller units (e.g., units that would not be able to overcome room enthalpy leakage) to be tested [1, 4–8].

iv. *Australian/New Zealand Standards*

The Australian AU/NZ 3823.4.1 standard is the ISO 16,358 standard that applies to air-cooled air conditioners and air to air heat pumps. It also covers the scope of ISO 5151, ISO 13,253 and ISO 15,042. The ISO 3253 applies to ducted air conditioner and ducted heat pumps. The ISO 15,042 covers non-ducted multi-circuit and multi-split units [1, 9–17].

Recently, the test procedure was improved (AS/NZS 3823.1.1:1998, Appendix ZZ) to provide more details on testing, nozzle discharge coefficient calculation, equation addition and uncertainty addition in measurements [1, 17].

v. *Japanese Standards*

The Japanese test procedures, JIS B 9612 and JIS B 8615, are applicable to both split and package units up to 10 kW cooling capacity [1, 18–20].

These test procedures were revised in 2013 and updated standard adopts ISO 5151:2010 as the testing procedure with country-specific adjustments to the testing conditions [1].

vi. *Chinese Standards*

The Chinese standard, GB 12,021.3–89 and GB/T7725 cover both air or water-cooled non-ducted air-conditioning units with a cooling capacity lower than 14 kW [1, 21–25].

vii. *Korean Standards*

The Korean test procedure (KS C 9306) covers only split and packaged units of capacity up to 35 kW. They do not cover multiple indoor split units [1, 26]. This standard was revised in 2017 to include greater scope in test procedure [27].

To evaluate the energy efficiency of an air conditioner, different parameters are used in different counties, which will be explained in the following sections.

8.2 Energy Efficiency Measurement

Energy input to any thermodynamically closed system that is being maintained at a set temperature (standard operation of air conditioners) requires that the energy removal rate from the air conditioner to increase. Consequently, for each unit of input energy to the system, this requires the air conditioner to remove that energy in addition to heat removal. To handle the overall load, the air conditioner energy input increases, computed by the inverse of its efficiency times the input of energy [28].

The energy efficiency rating of an air conditioner is defined by how many Btu per hour are removed for each watt of power consumed. For room air conditioners, this efficiency rating is the Energy Efficiency Ratio (EER), while for central air conditioners, it is the Seasonal Energy Efficiency Ratio (SEER) [28].

The EER is a measure of how efficiently an air-conditioning system will operate when the outdoor air temperature is maintained at a specific level (95°F). Technically, EER is the steady-state rate of heat removal (cooling capacity) in Btuh divided by steady-state rate of energy input in watts. This ratio is expressed in Btuh/watt as shown in Eq. 8.1 [28].

$$\text{EnergyEfficiencyRatio(EER)} = \frac{Coolingcapcity(Btuh)}{Energyinput(watt)} \tag{8.1}$$

EER is a direct measurement of system efficiency, for example, a unit with EER 7 is 18% less efficient than the unit with EER 8.5. That means it will spend 18% more electricity for the same effect. Figure 8.1 shows the energy efficiency of a portable air conditioner compares to a unit of EER 8.5. Theoretical EER derives from the COP ratio as: EER = 3.41 × COP [29].

On the other hand, the SEER is the measurement of system efficiency over a certain period of time (entire season). Technically, SEER measures the total cooling

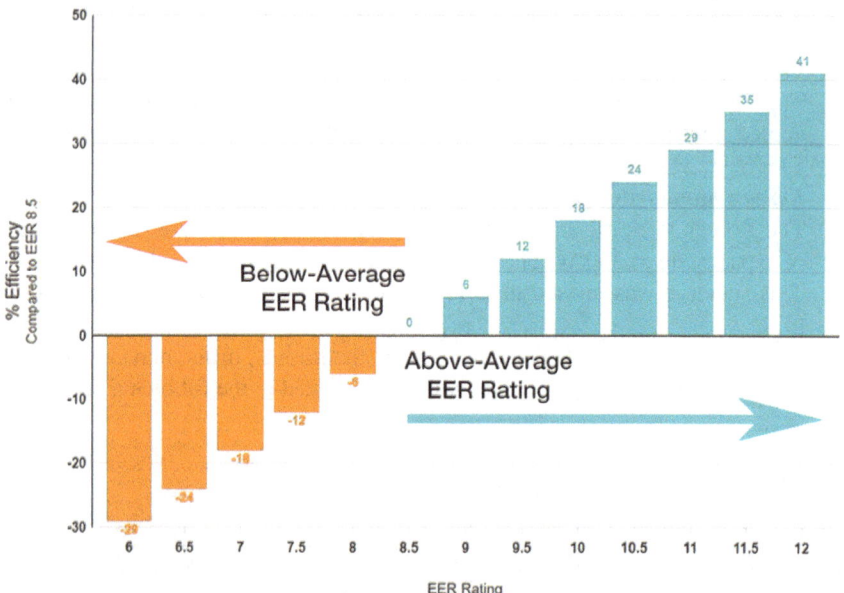

Fig. 8.1 Efficiency evaluation of an air conditioner based on EER value

of a central air conditioner or heat pump (in Btu) during the entire season divided by total energy (in watt-hours) consumed during the same period. A higher SEER value means that the system is more efficient [28].

$$SeasonalEnergyEfficiencyRatio(SEER) = \frac{Totalcooling(Btu)}{Totalenergyinput(watt - hour)} \quad (8.2)$$

Another direct efficiency measurement parameter is the coefficient of performance (COP), which demonstrates the cooling power of an air conditioning system divided by fan and compressor power [28].

$$Coefficientofperformance(COP) = \frac{Cooling(Watt)}{Energyinput(watt)} \quad (8.3)$$

The SCOP, or Seasonal Coefficient of Performance, describes the average COP during a heating season. Depending on definition, the SCOP value could also include other parts of the heating system than the heat pump only. For cooling and air-conditioning applications SEER, Seasonal Energy Efficiency Ratio, is normally used [28].

Heating seasonal performance factor (HSPF) is specifically used to measure the efficiency of air source heat pumps. HSPF is defined as the ratio of heat output (measured in BTUs) over the heating season to electricity used (measured in watt-hours). It, therefore, has units of BTU/watt-hr [30].

Every country has its national energy efficiency program and corresponding standards and labelling programs. The minimum efficiency performance standards vary in different economies. The energy efficiency programs along with corresponding efficiency metrics are shown in Table 8.2 [1].

There are two main methods to test air-conditioner units namely, the calorimeter room method and the indoor air enthalpy (or psychrometric) method. The calorimeter room method measures the energy input to a room that is being temperature controlled with an air-conditioner unit. The cooling capacity is considered equivalent to the energy used to maintain the air temperature at a constant value. On the other hand, the indoor air enthalpy method measures the air enthalpy as it enters and exits the AC's indoor unit. To calculate the cooling capacity of air-conditioning unit, the change in enthalpy is multiplied by the airflow rates. Table 8.3 summarized the test method adopted by different countries [1, 31].

8.3 Energy Classification of Air Conditioners

The classification of air conditioners in a certain class (A, B, C, etc.) is largely based on energy efficiency. In accordance to the instructions of the CE, the entire electrical devices are classified into seven energy classes (from A to G), depending on the energy they consume towards the energy they attribute. Concerning the air

Table 8.2 National energy efficiency programs for air conditioners

Country	Implementing body	Energy efficiency program	Energy efficiency parameter	Program goal
EU	European commission	Ecodesign/Energy label	SEER (cooling) SCOP (heating)	Improving efficiency through informed customer choice
Australia	Department of environment and energy	Equipment energy efficiency program	SEER (cooling) SCOP (heating)	Reduce energy bills and GHG emissions
Japan	Ministry of economy trade and industry	Top runner program	CSPF(cooling) HCSPF(heating)	Sets weighted average efficiency targets for manufacturers
China	Administration of quality supervision, inspection and quarantine	China energy label	EER (cooling) SEER(seasonal cooling) HSPF(seasonal heating)	Encourage customer to buy efficient product
Korea	Korea energy agency	High-Efficiency Appliance Cert Program E-Standby Program	EER (Cooling) CSPF (Seasonal Cooling) HSPF (Seasonal Heating)	Uses both comparative and endorsement efficiency labels

conditioning units, it depends on the rate of EER (Energy Efficiency Rate) for summer and of COP (Coefficient Of Performance) for winter. The higher these factors are, the better energy classifications the units have entailed more economical operations. The energy labels classify air conditioners units based on SEER and SCOP as shown in Fig. 8.2 [32].

Energy Stars is another rating criterion for efficiency evaluation of electrical appliances. It has been used in Australia for more than 30 years. Energy Rating Labels appear on all sorts of appliances, from air conditioners to fridges, washing machines and dryers.

For an air conditioner, they can be rated between 1 and 6 stars, but as technology is improving, so too is energy efficiency. Now, it is possible for air conditioners to be rated as many as 10 stars, thanks to new 'Super Efficiency Ratings'—an additional band of 4 stars that appears when an appliance is extra efficient. Figure 8.3 shows the energy star rating for cooling and heating [33].

Table 8.3 Air-conditioner units test procedure comparison summary

Country/Standard	Test method	Energy efficiency parameter	Type of test	Test conditions
ISO	ISO5151	EER	Calorimetric, Air enthalpy	Entering indoor 27(DB)/19(WB) Entering outdoor 35 (DB)
US	10 CFR 430 subpart B Appendix M & M1	EER, SEER (Cooling) HSPF (Heating) SEER2 (Cooling) HSPF2 (Heating)	Air enthalpy	Entering indoor: 26.7(DB)/19.4 (WB) Entering outdoor: 35 (DB) Entering indoor: 27 (DB)/19 (WB) Entering outdoor: 28 (DB)
EU	BS EN 14,511:2018	SEER (Cooling) SCOP (Heating)	Calorimetric, Air enthalpy	Entering indoor: 27 (DB)/19 (WB) Entering outdoor: 35 (DB)
Australia/New Zealand	AU/NZS 3823.4.1:2014	EER, CSPF (Cooling) HSPF (Heating)	Calorimetric, Air enthalpy	Entering Indoor: 27 (DB) / 19 (WB) Entering Outdoor: 35 (DB)
Japan	JIS B 8615–1:2013 JIS B 9612:2013	CSPF (Cooling) HSPF (Heating)	Calorimetric, Air enthalpy	Entering indoor: 27 (DB)/19 (WB) Entering outdoor: 35 (DB)
China	GB/T 7725–2004	EER, SEER (Cooling) HSPF (Heating)	Calorimetric, Air enthalpy	Entering indoor: 27 (DB)/19 (WB) Entering outdoor: 35 (DB)
Korea	KSC 9306 2017	CSPF, EER (Cooling) HSPF, COP (Heating)	Calorimetric, Air enthalpy	Entering indoor: 27 (DB)/19 (WB) Entering outdoor: 35 (DB)

8.4 Sample Calculations for Conventional Chiller Performance

The coefficient of performance (COP) is the well-accepted parameter to evaluate the efficiency of conventional chillers. In conventional chillers, there are two main components namely chiller unit and heat rejection unit. The heat rejection unit can be water-based like cooling tower (CT) or air-based like radiator.

In these sample calculations, we considered a water-cooled chiller of nominal capacity 20 Rton as shown in Fig. 8.4. The detail of the calculation is in Table 8.4.

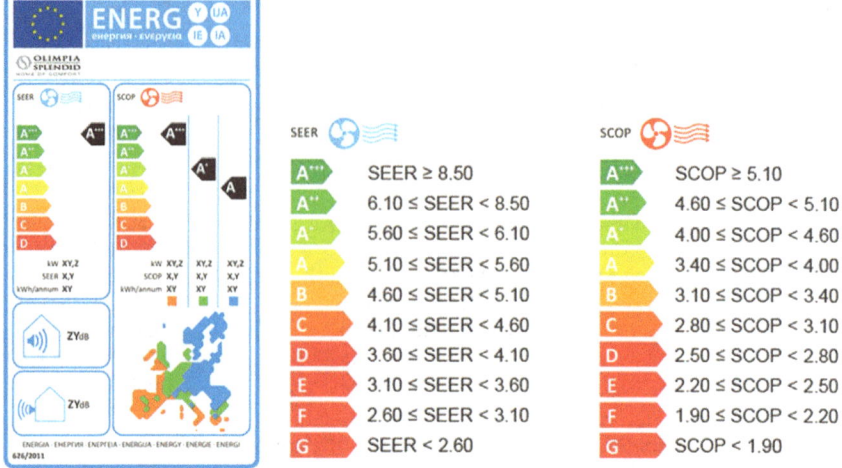

Fig. 8.2 Air conditioner classifications based on energy labels

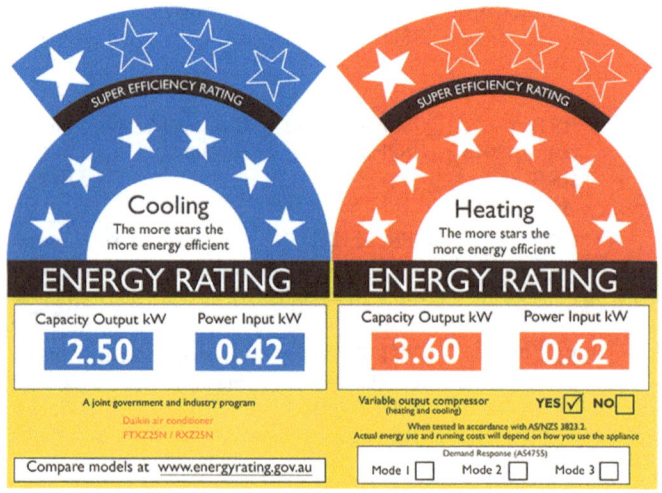

Fig. 8.3 Air conditioner classifications based on energy stars

8.5   Performance Evaluation of Indirect Evaporative Coolers

Unlike the conventional chillers, there is no international standard available for Indirect Evaporative Cooler (IEC) testing. Many national and regional standards are developed based on local conditions to evaluate the IEC performance. The most

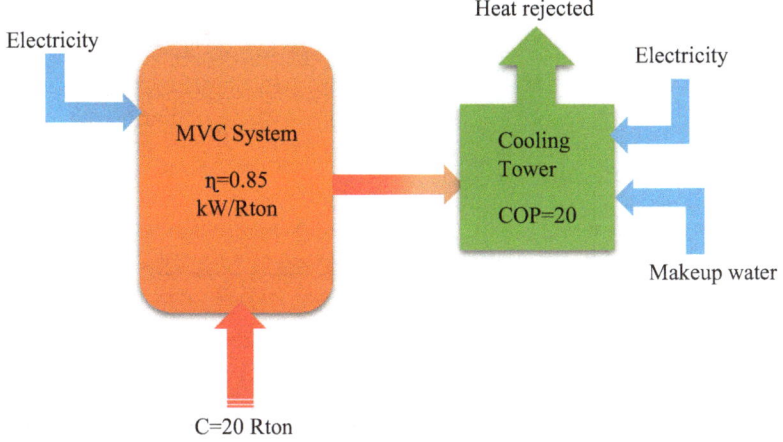

Fig. 8.4 Representative water-cooled chiller of nominal capacity 20Rton

Table 8.4 Sample calculations for MVC system

Sr#	Description	Equation	Calculation	Units
1	Nominal capacity	C	20 70.40	Rton kW_{th}
2	Chiller efficiency	η_{MVC}	0.85	kW/Rton
3	Electricity consumed by chiller	E_{MVC}	17	kW_e
4	CT performance	COP_{CT}	20	
5	Total heat rejected to CT	$HR_{Total} =$ $(C + E_{mvc}) + \left(\frac{C+E_{mvc}}{COP_{CT}-1}\right)$	92	kW_{th}
6	Saturation enthalpy	h_{fg}	2300	kJ/kg
7	Water consumed in CT	$\dot{m} = \frac{HR_{Total}}{h_{fg}}$	0.04 3.38	kg/s m^3/day
8	Electricity-related to water (assuming 3.5 kWh/m^3)	$E_{water\_consumed} =$ $\dot{m}_w e_{\_desalination}$	11.84	kWh_e/day
10	COP	$COP_{MVC} = \dfrac{C}{E_{MVC}+\frac{HR_{Total}}{COP_{CT}}}$	3.26	

common standards are (i) ANSI/ASHRAE Standards 133-2008/143-2007 and California Appliance Efficiency Regulations, USA, (ii) SASO35/36, Saudi Arabia, (iii) GB/T 25,860–2010, China, (iv) IS 3315-1974, India, (v) C22.2 No. 104, Canada and (vi) AS/NZS 2913–2000, Australia [31].

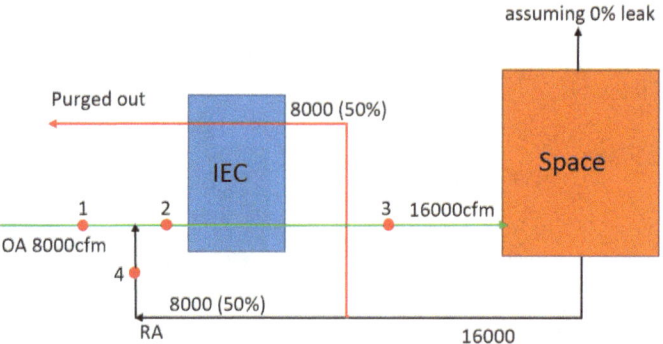

Fig. 8.5 Representative IEC unit of nominal capacity 20Rton

All these standards take into account different operating conditions and assumptions. To demonstrate the performance calculation procedure, we considered an IEC unit of nominal capacity of 20 Rton as shown in Fig. 8.5. The detailed calculations are presented in Table 8.5.

8.6 Performance Comparison

It can be seen clearly that the IEC unit performance is better than conventional chillers. In addition, it does not utilize chemical-based refrigerants, so it helps to reduce global warming effect. The comparison of performance is presented in Table 8.6.

8.7 Conclusions

There are well-established standards for conventional chillers testing and performance evaluation. Many international standards are modified for national use at different locations. Thus far, there is no known international standard for IEC unit testing even though they have great potential for future cooling applications. A simple mathematical model was used to compare the performance of conventional chillers and IEC. Key results revealed that the IEC has higher COPs while consuming less water when compared with conventional chillers.

Table 8.5 Sample calculations for IEC system

Indirect Evaporative Cooler

Sr#	Description	Equation	Calculation	Units
1	Nominal capacity	C	20 70.40	Rton kW_{th}
2	Air flow rate	$\dot{m}_{air}$	800 16,000	cfm/Rton cfm
	Purge air (PA) ratio		50	%
	Return air (RA) ratio		50	%
	Outdoor air (OA) ratio		50	%
	Outdoor air temperature (1)	T_{OA_1}	45	°C
	Mixed air temperature (2)	T_{MA_2}	35	°C
	Return air temperature (4)	T_{RA_4}	25	°C
	Supply air temperature (3)	T_{SA_3}	20	°C
3	Enthalpy of OA (45 °C and 12 g/kg)	$h_{OA\_45\_12}$	70	kJ/kg
4	Enthalpy of MA (35°C and 12 g/kg)	$h_{MA\_35\_12}$	65	kJ/kg
	Enthalpy of SA (20°C and 12 g/kg)	$h_{SA\_20\_12}$	17	kJ/kg
15	Blower power (η = 60%)	$P_{blower} = \dfrac{[\rho \cdot Q] \cdot [g \cdot H_{loss\_across\_MD.IEC}]}{\eta_{blower}}$ $P_{blower} = \dfrac{\dot{m}_{air\_total} \cdot [g \cdot H_{loss\_across\_MD.IEC}]}{\eta_{blower}}$	200 5.0	Watt/Rton kW
16	Water pump power (2 bar pressure)		100 2.0	Watt/Rton kW
17	Total heat rejected by IEC	$HR_{Total} = (C + P_{blower})$	75.40	kg/s
22	Total water consumption	$\dot{m}_{water\_IEC} = \dfrac{HR_{total}}{h_{fg\_vapor}}$	0.03 2.77	kg/s m^3/day
23	Water related electricity	$E_{water\_consumed} = \dot{m}_w \cdot e_{\_desalination}$	9.70	kWh$_e$/day

(continued)

Table 8.5 (continued)

Indirect Evaporative Cooler

Sr#	Description	Equation	Calculation	Units
25	Overall COP (cooling only)	$COP_{cooling\_only} = \frac{\dot{m}_{air}(h_{OA}-h_{SA})}{P_{blower}+P_{pump}}$	21.52	

Table 8.6 Conventional chiller and IEC performance comparison

Description	Water-cooled chiller	IEC
Nominal capacity (Rton)	20	20
COP	3.26	21.25
Water consumption (m^3/day)	3.38	2.77
Global warming potential	high	zero

References

1. Carmichael R (2020) Domestic air conditioner test standards and harmonization. Final Report. Cadeo Group, Washington, DC
2. ISO 5151:2017(en) Non-ducted air conditioners and heat pumps—testing and rating for performance
3. 10 CFR Appendix M to Subpart B of Part 430—Uniform Test Method for Measuring the Energy Consumption of Central Air Conditioners and Heat Pumps https://www.govinfo.gov/content/pkg/CFR-2017-title10-vol3/pdf/CFR-2017-title10-vol3-part430-subpartB-appM.pdf
4. BS EN 14511-1 2018 Edition (2018) Air conditioners, liquid chilling packages and heat pumps for space heating and cooling and process chillers, with electrically driven compressors Part 1: Terms and definitions. British Standards Institution
5. BS EN 14511-1:2018 (n.d.) (2018) Air Conditioners, liquid chilling packages and heat pumps for space heating and cooling and process chillers, with electrically driven compressors (Part 1)
6. BS EN 14511-2:2018 (n.d.) (2018) Air Conditioners, liquid chilling packages and heat pumps for space heating and cooling and process chillers, with electrically driven compressors (Part 2)
7. BS EN 14511-3:2018 (n.d.) (2018) Air Conditioners, liquid chilling packages and heat pumps for space heating and cooling and process chillers, with electrically driven compressors (Part 3)
8. BS EN 14511-4:2018 (n.d.) (2018) Air Conditioners, liquid chilling packages and heat pumps for space heating and cooling and process chillers, with electrically driven compressors (Part 4)
9. AS/NZ 3823.4.2:2014 (n.d.) (2014) Performance of electrical appliances - Air conditioners and heat pumps
10. AS/NZS 3823.1.1:2012 (n.d.) (2012) Performance of electrical appliances - Air conditioners and heat pumps (Part 1.1)
11. AS/NZS 3823.4.1:2014 (n.d.) (2014) Performance of electrical appliances - Air conditioners and heat pumps (Part 4.1)
12. AS/NZS 3823.4.1:2014 (2014) Performance of electrical appliances - Air conditioners and heat pumps, Part 4: Air-cooled air conditioners and air-to-air heat pumps -Testing and calculating methods for seasonal performance factors - Cooling seasonal performance factor (ISO 16358-1:2013, MOD)

13. ISO 16358-1:2013 (2013) Air-cooled air conditioners and air-to-air heat pumps—testing and calculating methods for seasonal performance factors — Part 1: Cooling seasonal performance factor
14. ISO 16358-1:2013/AMD 1:2019 (2019) Air-cooled air conditioners and air-to-air heat pumps—testing and calculating methods for seasonal performance factors — Part 1: Cooling seasonal performance factor—Amendment 1
15. ISO 13253:2017(en) (2017) Ducted air-conditioners and air-to-air heat pumps—testing and rating for performance
16. ISO 15042:2017 (2017) Multiple split-system air conditioners and air-to-air heat pumps—testing and rating for performance
17. AS/NZS 3823.1.1:1998, Appendix ZZ (2010) Performance of electrical appliances—airconditioners and heat pumps Non-ducted airconditioners and heat pumps—testing and rating for performance (ISO 5151:2010, MOD)
18. JIS C 9335-2-40:2004 (2004) Household And Similar Electrical Appliances—Part 2-40: Particular requirements For Electrical Heat Pumps, Air-conditioners And Dehumidifiers. Japanese Standards Association
19. JIS B 8615-1 (2013) Non-ducted air conditioners and heat pumps—Testing and rating for performance. Japanese Standards Association
20. JIS C 9612:2005 (2005) Room Air Conditioners. Japanese Standards Association
21. SBTS, State Bureau of Technical Supervision (1989) Limited values of energy consumption and method of testing for room air conditioners. GB 12021.3-89. China
22. SBTS, State Bureau of Technical Supervision (1996) Room Air Conditioners. GB/T7725. China
23. SSB, State Statistical Bureau (1998) ZhongguoGongyeJingjiTongjiNianjian (China Industrial Economic Statistics Yearbook). China
24. SSB, State Statistical Bureau (1998) Zhongguo Nengyuan TongjiNianjian (China Energy Statistical Yearbook)
25. SSB, State Statistical Bureau (1999) Zhongguo Tongji Nianjian 1999 (China Statistical Yearbook, 1999). China
26. KS C 9306 (2011) Air Conditioners. Korean Standards Association
27. KS C 9306 (2017) Air Conditioners. Korean Standards Association
28. Climate Technology Center & Network. https://www.ctc-n.org/technologies/efficient-air-conditioning-systems
29. https://learnmetrics.com/eer-rating/
30. ANSI/AHRI 210/240–2008: 2008 Standard for Performance Rating of Unitary Air-Conditioning & Air-Source Heat Pump Equipment.
31. Duan Z, Zhan C, Zhang X, Mustafa M, Zhao X, Alimohammadisagvand B, Hasan A (2012) Indirect evaporative cooling: Past, present and future potentials. Renew Sustain Energy Rev 16:6823–6850. https://doi.org/10.1016/j.rser.2012.07.007
32. https://www.olimpiasplendid.com/seer-scop.
33. https://www.daikin.com.au/articles/buyers-guide/understanding-energy-stars-when-buying-air-conditioner

Chapter 9
Thermo-Economic Analysis for Cooling Cycles

Abbreviations

c	Cost, $
CRF	Capital recovery factor
i	Interest rate, %
n	Lifespan, year
P	Power, W
Q	Heating/cooling rate, W
t	Operation time, hour

Subscript

cool	Cooling
elec	Electricity
h	Heat
inv	Initial investment
O&M	Operational and maintenance
th	Thermal

9.1 Introduction

With the increase in global cooling demand, air conditioning is becoming increasingly crucial, and growing interests have been sharply focused on developing novel cooling processes that are energy-efficient, cost-effective, and environmentally friendly.

© Springer Nature Singapore Pte Ltd. 2021

C. Kian Jon et al., *Advances in Air Conditioning Technologies*, Green Energy and Technology, https://doi.org/10.1007/978-981-15-8477-0_9

Currently, the air-conditioning market is dominated by the mechanical vapour compression refrigeration system. Due to its long history and massive scale production, this technology has many advantages, such as high technology maturity, good stability, low cost, and a high coefficient of performance (COP). However, mechanical chillers can only be driven by electricity, which consumes fossil fuels and leads to greenhouse gas emissions. Therefore, mechanical chillers are regarded to be unsustainable.

The absorption and adsorption cooling systems are potential alternatives to conventional mechanical vapour compression systems. They employ thermal compressors, i.e. liquid desiccant and solid absorbent, thus eliminating the need for a mechanical compressor and significantly reducing electricity consumption. However, these systems have a lower COP and require a substantial amount of heat to support its operation. Therefore, they are more suitable for applications where cost-effective thermal energy is available. Moreover, the absorption and adsorption systems have complex system configurations and high system costs, which further reduce their attraction to users.

The indirect evaporative cooling technology works in a different principle. Instead of relying on vapour compression for heat removal, it employs the evaporative potential of unsaturated air as the driving force for cooling. The system is energy efficient since it consumes only a small amount of electricity for air and water pumping. The system configuration is also simple, making it economically competitive.

To allow a direct comparison of the existing cooling cycles, this chapter conducts an economic analysis of different technologies to achieve their life-cycle costs. Thermodynamic and economic data will be collected first, based on which the levelized cost of cooling, expressed in $/kWh, will be calculated under different scenarios. The effects of key parameters on the final cost will also be evaluated. The results will provide a common platform to compare the existing cooling cycles and enable quick decision-making.

9.2 Economic Model

The levelized cost of cooling over the lifetime of a system consists of initial equipment costs, annual maintenance and operation (O&M), and energy costs. It can be expressed as

$$c_{cool} = \frac{C_{inv} \times CRF + C_{O\&M} + Q_h \times c_{th} + P \times c_{elec}}{\dot{Q}_{cool} \times t} \tag{9.1}$$

where C_{inv} is the initial investment and $C_{O\&M}$ is the O&M costs. Q_h, P, Q_{cool}, and t are the annual heat consumption, electricity consumption, cooling power, and operation time, respectively. c_{th} and c_{elec} represent the price of electricity and heat, respectively. **CRF** is the capital recovery factor, which is calculated as

$$CRF = \frac{i \times (1+i)^n}{(1+i)^n - 1} \tag{9.2}$$

where i is the interest rate and n is the plant lifespan.

9.3 Economic Data

To enable thermo-economic analysis on the cooling systems, economic and thermodynamic data are firstly collected from the open literature. As highlighted in the previous section, the required data include initial system costs, annual maintenance and operational costs, and energy consumption of the system.

Table 9.1 summarizes the economic and performance data for a mechanical vapour compression chiller with a rated cooling capacity of 34.3 kW [1]. The initial cost includes the purchase and installation costs of the chiller as well as the circulation pumps and the fan coil. The overall initial cost is $13,086, yielding a unit initial cost of $381/kW. The annual O&M costs are reported to be $2093, and the coefficient-of-performance (COP) is 3.5.

Table 9.2 shows the thermo-economic data for a LiBr/H_2O absorption chiller [2]. In addition to evaporator, condenser, absorber, and generator, several heat exchangers are included to allow internal heat recovery, and the overall initial cost is $45,564.6. The annual O&M cost is not directly available in the literature. It is assumed to be 5% of the initial cost due to small requirements for maintenance and operation for absorption chillers. The COP of the chiller is 0.72, which represents the thermal energy consumption. A small amount of electricity is required for pumping the fluids, and its cost is also considered for the sake of completeness.

Table 9.3 gives the costs and performance data for a silica gel–water adsorption chiller [3]. The chiller consists of two absorbers, an evaporator, a condenser, and several pumps and valves. A controller is also required to enable automatic switch of adsorption/desorption mode for two absorbers. The unit cost is $541.2/kW, which is higher than the mechanical chillers and absorption chillers. On the other hand, its COP is much lower.

	Variable	Value
Costs	Purchase cost ($)	9840
	Installation cost ($)	1680
	Circulation pump ($)	234
	Fan coil ($)	1332
Performance	O&M costs (exclude energy cost)	2093
	Capacity (kW)	70
	COP	3.5

Table 9.1 Economic and performance data for a mechanical vapour compression chiller [1]

Table 9.2 Economic and performance data for a LiBr/H$_2$O absorption chiller [2]

	Variable	Value
Costs	Generator ($)	8458.8
	Condenser ($)	13,838
	Evaporator ($)	10,047.1
	Absorber ($)	8612.4
	Heat exchanger ($)	4608.3
	O&M costs (exclude energy cost)	5% of initial cost
Performance	Capacity (kW)	201.29
	COP (thermal only)	0.72
	Electricity (kW/kW)	0.05

Table 9.3 Economic and performance data for a silica gel–water adsorption chiller [3]

	Variable	Value
Costs	Absorbers ($)	1752
	Evaporators and condensers ($)	1569
	Pumps and valves ($)	458
	Frame ($)	157.8
	Controller ($)	30
	O&M costs (exclude energy cost)	5% of initial cost
Performance	Capacity (kW)	7.33
	COP (thermal only)	0.4
	Electricity consumption (kW/kW)	0.04

The indirect evaporative cooler uses the evaporative potential of unsaturated air as the driving force for cooling. Therefore, it has a simple design and high energy efficiency. Also, the requirement for maintenance and operation is low. Table 9.4 provides the economic and performance data for an indirect evaporative cooler. Following these data reported in the open literature [4], the initial cost for a 10 kW

Table 9.4 Economic and performance data for an indirect evaporative cooler [4]

	Variable	Value
Costs	Purchase cost ($)	2600
	Capacity (kW)	10
	O&M costs	2% of initial cost
Performance	Capacity (kW)	10
	COP	10

Table 9.5 Costs of electricity and thermal energy

Energy	Source	Cost ($/kWh)
Electricity [5]	Solar photovoltaic	0.151
	Combined gas cycle	0.074
Thermal energy [6]	Solar thermal collector	0.032
	Gas heater	0.059
	Waste heat	0

unit is estimated to be $2600, and the annual O&M cost is considered as 2% of the initial cost.

The costs of cooling systems are also sensitive to the prices of electricity and thermal energy. We consider several sources of electricity and thermal energy to represent different applications. Table 9.5 provides the prices for electricity and heat from different sources.

Other input data for the economic model include plant lifespan, interest rate, and annual operating hours. Without loss of generality, the plant lifespan is assumed to be 20 years for all the cooling systems, while the annual interest rate and operating hours are 5% and 3000, respectively. A sensitivity analysis will be conducted in the ensuing sections to evaluate the effects of these parameters.

9.4   Levelized Costs of Cooling Systems

Employing the model and data presented in the previous sections, this section evaluates the levelized costs for different cooling systems. Considering the sources of energy, three scenarios are defined to represent different applications. Scenario 1 is the regular case that electricity is obtained from a power plant, and heat is provided by a gas heater. Scenario 2 considers the solar application, which uses PV panels and solar thermal collectors for electricity and heat generation, respectively. Scenario 3 represents waste heat utilization, which assumes that abundant waste heat is available. Table 9.6 summarizes the energy sources for the different scenarios.

Table 9.6 Energy sources for different scenarios

Scenario	Electricity	Thermal energy
1	Combined gas cycle	Gas heater
2	Photovoltaic	Solar thermal collector
3	Combined gas cycle	Waste heat

9.4.1 Base Case

Figure 9.1a shows the specific costs of different systems under Scenario 1. It is clear that energy costs constitute greater than 50% of the overall cost for all the cycles. The costs for the absorption and adsorption chillers are several times higher than the other two systems due to a significant amount of thermal energy consumption. On the other hand, the indirect evaporative cooler is much cheaper as a result of low initial and O&M costs coupled with high energy efficiency.

When solar energy is employed to drive cooling systems, the price of electricity is slightly higher, while thermal energy becomes cheaper. Therefore, it favours heat-driven systems. Figure 9.1b shows the costs of different systems under Scenario 2. The costs for the mechanical chiller and the indirect evaporative cooler are increased due to more remarkable electricity costs. On the other hand, the absorption and adsorption chillers benefit from cost-effective solar thermal energy. The indirect evaporative cooler remains the most economic cooling option under such a scenario, while the cost of absorption chiller becomes comparable with that of the mechanical chiller.

The costs of the thermally driven systems can be further reduced when waste heat is available. The actual cases include combustion gasoline engines, power plants, as well as various industrial processes. Low-grade heat from these processes would otherwise be disposed of if unrecovered. As shown in Fig. 9.1c, the absorption chiller overtakes the indirect evaporative cooler and becomes the most cost-effective under Scenario 3. The cost of the adsorption chiller is also significantly reduced. The mechanical chiller is now the most expensive because of high electricity costs.

9.4.2 Effect of Initial Costs

The above results are based on an annual operation time of 3000 h. If the operation time is increased, the initial cost is allocated to larger cooling capacity, and thus the cost will decrease. Figure 9.2 shows the impact of annual operating hours on cooling costs under different scenarios. The overall trend is similar, but the magnitude of change is different.

For the heat-driven systems, the costs are more sensitive to the change of energy costs, and the effects of the operating hour are different under different scenarios. Under scenarios 1 and 2, the costs are mainly attributed to the energy cost due to both absorption chiller and adsorption chiller, while the contribution of the initial costs is relatively small. Consequently, an increase in the operation hour has a minor impact on the final cost. When the annual operation hour is increased from 3000 to 6000, the costs for absorption and adsorption chillers decrease by approximately 10%, as shown in Fig. 9.2a, b. On the other hand, since thermal energy is free under Scenario 3, initial costs become more significant, and increasing the operation hour leads to

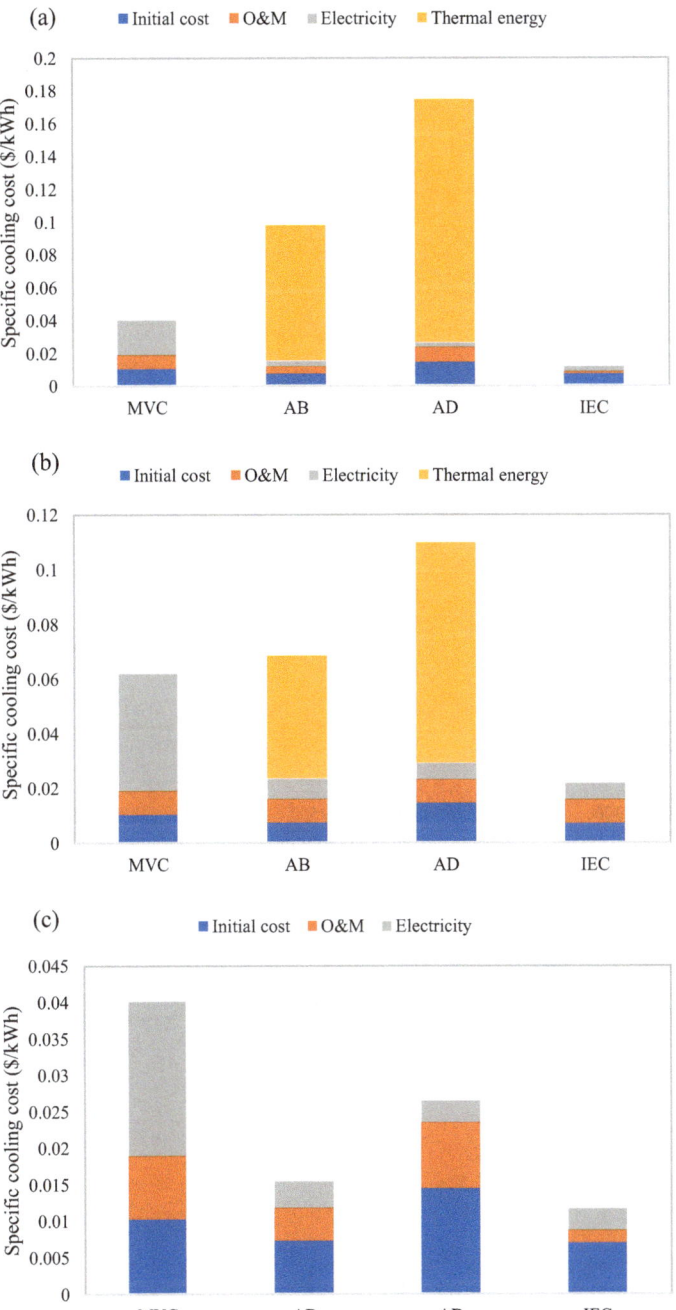

Fig. 9.1 Specific cooling costs for different systems under **a** Scenario 1, **b** Scenario 2, and **c** Scenario 3

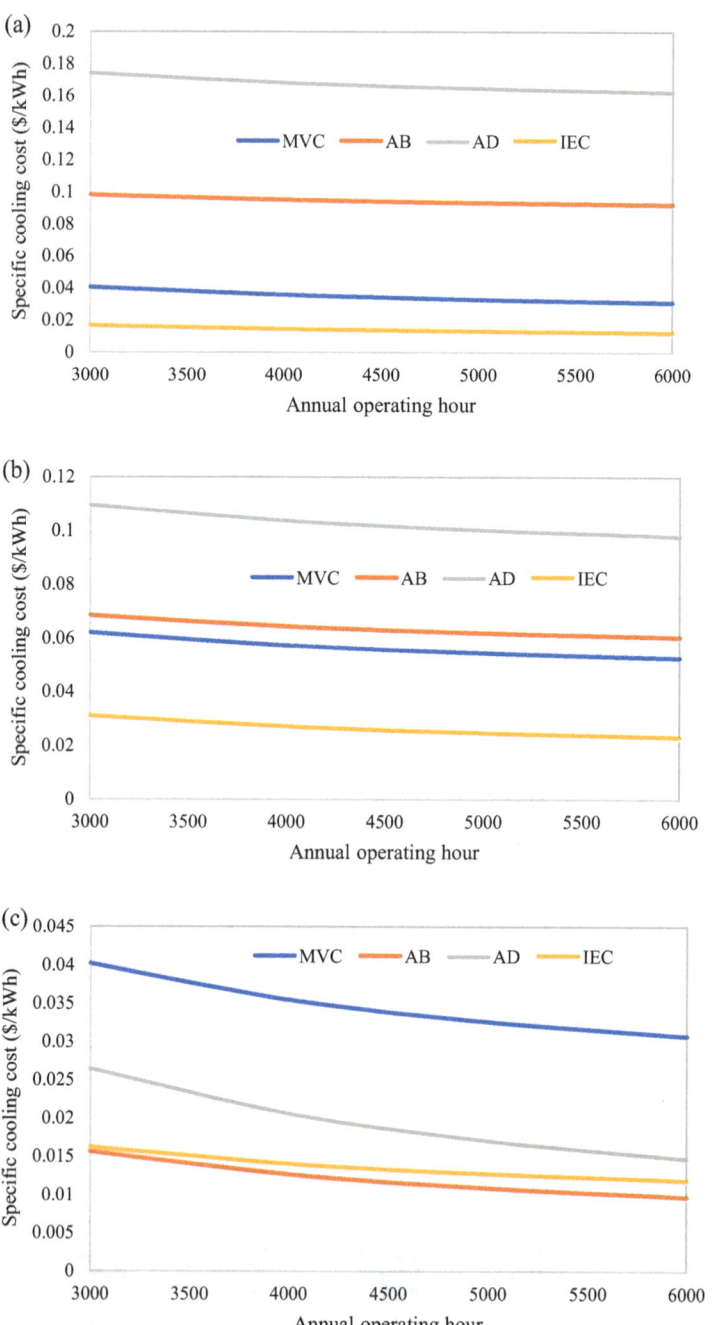

Fig. 9.2 Impact of annual operating hour on cooling costs under **a** Scenario 1, **b** Scenario 2, and **c** Scenario 3

40–30% of cost reduction for absorption and adsorption chillers, respectively, as can be seen in Fig. 9.2c.

The trend for electricity-driven systems, on the other hand, is similar under three scenarios due to a small change in electricity price. Under the range of operation hours considered, the cost for the mechanical chiller is reduced by 24% under Scenario 1 and Scenario 3, while the drop is 12% under Scenario 2. For the indirect evaporative cooler, the variation of the final cost is approximately 25% for all three scenarios.

The interest rate also impacts the contribution of the initial cost. When the interest rate is higher, the value of the initial cost that is allocated to the unit cooling effect is also higher, leading to a higher final cost for cooling, as revealed in Fig. 9.3. Similar to the effect of the annual operating hour, the interest rate has the most impact on the costs of the thermally driven chillers under Scenario 3. When the interest rate increases from 1 to 9%, the costs for absorption and adsorption chillers appreciate by 37% and 44%, respectively, under Scenario 3, while the changes are less than 10% under the other two scenarios. For the mechanical chiller, the variations of costs are 18, 12, and 18% for the three scenarios, while for the indirect evaporative cooler the changes are 33, 16, and 33%.

9.4.3 Effect of Energy Efficiency

The costs for cooling are also subjected to energy efficiency, which affects the annual energy costs. Figure 9.4a shows the final cost of mechanical chiller under different COPs. The COP varies from 2 to 5 considering factors like different weather conditions, partial-load operation, and chiller performance degradation. As shown in the figure, cost reduction with increasing COP is more obvious under Scenario 2 which has a higher electricity price. The results quantify the benefits of selecting efficient chillers, optimal operation, and regular maintenance (e.g. wash the condenser).

Figure 9.4b shows the costs of the absorption chiller under different energy efficiencies. The cost reduction is most significant under Scenario 1 which has the highest thermal price, while no variation is observed under Scenario 3 that offers free thermal energy. Under the range of COP considered, cooling cost can be reduced by 0.082 and 0.045 $/kWh for Scenario 1 and Scenario 2, respectively

In actual operation, the common ways of increasing the COP include increasing the regeneration temperature and reducing the cooling water temperature [7]. However, these methods are subjected to the availability of heat source and cooling medium. Another effective way is to add heat exchangers to enable internal heat recovery [8], but it induces additional initial investment, and its economic viability needs to be further evaluated.

Figure 9.4c shows the effect of COP on the cost of the adsorption chiller. Similar to the absorption chiller, the cost for the adsorption chiller can be reduced remarkably when increasing the COP under Scenarios 1 and 2, while the cost remains constant under Scenario 3. When the COP is increased from 0.2 to 0.7, the cost reduction is 0.21 and 0.11 $/kWh, for Scenario 1 and Scenario 2, respectively. They represent 66

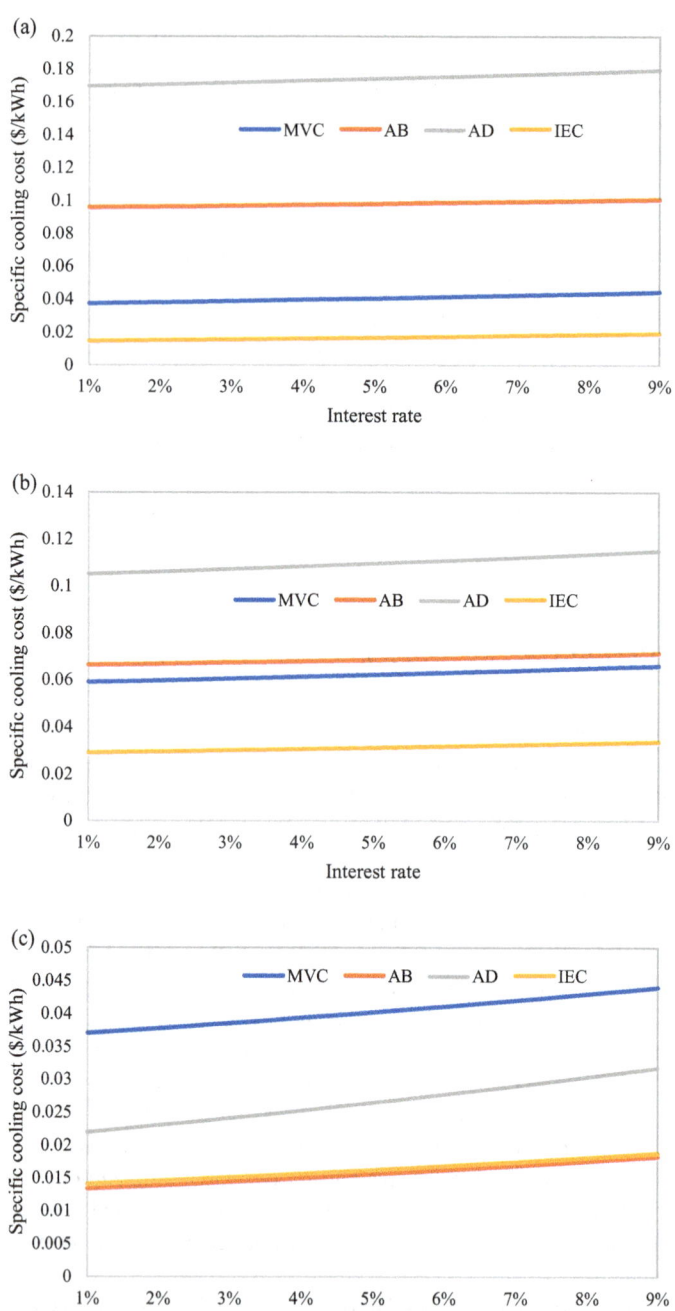

Fig. 9.3 Impact of interest rate on cooling costs under **a** Scenario 1, **b** Scenario 2, and **c** Scenario 3

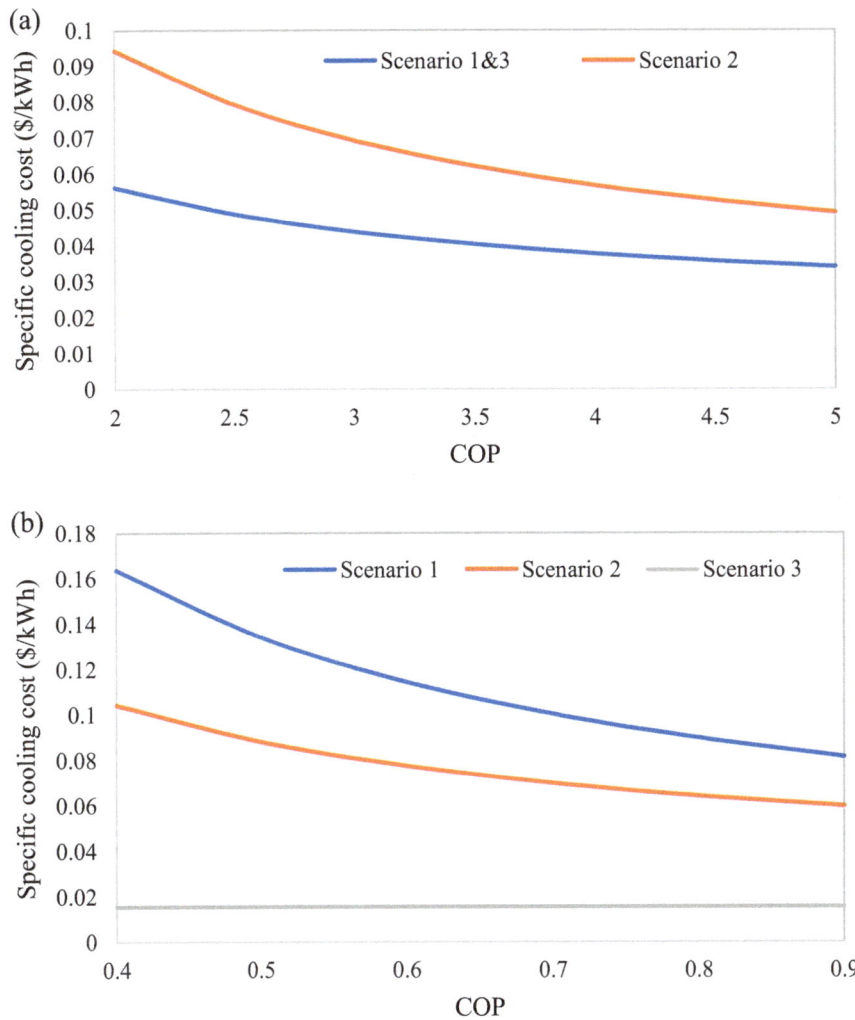

Fig. 9.4 Impact of COP on cooling costs for **a** mechanical vapour compression chiller, **b** absorption chiller, **c** adsorption chiller, and **d** indirect evaporative cooler

and 60% of the initial costs with COP of 0.2. Such significant changes are attributed to a high fraction of energy cost

From the design point of view, the energy efficiency of the adsorption chiller can be improved by enhancing heat transfer in the adsorbent beds [9], conducting internal heat and mass recovery [10, 11] and optimizing the system configuration [12]. The COP can also be improved during operation by increasing the regeneration temperature, reducing the cooling water temperature, and optimizing the cycle time [13].

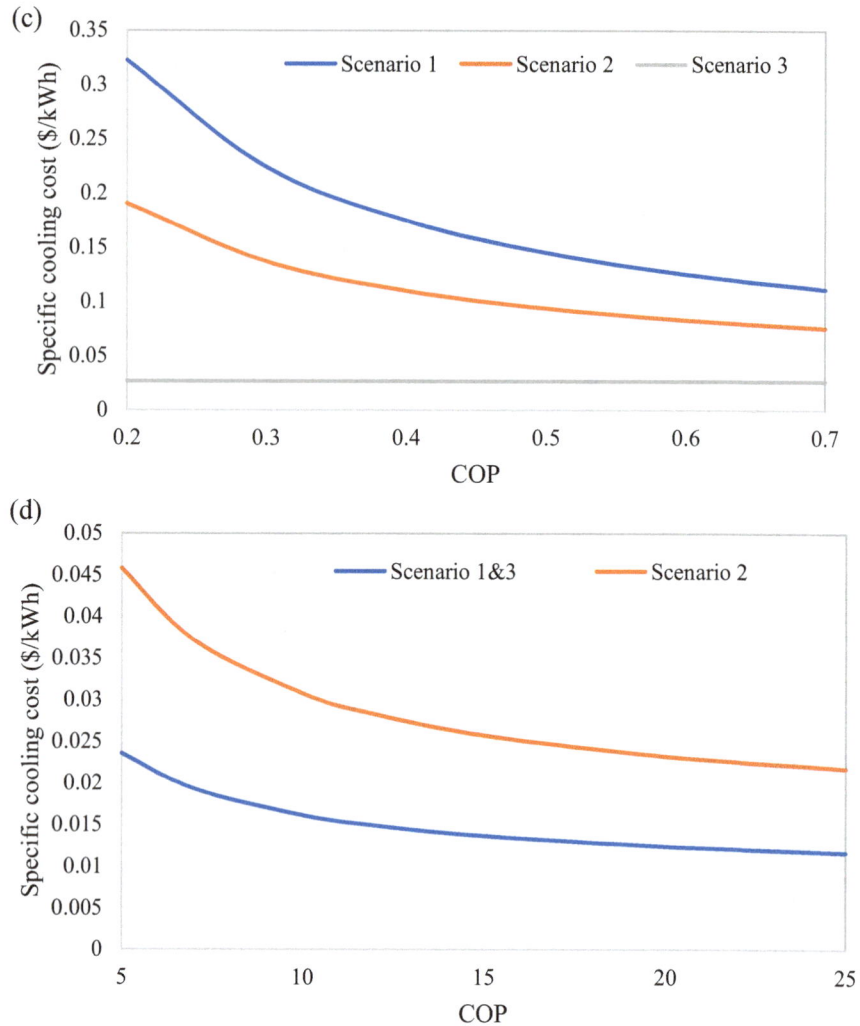

Fig. 9.4 (continued)

Figure 9.4d shows the cost of the indirect evaporative cooler under different COPs. Under all scenarios, the cost reduces significantly when the COP is increased from 5 to 10, after which the trend slows down. When the cooler COP is below 10, the energy cost is significant compared with other costs, and increasing COP is able to effectively reduce the cost. After COP gets higher than 10, the energy cost becomes insignificant, and further increase of the COP has little effect on the final cost

Due to the simplicity of the configuration, achieving the optimal design for an indirect evaporative cooler is not difficult. Therefore, the factors that may limit its energy efficiency come from the operation period. The most critical parameter that determines the performance of the cooler is the inlet air humidity, as it defines the driving force for cooling. A common sense is that the indirect evaporative cooler is suitable only for tropical areas. When the humidity ratio of the air is high, a dehumidifier is usually required to remove the moisture in the air before supplied to the indirect evaporative cooler [14–16]. Such a system integration has higher initial costs. Also, the overall system efficiency degrades severely since the COP of dehumidifiers is usually low [17–19]. As a result, the indirect evaporative cooler may lose its economic competitiveness in humid regions.

9.4.4 Chiller Selection

From the above results, it is clear that the indirect evaporative cooler is the most economic choice in most scenarios, and the absorption becomes competitive when waste heat is available. However, in actual applications, price is not the only consideration. For example, the indirect evaporative cooler requires a low outdoor air humidity, and its performance degrades when the air humidity is high. On the other hand, although the adsorption chiller is expensive under most scenarios, it has no moving parts and is more stable under harsh environments. Therefore, it is a more viable choice than absorption chillers for application on the ship, which is subjected to severe vibrations. Figure 9.5 depicts the scheme for selecting the appropriate chillers by considering their features, applicability, and costs of different technologies.

9.5 Conclusions

This chapter conducts an economic analysis on different cooling cycles considering the initial investments, the operation and maintenance costs, and the energy costs. Different sources of energy are considered to cover various applications. Under the regular scenario, i.e. electricity and heat are obtained from the power plant and the gas heater, respectively, the mechanical chiller and the indirect evaporative cooler are cost-effective, while heat-driven processes like absorption and adsorption cooling cycles are expensive due to high thermal energy costs. When solar energy is employed as the heat source, the price for thermal heat gets lower, and the cost of the absorption chiller becomes comparable with that of the mechanical chiller. Costs of thermally driven cycles can be further reduced if waste heat is available, under which scenario the absorption chiller becomes the most cost-effective option. The costs of different systems are also subjected to the variation of the interest rate, the annual operation hour, and the cooling system's energy efficiency, which emphasize their importance for optimal design, selection, and operation. The derived results offer a robust and convenient basis for the selection of cooling systems in real applications.

Fig. 9.5 Decision tree for the selection of cooling cycles

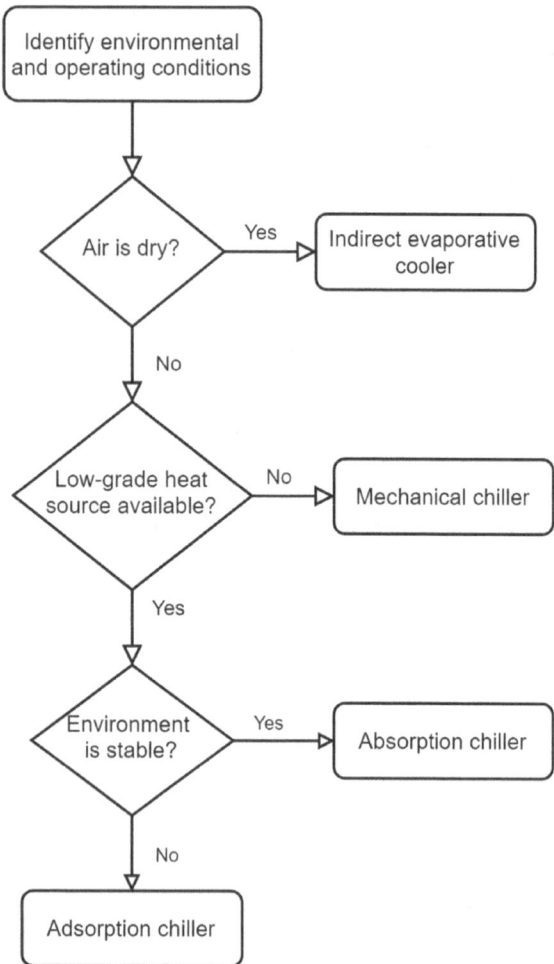

References

1. Imal M, Yılmaz K, Pınarbaşı A (2015) Energy efficiency evaluation and economic feasibility analysis of a geothermal heating and cooling system with a vapor-compression chiller system. Sustainability 7(9):12926–12946
2. Misra R, Sahoo P, Gupta A (2005) Thermoeconomic optimization of a LiBr/H$_2$O absorption chiller using structural method. J Energy Resour Technol 127(2):119–124
3. Lambert MA, Beyene A (2007) Thermo-economic analysis of solar powered adsorption heat pump. Appl Therm Eng 27(8–9):1593–1611
4. Camargo J, Ebinuma C, Silveira J (2003) Thermoeconomic analysis of an evaporative desiccant air conditioning system. Appl Therm Eng 23(12):1537–1549
5. Ray D (2019) Lazard's levelised cost of energy analysis—version 13.0. Lazard, New York, NY, USA, p 20
6. Wang, R. and X. Zhai, Handbook of energy systems in green buildings. 2018: Springer.

7. Şencan A, Yakut KA, Kalogirou SA (2005) Exergy analysis of lithium bromide/water absorption systems. Renewable Energy 30(5):645–657

8. Ng K, Tu K, Chua H, Gordon J, Kashiwagi T, Akisawa A, Saha BB (1998) Thermodynamic analysis of absorption chillers: internal dissipation and process average temperature. Appl Therm Eng 18(8):671–682

9. Rezk A, Al-Dadah R, Mahmoud S, Elsayed A (2013) Effects of contact resistance and metal additives in finned-tube adsorbent beds on the performance of silica gel/water adsorption chiller. Appl Therm Eng 53(2):278–284

10. Uyun AS, Miyazaki T, Ueda Y, Akisawa A (2009) Experimental investigation of a three-bed adsorption refrigeration chiller employing an advanced mass recovery cycle. Energies 2(3):531–544

11. Myat A, Choon NK, Thu K, Kim Y-D (2013) Experimental investigation on the optimal performance of Zeolite–water adsorption chiller. Appl Energy 102:582–590

12. Chua H, Ng K, Malek A, Kashiwagi T, Akisawa A, Saha BB (2001) Multi-bed regenerative adsorption chiller—improving the utilization of waste heat and reducing the chilled water outlet temperature fluctuation. Int J Refrig 24(2):124–136

13. Li A, Ismail AB, Thu K, Ng KC, Loh WS (2014) Performance evaluation of a zeolite–water adsorption chiller with entropy analysis of thermodynamic insight. Appl Energy 130:702–711

14. Cui X, Islam M, Mohan B, Chua K (2016) Theoretical analysis of a liquid desiccant based indirect evaporative cooling system. Energy 95:303–312

15. Woods J, Kozubal E (2013) A desiccant-enhanced evaporative air conditioner: Numerical model and experiments. Energy Convers Manage 65:208–220

16. Pandelidis D, Anisimov S, Worek WM, Drąg P (2016) Comparison of desiccant air conditioning systems with different indirect evaporative air coolers. Energy Convers Manage 117:375–392

17. La D, Dai Y, Li Y, Wang R, Ge T (2010) Technical development of rotary desiccant dehumidification and air conditioning: A review. Renew Sustain Energy Rev 14(1):130–147

18. Mei L, Dai Y (2008) A technical review on use of liquid-desiccant dehumidification for air-conditioning application. Renew Sustain Energy Rev 12(3):662–689

19. Parekh S, Farid M, Selman J, Al-Hallaj S (2004) Solar desalination with a humidification-dehumidification technique—a comprehensive technical review. Desalination 160(2):167–186